Superluminal Radio Sources

Superluminal Radio Sources

Proceedings of a Workshop
in honor of Professor Marshall H. Cohen,
held at Big Bear Solar Observatory, California,
October 28-30, 1986.

Edited by

J. ANTON ZENSUS AND TIMOTHY J. PEARSON
California Institute of Technology

CAMBRIDGE UNIVERSITY PRESS
Cambridge
New York   New Rochelle
Melbourne   Sydney

Published by the Press Syndicate of the University of Cambridge
The Pitt Building, Trumpington Street, Cambridge CB2 1RP
32 East 57th Street, New York, NY 10022, USA
10 Stamford Street, Oakleigh, Melbourne 3166, Australia

© Cambridge University Press 1987

First published 1987

Printed in Great Britain by Unwin Brothers, Woking

British Library CIP data available

Library of Congress CIP data available

ISBN 0 521 34560 X

Contents

Foreword	ix
Preface	xi
Acknowledgments	xii
Workshop Participants	xiii
Introduction Timothy J. Pearson and J. Anton Zensus	1
Summary of Known Superluminal Sources Richard W. Porcas	12
3C 273: Archetype of Superluminal Sources J. Anton Zensus	26
Observations of 3C 273 at 3 mm Wavelength Alan T. Moffet and Anthony C. S. Readhead	32
Superluminal Motion in the Quasar 3C 279 Stephen C. Unwin	34
Investigations of 3C 345 John A. Biretta and Marshall H. Cohen	40

3C 120
    R. Craig Walker, John M. Benson, and
    Stephen C. Unwin
	48

Structural Variations in the Quasar 3C 454.3 Ivan I. K. Pauliny-Toth	55
Superluminal Motion in BL Lac: Evidence for Deceleration in Two Events Robert L. Mutel and Robert B. Phillips	60
4C 39.25: Superluminal Motion Between Stationary Components David B. Shaffer and Alan P. Marscher	67

Superluminal Motion Towards a Stationary Component in Quasar 3C 395
    R. S. Simon, K. J. Johnston, J. Hall,
    J. H. Spencer, and J. A. Waak
	72

Subluminal Expansion in NGC 1275 *Donald C. Backer*	76
Superluminal Motion and Other Indications of Bulk Relativistic Motion in a Complete Sample of Radio Sources from the S5 Survey *Arno Witzel*	83
The Quest for Superluminal Sources *Timothy J. Pearson, Anthony C. S. Readhead, and Peter D. Barthel*	94
Tests of Beaming Models *Peter A. G. Scheuer*	104
Relativistic Beaming and the Nuclei of Double-Lobed Quasars *David H. Hough and Anthony C. S. Readhead*	114
Superluminal Motion in a Randomly Oriented Quasar Sample *J. Anton Zensus and Richard W. Porcas*	126
Extended Structure of Superluminal Radio Sources *Ian W. A. Browne*	129
Feeling Uncomfortable *Peter D. Barthel*	148
The Arcminute Structure of 1928+738 *R. S. Simon, K. J. Johnston, A. Eckart, P. Biermann, C. Schalinski, A. Witzel, and R. G. Strom*	155
Intrinsic Asymmetry in NGC 6251 *Dayton L. Jones*	162
Are Compact Doubles Misaligned Superluminals? *Mark W. Hodges and Robert L. Mutel*	168
VLBI Observations of Compact Steep-Spectrum Radio Sources *Carla Fanti and Roberto Fanti*	174
VLBI Observations of the Suspected Superluminal 3C 371 *Kevin R. Lind*	180
VLA Polarimetry of the Active Galaxy 3C 371 *J. M. Wrobel*	186
Milliarcsecond Polarization of Superluminal Sources *David H. Roberts and John F. C. Wardle*	193

The Low Frequency Variability of Extragalactic Radio Sources: a Relativistic Effect or Galactic Scintillation? *R. Fanti, L. Gregorini, L. Padrielli, and S. Spangler*	200
Superluminal Motion in CTA 102 *Lars B. Bååth*	206
Imaging Superluminal Sources: Prospects for the Next Decade *Peter N. Wilkinson*	211
A Different Perspective on Superluminal Sources *Lawrence Rudnick*	217
Infrared, Optical, UV, and X-Ray Properties of Superluminal Radio Sources *Chris Impey*	233
Superluminal Radio Sources: What Does X-ray Emission Tell Us? *Diana M. Worrall*	251
Optical Spectra of Superluminal Sources *Charles R. Lawrence, Anthony C. S. Readhead, Timothy J. Pearson, and Stephen C. Unwin*	260
Emission-Line Profile Changes in 3C 390.3 *J. Beverley Oke*	267
Evidence for Shocks in Relativistic Jets *Hugh D. Aller, Philip A. Hughes, and Margo F. Aller*	273
Synchro-Compton Emission from Superluminal Sources *Alan P. Marscher*	280
How Fast Can a Blob Go? *E. S. Phinney*	301
The μ–z Diagram *Marshall H. Cohen*	306
Grand Unified Models *Roger D. Blandford*	310
Bibliography	328
Index of Authors	352
Index of Objects	354
Index of Subjects	359

Foreword

JESSE L. GREENSTEIN
Lee A. DuBridge Professor of Astrophysics, Emeritus
California Institute of Technology

The source of energy for the radio noise emitted by astronomical objects has remained a central question, one which I tackled unsuccessfully 50 years ago. Nature has provided examples, at many scales, of magnetic plasmoids moving at thousands of kilometers per second. Electrons or positrons, at energies of billions of electron volts, are produced in some violent event and radiate myriads of low-energy photons, as they spiral in a magnetic field. Such synchrotron continua may have flat or steeply decreasing photon fluxes, dependent on the particle energy distribution, linking radio sources with cosmic-ray physics. Cosmic violence, explosion apparently subsequent to implosion, occurs in the Crab nebula, a supernova remnant, and in the collapsed pulsar at its heart. The radiation process is an indirect consequence of the violence, and merely signals the very large total energy that must exist. In 1963 the quasars, strong, variable, radio-emitting events in remote galaxies, were recognized to be small, a light year or less, and to demand more input than stellar nuclear energy could provide. Gravitational energy from 10^5 to $10^9 M_\odot$ was invoked; soon black holes became a respectable explanation. Given the high space density of relativistic electrons and radio-frequency photons, the inverse Compton effect upgraded the photons to X-rays. X-rays have been found from the Crab nebula, interacting binaries with accretion disks around a collapsed star, from quasars, and from active galactic nuclei.

The superb technical developments of the Very Long Baseline Interferometry of quasars traced the glowing, devouring dragon to its nest in the center of galaxies. Data processing of tapes from receivers separated by nearly the Earth's diameter made possible the mapping of radio sources at resolution of a milliarcsecond, in international, well-coordinated observing campaigns. By 1971, a double source in 3C 279 was found to have increased in angular separation, at a rate corresponding to three times the velocity of light. Such rates, and the energies involved in many sources, required a more complex model. Beaming from a relativistically expanding core increases the flux observed in certain directions, and the rate of superluminal expansion. Still more detailed explanations include a massive black hole, an accretion disk, polar jets of relativistic plasma, blast waves.

The observations reported in this 1986 Workshop are intrinsically difficult, at the margin of the barely possible. Marshall H. Cohen, whose sixtieth birthday was celebrated, has been a pioneer and central figure in the international campaigns to obtain these data; he has, by his human warmth and scientific strength, been an ample energy source for understanding superluminal radio sources.

Preface

A workshop on "Superluminal Radio Sources" was held in honor of Professor Marshall H. Cohen on October 28–30, 1986, at Big Bear Solar Observatory, Big Bear, California. Forty astronomers from around the world gathered at this meeting to discuss observations and theories of these objects in an informal atmosphere.

This book contains the proceedings of the Workshop. We hope that it will not only be useful as a summary but also as a source of encouragement for future work. We thank all the participants for providing written versions of their contributions soon after the meeting. Special thanks are due to Alan Moffet who could not attend the meeting owing to ill health but has nevertheless allowed us to include his paper.

The astronomical observing technique of Very Long Baseline Interferometry and the study of superluminal radio sources have always been a "family affair": experiments are jointly planned by scientists in different countries and observations are carried out simultaneously at observatories thousands of kilometers apart. Naturally there is a need for occasional "family gatherings" to discuss the latest results, and we felt that it was appropriate to combine such a meeting with the celebration of Marshall Cohen's sixtieth birthday.

Marshall Cohen is a pioneer of VLBI and the research of superluminal sources. He is also mentor and friend of many young scientists in this field. It is with pleasure that we dedicate this book to him.

J. ANTON ZENSUS
TIMOTHY J. PEARSON

Pasadena
March 1987

Acknowledgments

We thank all those who assisted in organizing the Workshop, especially Jo Ann Wolpert. In particular, we acknowledge with gratitude the enthusiasm shown by Terri Arnold both during the Workshop and in preparing the proceedings for publication. We are grateful to Harold Zirin and the staff of the Big Bear Solar Observatory for their hospitality, and to E. C. Stone, A. T. Moffet, A. C. S. Readhead, and N. Z. Scoville for making the Workshop possible. The Workshop was supported in part by the California Institute of Technology and the National Science Foundation.

We thank the editors and publishers of the following works for permission to reproduce copyright material.

The Astrophysical Journal: pages 21, 29, 36, 41, 43, 78, 98, 100, 156, 157, 158, 163, 164, 170, 195, 221, 222; copyright © 1983–1987 by the American Astronomical Society.

Nature: pages 20, 28, 119; copyright © 1986, 1987 by Macmillan Journals Ltd., London.

Monthly Notices of the Royal Astronomical Society: page 56, copyright © 1982 by the Royal Astronomical Society, London.

IAU Symposia No. 110 (VLBI and compact radio sources), No. 119 (Quasars), and No. 121 (Observational evidences of activity in galaxies): pages 57, 79, 227, 325; copyright © 1984, 1986, 1987 by the International Astronomical Union and D. Reidel Publishing Company, Dordrecht, Holland.

Science: page 15, copyright © 1971 by the American Association for the Advancement of Science, Washington.

Workshop Participants

Hugh D. Aller	Department of Astronomy, University of Michigan, Ann Arbor, Michigan, USA
Lars B. Bååth	Onsala Space Observatory, Onsala, Sweden
Donald C. Backer	Radio Astronomy Laboratory, University of California, Berkeley, California, USA
Peter D. Barthel	Owens Valley Radio Observatory, California Institute of Technology, Pasadena, California, USA
John A. Biretta	Harvard-Smithsonian Center for Astrophysics, Cambridge, Massachusetts, USA
Roger D. Blandford	California Institute of Technology, Pasadena, California, USA
Ian W. Browne	Nuffield Radio Astronomy Laboratories, Jodrell Bank, United Kingdom
Marshall H. Cohen	Owens Valley Radio Observatory, California Institute of Technology, Pasadena, California, USA
Carla Fanti	Istituto di Radioastronomia CNR, Bologna, Italy
Roberto Fanti	Istituto di Radioastronomia CNR, Bologna, Italy
Jesse L. Greenstein	California Institute of Technology, Pasadena, California, USA
Mark W. Hodges	Owens Valley Radio Observatory, California Institute of Technology, Pasadena, California, USA
David H. Hough	Jet Propulsion Laboratory, California Institute of Technology, Pasadena, California, USA
Chris Impey	Steward Observatory, University of Arizona, Tucson, Arizona, USA
Dayton L. Jones	Jet Propulsion Laboratory, California Institute of Technology, Pasadena, California, USA
Andrzej J. Kus	Katedra Radioastronomii, Uniwersytet Mikolaja Kopernika, Toruń, Poland
Charles R. Lawrence	Owens Valley Radio Observatory, California Institute of Technology, Pasadena, California, USA
Kevin R. Lind	National Radio Astronomy Observatory, Charlottesville, Virginia, USA

Front row, left to right: Porcas, Shaffer, Unwin, Cohen, Lawrence, Phinney, Biretta, Hodges. *Second row:* Readhead, Pearson, Barthel, Worrall, Bååth, Wilkinson, Lind, Browne, Impey, Rudnick, Walker, Terri Arnold. *Interstitial:* Zensus. *Back row:* Kus, Pauliny-Toth, Scheuer, Wrobel, Aller, Oke *(partially occulted)*, Marscher, Mutel, Roberts, Simon, Hough, Schmidt, Jones, Blandford, Backer, C. Fanti, Witzel, R. Fanti, Nancy.

Participants

Alan P. Marscher	Department of Astronomy, Boston University, Boston, Massachusetts, USA
Robert L. Mutel	Physics and Astronomy Department, University of Iowa, Iowa City, Iowa, USA
J. Beverley Oke	Palomar Observatory, California Institute of Technology, Pasadena, California, USA
Ivan I. K. Pauliny-Toth	Max-Planck Institut für Radioastronomie, Bonn, Federal Republic of Germany
Timothy J. Pearson	Owens Valley Radio Observatory, California Institute of Technology, Pasadena, California, USA
E. Sterl Phinney	California Institute of Technology, Pasadena, California, USA
Richard W. Porcas	Max-Planck Institut für Radioastronomie, Bonn, Federal Republic of Germany
Anthony C. S. Readhead	Owens Valley Radio Observatory, California Institute of Technology, Pasadena, California, USA
David H. Roberts	Department of Physics, Brandeis University, Waltham, Massachusetts, USA
Lawrence Rudnick	School of Physics and Astronomy, University of Minnesota, Minneapolis, Minnesota, USA
Edwin E. Salpeter	Newman Laboratory, Cornell University, Ithaca, New York, USA
Peter A. G. Scheuer	Mullard Radio Astronomy Observatory, Cavendish Laboratory, Cambridge, United Kingdom
Maarten Schmidt	California Institute of Technology, Pasadena, California, USA
David B. Shaffer	NASA Goddard Space Flight Center, Greenbelt, Maryland, USA
Richard S. Simon	E. O. Hulburt Center for Space Research, Naval Research Laboratory, Washington, D.C., USA
Stephen C. Unwin	Owens Valley Radio Observatory, California Institute of Technology, Pasadena, California, USA
R. Craig Walker	National Radio Astronomy Observatory, Socorro, New Mexico, USA
Peter N. Wilkinson	Nuffield Radio Astronomy Laboratories, Jodrell Bank, United Kingdom
Arno Witzel	Max-Planck Institut für Radioastronomie, Bonn, Federal Republic of Germany
Diana M. Worrall	Harvard-Smithsonian Center for Astrophysics, Cambridge, Massachusetts, USA
Joan M. Wrobel	Astrophysics Research Center, New Mexico Institute of Mining and Technology, Socorro, New Mexico, USA
J. Anton Zensus	Owens Valley Radio Observatory, California Institute of Technology, Pasadena, California, USA

Superluminal Radio Sources: Introduction

TIMOTHY J. PEARSON AND J. ANTON ZENSUS
California Institute of Technology

1. The Discovery

The discovery in 1970–1971 of radio sources that appeared to be expanding faster than the speed of light was a triumph for the new technique of Very Long Baseline Interferometry, and a remarkable confirmation of a prediction made several years earlier. It provides an interesting example of the parallel and complementary development of observational methods and theoretical understanding in astronomy.

The theory of radio galaxies and quasars in vogue in the mid-1960s was based on the hypothesis that their emission was synchrotron radiation from expanding clouds of electrons moving with relativistic speeds in weak magnetic fields (e.g., van der Laan 1966). This model could explain the characteristic radio spectra of the sources and the intensity fluctuations on time scales of years that were observed in some of them. If it was assumed that the quasar redshifts were indicators of their cosmological distances, then the theory implied that some sources had angular dimensions smaller than 1–10 milliarcseconds. The origin of the redshifts was, of course, a subject of controversy (e.g., Hoyle and Burbidge 1966).

The discovery of rapid variations in the quasar 3C 273 (Dent 1965) and a number of other sources presented considerable problems with the synchrotron theory, however. The short variability time scales indicated physical dimensions as small as light-months, implying higher energy densities in the sources than could be allowed by the synchrotron model. Under such conditions, the radio emission would be rapidly quenched by the "Compton catastrophe", in which the radio photons would be raised to X-ray energies by inverse Compton scattering off the relativistic electrons. One solution to this problem was to retract the assumption of cosmological redshifts and to postulate instead that quasars are located at "local" distances of 10 Mpc or less (Hoyle, Burbidge, and Sargent 1966). Other solutions included the possibility that different radiation processes, more "exotic" than synchrotron radiation, were active, or that the variability was not intrinsic to the sources but arose from propagation effects.

An alternative explanation was presented by Rees (1966, 1967), who suggested that the inverse Compton problem would be alleviated if the rapidly varying sources were expanding with relativistic velocities. He showed that this would yield shorter apparent time scales for the variations and permit larger source dimensions. The confirmation of this suggestion by direct observation of expansion, however, had to await the development of observational techniques to measure the angular sizes of the compact sources.

During the 1960s, radio astronomers gradually extended the resolution of their instruments from arcseconds to milliarcseconds. The study of interplanetary scintillations gave size estimates smaller than about a tenth of an arcsecond for some quasars (e.g., Cohen and Gundermann 1967; Cohen 1969). At Jodrell Bank, Palmer and his colleagues measured similar diameters with their radio-linked interferometers and found numerous objects still unresolved, in spite of their steadily increasing baselines (Palmer et al. 1967; see also Lovell 1973). The need for yet higher resolution motivated groups in Canada, the U.S., England, and the U.S.S.R. to develop Very Long Baseline Interferometry (VLBI), in which independent local oscillators are used and the signals are recorded on magnetic tape for later correlation.

Two such tape-recording systems were implemented by groups in Canada and the U.S.; and in 1967, the Canadian team won—by only a few weeks—the race for the first fringes (Broten et al. 1967; Bare et al. 1967). The early experiments employed baselines across Canada and the U.S. and confirmed that many compact sources were as small as a milliarcsecond, but a number of sources were still unresolved. In the quest for ever more angular resolution, experiments with longer baselines and at higher frequencies were performed between the U.S. and Australia (Gubbay et al. 1969b), Sweden (Kellermann et al. 1968), and the U.S.S.R. (Broderick et al. 1970), sometimes in the face of considerable logistic difficulties (e.g., Kellermann 1970). These experiments yielded information about the structure of many sources that could be compared with the spectra and variability time scales to test the synchrotron model.

The first indications of changes in the structure of some sources were obtained by an American-Australian team in a series of transpacific VLBI observations between 1968 and 1970 (Gubbay et al. 1969a). Following the early experiments, they had realized the potential of the NASA tracking antennas for VLBI measurements and set up an interferometer operating between California and Australia. The changes in the source visibility that they measured for 3C 279, combined with changes in total flux density, indicated that a component first seen in 1969 had reached a diameter of about 1 milliarcsecond, implying expansion at an apparent velocity of at least twice the speed of light. Aware of Rees's model, Moffet et al. (1972) concluded that their measurements presented evidence for relativistic expansion of this component. This interpretation, although by no means unique, was later confirmed, and in hindsight it seems fair to say that their experiment was the first interferometric measurement of superluminal expansion.

The result of Moffet et al. and its interpretation attracted much attention (e.g., Kellermann 1972), and several programs were begun to study the variations of more sources. However, the first detailed measurement of the compact structure in 3C 279 that allowed the exact determination of the structural changes in this source were obtained by a team that was primarily interested in another project (Knight et al. 1971). Their goal was to measure the gravitational bending of radio signals by the sun, one of the critical tests of the general theory of relativity (e.g., Shapiro 1967). The experiment involved the measurement of the change in the angular separation of 3C 279 from a nearby quasar, 3C 273, as the signal from 3C 279 passed close to the Sun in October 1970. Surprisingly, this experiment showed that both quasars were not unresolved point sources but had a dependence of visibility on baseline orientation characteristic of a double structure (Knight et al. 1971). It was clear that these double sources were perfect test cases for Rees's prediction of superluminal expansion, and independent observations in February 1971 by two teams did reveal changes in the visibility of both sources (Whitney et al. 1971; Cohen et al. 1971). In the case of 3C 279, the changes were most simply interpreted as due to an increase in the separation of the double with a speed of about $3c$, confirming the results of Moffet et al.†

The discovery of apparent superluminal motion generated considerable public attention and controversy and it was welcomed by some as a refutation of special relativity or of cosmological redshifts (e.g., Stubbs 1971). One problem was that the inference of superluminal expansion from the observations was model-dependent: the visibility changes could be produced, for example, by intensity fluctuations in a complex, stationary system. In the subsequent years, however, extensive observing programs and the development of more sophisticated analysis methods for VLBI have enabled observers to make true images of the superluminal sources with milliarcsecond resolution, allowing them to follow

---

† These results were discussed at a symposium of the American Academy of Arts and Sciences on the occasion of the presentation of the Rumford Medal to three groups of scientists for their pioneering work in long-baseline interferometry (Rogers and Morrison 1972; Rossi 1971).

the birth and evolution of superluminally moving components in many sources. These observations and their interpretation are the subject of this Workshop.

2. Very Long Baseline Interferometry

The superluminal radio sources were discovered by Very Long Baseline Interferometry, and it is still the only available technique for imaging them with milliarcsecond resolution. VLBI images can now be made routinely at wavelengths $\lambda = 50$, 18, 6, 2.8, and 1.3 cm, using as many as 18 antennas. The maximum resolution (FWHM) is $\sim 0.5\lambda/D$; for baseline length $D = 8200$ km (Europe–California) and $\lambda = 1.3$ cm this is 0.2 milliarcseconds (mas). Most of the available images have been made at 6 cm and have resolution ~ 1 mas. Shorter wavelengths (7 mm, 3 mm) have been used successfully but are not yet routine, and image-forming capability is crude (see the papers by Moffet and Readhead, page 32, and Backer, page 76). World-wide millimeter arrays will increase the resolution to 0.05 mas, and even higher resolution will eventually be available with antennas in space.

In terms of linear resolution, 1 mas ranges from about 0.25 pc in nearby active galactic nuclei to about 4 pc in most quasars. For comparison, the Schwarzschild radius of a $10^9 M_\odot$ black hole is $\sim 10^{-4}$ pc, the size of the "broad emission line region" in active nuclei is typically ~ 1 pc, and the size of the "narrow emission line region" is typically ~ 100 pc. Thus VLBI can resolve the broad-line region, but it is still some way from resolving the "central engine".

When making VLBI images, self-calibration techniques are used to circumvent atmospheric and oscillator phase and amplitude instability (e.g., Pearson and Readhead 1984). These techniques require a source to be detectable in the coherence time (in the range 30 s to 10 min, depending on frequency and weather conditions), so that observations are restricted to strong sources, brighter than ~ 0.5 Jy for the Mark II recording system (bandwidth 2 MHz), or ~ 0.1 Jy for the Mark III system (bandwidth up to 112 MHz). When self-calibration is used, absolute position information is lost, which makes it difficult to register images made at different wavelengths or epochs.

The dynamic range achieved in VLBI images (the ratio of the strongest feature in the image to the noise level) ranges from about 20 : 1 to 1000 : 1, and is limited by (a) sensitivity (a few millijanskys per beam area), (b) "non-closing" calibration errors which cannot be removed by self-calibration, and (c) inadequate coverage of the (u, v) plane (i.e., missing Fourier components of the image). All of these problems should be alleviated when the Very Long Baseline Array, currently under construction, is completed in 1992 (Wilkinson, this Workshop, page 211).

Images of the linear and circular polarization of extended radio sources are made routinely with connected-element interferometers such as the NRAO Very Large Array, and they provide valuable information about the distribution of magnetic fields in the sources. The extension of these techniques to VLBI is desirable, but is as yet very difficult: only a few VLBI polarization images have been made (Roberts and Wardle, this Workshop, page 193).

3. Cosmology

Observers of superluminal sources measure an apparent internal *proper motion* μ, that is, the rate of change of the apparent separation of two features in the brightness distribution. To convert the angular velocity into an apparent linear velocity $v_{\rm app} = c\beta_{\rm app}$ (projected on the sky) the distance to the source is needed. For almost all the known superluminal sources, the only available distance indicator is the *redshift* z; to obtain the distance, we need in addition some knowledge of the geometry of the universe. We shall assume that the universe is described by the standard Friedmann–Robertson–Walker model, parameterized by the *Hubble constant* H_0 and the *deceleration parameter* q_0. The deceleration parameter is equal to one-half the *density parameter* Ω, the ratio of the mean density to the critical density required to close the universe.

The *luminosity distance* D_L of an object with redshift z is

$$D_L = \frac{c}{H_0 q_0^2} \left(q_0 z + (q_0 - 1)(\sqrt{1 + 2q_0 z} - 1) \right), \tag{1}$$

where c is the speed of light. It is more convenient for computation to write this as

$$D_L = \frac{cz}{H_0} \left(\frac{1 + \sqrt{1 + 2q_0 z} + z}{1 + \sqrt{1 + 2q_0 z} + q_0 z} \right), \tag{2}$$

The *angular size distance* D_θ used to derive projected linear size l from observed angular size θ is

$$D_\theta = D_L/(1+z)^2. \tag{3}$$

To derive the apparent transverse velocity $\beta_{\rm app}$ from an observed proper motion μ it is necessary to take account of the time-dilation between the observer's frame and the frame of the object:

$$\begin{aligned}\beta_{\rm app} &= \mu D_\theta (1+z)/c \\ &= \frac{\mu z}{H_0(1+z)} \left(\frac{1 + \sqrt{1 + 2q_0 z} + z}{1 + \sqrt{1 + 2q_0 z} + q_0 z} \right).\end{aligned} \tag{4}$$

The value of the Hubble constant is uncertain; values in the range 50–100 km s^{-1} Mpc^{-1} are advocated by various authors. We shall hide the uncertainty in a dimensionless parameter h, the Hubble constant in units of 100 km s^{-1} Mpc^{-1}. Proper motion μ is usually measured in milliarcseconds per year (1 mas = 4.85 × 10^{-9} rad); in this case, the appropriate numerical factor in equation (4) is

$$H_0 = 2.11 \times 10^{-2} h \text{ mas yr}^{-1}. \tag{5}$$

The value of q_0 seems to be entirely unknown. We shall use the "critical" value $q_0 = 0.5$ throughout this book, although this should not be taken as an endorsement of this particular value.

4. The "Standard Model" of Superluminal Sources

The various explanations of superluminal motion have been reviewed by Blandford, McKee, and Rees (1977), Marscher and Scott (1980), and Scheuer (1984). The models generally fall into one of the following classes. (*a*) *Non-cosmological redshifts.* If the objects are much closer than we think, the apparent speeds are no longer superluminal. (*b*) *Gravitational lenses.* Gravitational lenses can magnify both flux densities and apparent motions, but such models have difficulties in producing the observed morphologies of the superluminal sources. (*c*) *Screen or light-house models.* In these a fixed or moving screen is illuminated or excited to radiate by a stationary source. With suitable geometry, apparent superluminal motion of the illuminated spot is readily obtained. The simplest is the "light echo" theory that was introduced to explain the apparent superluminal expansion of a shell around Nova Persei 1901 (e.g., Couderc 1939), and applied to the superluminal radio sources by Lynden-Bell (1977). De Waard (1986) has recently reviewed these models and presented some variants in which the screen is excited by a relativistic jet from the precessing central source. (*d*) A class of models that Scheuer (1984) calls *computer-controlled Christmas trees with beamed lights*. These include the magnetic-dipole model (Sanders 1974; Bahcall and Milgrom 1980), which appears to be no longer viable, but the version proposed by Scheuer retains some of the advantages of the relativistic beaming model while avoiding some of its problems. (*e*) The *relativistic beaming model* first proposed by Rees (1966, 1967; see also Ozernoj and Sazonov 1969) and elaborated for a relativistic jet by Blandford and Königl (1979). This is the most generally accepted explanation; indeed, few of the contributors to this Workshop consider any other possibilities. In the remainder of this introduction we shall outline the basic features of this "standard model".

Kinematics of Relativistic Expansion

The "simple fact that an object moving relativistically in suitable directions may appear to a distant observer to have a transverse velocity much greater than c" was pointed out by Rees (1966). Rees considered primarily the case of an expanding spherical shell. Because radiation from different parts of the shell takes different times to reach a distant observer, the observer does not see a spherical source; rather, the locus of points from which radiation reaches the observer at a given instant is a spheroid, and the apparent transverse expansion speed of the limb of the spheroid is γv, where γ is the Lorentz factor corresponding to the radial expansion velocity v,

$$\gamma = (1 - v^2/c^2)^{-1/2}. \tag{6}$$

The observed superluminal sources are better modeled by one or more radiating "blobs" or "plasmoids" moving at relativistic velocity v away from a stationary "core". In this case, the observed transverse velocity of separation of the blob from the core, $v_{\text{app}} = \beta_{\text{app}} c$, is related to the true velocity, $v = \beta c$, and the angle to the line of sight, θ, by

$$\beta_{\text{app}} = \frac{\beta \sin\theta}{1 - \beta \cos\theta}. \tag{7}$$

Angle $\theta = 0$ corresponds to motion directly towards the observer, $\theta = \pi$ to motion directly away from the observer.

If we consider β as a fixed parameter, then equation (7) shows that β_{app} has a maximum value of $\gamma\beta = \sqrt{\gamma^2 - 1}$ when $\cos\theta = \beta$ or $\sin\theta = 1/\gamma$. This maximum exceeds unity, and apparent superluminal motion will be observed at some angles, when $\beta > 1/\sqrt{2}$ or $\gamma > \sqrt{2}$. The apparent speed β_{app} exceeds the true speed β for all angles in the range

$$\frac{2\beta}{1 + \beta^2} > \cos\theta > 0.$$

Alternatively, we can consider β_{app} as a fixed parameter. Then β and γ are minimized when $\cot\theta = \beta_{\text{app}}$, with minimum values

$$\beta_{\min} = \cos\theta = \beta_{\text{app}}(1 + \beta_{\text{app}}^2)^{-1/2},$$

$$\gamma_{\min} = \csc\theta = \sqrt{1 + \beta_{\text{app}}^2}.$$

The maximum possible value of θ occurs when $\beta \to 1$ and $\gamma \to \infty$, and is given by

$$\cot(\theta_{\max}/2) = \beta_{\text{app}};$$

i.e., the maximum value of θ is twice the value that minimizes γ.

Finally, we can consider θ as a fixed parameter. For a given θ, β_{app} approaches a maximum value as $\beta \to 1$,

$$(\beta_{\text{app}})_{\max} = \frac{\sin\theta}{1 - \cos\theta}.$$

In a sample of objects that emit blobs with speed β in random directions, the fraction of objects with apparent speed greater than β_{app} is $(\cos\theta_2 - \cos\theta_1)/2$ where θ_2 and θ_1 are the solutions of equation (7) for the given β and β_{app}; this fraction (expressed as a function of γ) is

$$p(> \beta_{\text{app}}) = \frac{\sqrt{1 - \beta_{\text{app}}^2/(\gamma^2 - 1)}}{1 + \beta_{\text{app}}^2}$$

(Cawthorne et al. 1986). Note that even if β (or γ) is unknown, there is an upper limit on the fraction of objects with apparent speed greater than β_{app}:

$$p(> \beta_{\text{app}}) < \frac{1}{1 + \beta_{\text{app}}^2}.$$

For example, not more than $\frac{1}{17} = 5.9\%$ of a randomly oriented sample can have $\beta_{\text{app}} > 4$ (or 11.8% if the objects emit blobs in two opposed directions). This provides a powerful test of the model (Scheuer, this Workshop, page 104).

Relativistic Beaming

If the radiating material is moving relativistically then the apparent surface brightness of the source will be affected by relativistic aberration. In the simplest case of an optically thin blob moving at angle θ towards the observer with speed βc (Lorentz factor γ), if the blob emits an isotropic flux density $S_0(\nu)$ at frequency ν, the observer will measure a flux density

$$S(\nu) = S_0(\nu/\delta)\delta^3 = S_0(\nu)\delta^{3-\alpha}$$

if the source has a power-law spectrum, $S(\nu) \propto \nu^\alpha$;† where δ is the Doppler factor, the ratio of observed to emitted frequency:

$$\delta = \gamma^{-1}(1 - \beta\cos\theta)^{-1}.$$

† There is no standard convention for the sign of the *spectral index* α. In most of the contributions to this Workshop, $S(\nu) \propto \nu^{+\alpha}$ has been used, but some authors have preferred the opposite convention.

With more complicated models, the "Doppler boosting" factor can be different; e.g., for a "jet" formed out of a series of blobs of finite lifetime, the number of blobs that can be seen at once scales as δ^{-1} and the observed flux density is proportional to $\delta^{2-\alpha}$. The boosting effect is very strong, and as it selectively amplifies radiation from parts of a source moving towards the observer, it can dramatically change the appearance of a source. Quantitative interpretation is difficult, however, because the true velocity of the radiating material is unknown—it may not be the same as that inferred from the apparent superluminal motion—and it is likely to be different in different parts of the source.

Unified Beaming Models

If the cores of the superluminal sources are directed towards us and amplified by relativistic beaming, it is natural to ask what the sources would look like if they were viewed from a less privileged direction. In the simplest models, beaming is confined to a cone of half-angle $\sim 1/\gamma$, so for every beamed source, there should be $\sim \gamma^2$ similar sources beamed in other directions (Blandford and Rees 1978). These sources will of course be much fainter, but should presumably be detectable in some waveband. Much of the discussion at this Workshop addresses the question: where are the unbeamed sources? The most popular answer is provided by the "unified scheme" of Orr and Browne (1982): the unbeamed counterparts of the superluminal sources (and, indeed, of all core-dominated, flat-spectrum radio sources) are the normal extended double radio galaxies and quasars.

Scheuer and Readhead (1979) earlier suggested that the sources at large angles to the line of sight should be identified with some or all of the radio-quiet quasars. This hypothesis derived support from the similarity of the optical spectra of radio-emitting and radio-quiet quasars, and from the fact that radio-quiet quasars are more numerous than radio-emitting quasars by a factor ~ 100. They assumed that all the radio emission was beamed and made estimates of the numbers of "radio-quiet" quasars that should be detected at various flux density levels. These predictions have not been confirmed: a much larger fraction than expected of optically-selected quasars are found to be radio sources (Kellermann et al. 1986).

When examined with sufficient dynamic range, e.g., with the Very Large Array, the Jodrell Bank MERLIN array, and the Westerbork Synthesis Radio Telescope, most of the compact, flat-spectrum, radio sources are found to have weak, extended components ($\gtrsim 1''$) emitting a few per cent of the total flux density. In many cases, the extended structure is similar to that of the normal extended double radio sources, the only difference being the strength of the compact core. It is unlikely that a large part of the extended flux density can be beamed, and so the unbeamed counterparts of these sources should still be moderately strong radio sources—much stronger than the radio-quiet quasars. These observations provide good evidence that the simple Scheuer–Readhead model is not correct; rather, they provide support for the revised "unified scheme" of Orr and Browne (1982), where the core-dominated and superluminal sources are normal double sources in which the core is beamed towards us.

The unified schemes are susceptible to statistical tests. Orr and Browne (1982) used their model to predict the proportion of flat-spectrum sources in flux-limited samples, and to derive the source counts of flat-spectrum quasars from the steep-spectrum counts; their results for a Lorentz factor $\gamma \sim 5$ were quite encouraging. There should also be correlations between aspect-dependent properties of the sources. Such properties include (a) the relative core strength (e.g., the ratio R of core flux density to extended flux density)—cores beamed towards the observer should be brighter; (b) the measured core expansion speed—the highest speeds should be seen in sources making a small angle to the line of sight; (c) the projected linear size—sources lying close to the line of sight should be smaller (assuming that the beaming axis of the core is roughly parallel to the largest dimension of the source); and (d) source curvature—a small bend in the source axis is amplified by projection if the axis lies close to the line of sight. Unfortunately, the correlations are likely to be masked by a spread of the intrinsic properties of the sources; but they appear to be present in the expected sense. One problem is that the projected sizes of the superluminal sources tend to be larger than expected (Schilizzi and de Bruyn 1983). It is too early to say that the unified scheme is correct, but the results are at least promising enough for most of the contributors to this Workshop to discuss their observations in this context.

References

Bahcall, J. N., and Milgrom, M. 1980, *Astrophys. J.*, **236**, 24.
Bare, C., Clark, B. G., Kellermann, K. I., Cohen, M. H., and Jauncey, D. L. 1967, *Science*, **157**, 189.
Blandford, R. D., and Königl, A. 1979, *Astrophys. J.*, **232**, 34.
Blandford, R. D., McKee, C. F., and Rees, M. J. 1977, *Nature*, **267**, 211.
Blandford, R. D., and Rees, M. J. 1978, in *Pittsburgh Conference on BL Lac Objects*, ed. A. M. Wolfe (Pittsburgh: University of Pittsburgh), p. 328.
Broderick, J. J., Vitkevich, V. V., Jauncey, D. L., Efanov, V. A., Kellermann, K. I., Clark, B. G., Kogan, L. R., Kostenko, V. I., Cohen, M. H., Matveenko, L. I., Moiseev, I. G., Payne, J., and Hansson, B. 1970, *Astron. Zh.*, **47**, 784. English translation: 1971, *Soviet Astron.*, **14**, 627.
Broten, N. W., Legg, T. H., Locke, J. L., McLeish, C. W., Richards, R. S., Chisholm, R. M., Gush, H. P., Yen, J. L., and Galt, J. A. 1967, *Nature*, **215**, 38.

Cawthorne, T. V., Scheuer, P. A. G., Morison, I., and Muxlow, T. W. B. 1986, *Monthly Notices Roy. Astron. Soc.*, **219**, 883; erratum *Monthly Notices Roy. Astron. Soc.*, **222**, 895.
Cohen, M. H. 1969, *Ann. Rev. Astron. Astrophys.*, **7**, 619.
Cohen, M. H., Cannon, W., Purcell, G. H., Shaffer, D. B., Broderick, J. J., Kellermann, K. I., and Jauncey, D. L. 1971, *Astrophys. J.*, **170**, 207.
Cohen, M. H., and Gundermann, E. J. 1967, *Astrophys. J. (Letters)*, **148**, L49.
Couderc, P. 1939, *Ann. Astrophys.*, **2**, 271.
Dent, W. A. 1965, *Science*, **148**, 1458.
de Waard, G. J. 1986, Ph. D. thesis, Rijksuniversiteit Leiden.
Gubbay, J., Legg, A. J., Robertson, D. S., Moffet, A. T., Ekers, R. D., and Seidel, B. 1969a, *Nature*, **224**, 1094.
Gubbay, J., Legg, A. J., Robertson, D. S., Moffet, A. T., and Seidel, B. 1969b, *Nature*, **222**, 730.
Hoyle, F., and Burbidge, G. R. 1966, *Astrophys. J.*, **144**, 534.
Hoyle, F., Burbidge, G. R., and Sargent, W. L. W. 1966, *Nature*, **209**, 751.
Kellermann, K. 1970, "The Russian-American VLB experiment (parts I–III)", *The Observer* (Green Bank: National Radio Astronomy Observatory), **10**, No. 1, p. 9; No. 2, p. 7; No. 4, p. 5.
Kellermann, K. I. 1972, in *IAU Symposium 44, External Galaxies and Quasi-Stellar Objects*, ed. D. S. Evans (Dordrecht: Reidel), p. 190.
Kellermann, K. I., Clark, B. G., Bare, C. C., Rydbeck, O., Ellder, J., Hansson, B., Kollberg, E., Hoglund, B., Cohen, M. H., and Jauncey, D. L. 1968, *Astrophys. J. (Letters)*, **153**, L209.
Kellermann, K., Sramek, R., Shaffer, D., Green, R., and Schmidt, M. 1986, in *IAU Symposium 119, Quasars*, ed. G. Swarup and V. K. Kapahi (Dordrecht: Reidel), p. 95.
Knight, C. A., Robertson, D. S., Rogers, A. E. E., Shapiro, I. I., Whitney, A. R., Clark, T. A., Goldstein, R. M., Marandino, G. E., and Vandenberg, N. R. 1971, *Science*, **172**, 52.
Lovell, B. 1973, *Out of the Zenith* (Oxford: Oxford University Press), chapter 6.
Lynden-Bell, D. 1977, *Nature*, **270**, 396.
Marscher, A. P., and Scott, J. S. 1980, *Publ. Astron. Soc. Pacific*, **92**, 127.
Moffet, A. T., Gubbay, J., Robertson, D. S., and Legg, A. J. 1972, in *IAU Symposium 44, External Galaxies and Quasi-Stellar Objects*, ed. D. S. Evans (Dordrecht: Reidel), p. 228.
Orr, M. J. L., and Browne, I. W. A. 1982, *Monthly Notices Roy. Astron. Soc.*, **200**, 1067.
Ozernoj, L. M., and Sazonov, V. N. 1969, *Astrophys. Space Sci.*, **3**, 365. English translation: *Astrophys. Space Sci.*, **3**, 395.
Palmer, H. P., Rowson, B., Anderson, B., Donaldson, W., Miley, G. K., Gent, H., Adgie, R. L., Slee, O. B., and Crowther, J. H. 1967, *Nature*, **213**, 789.
Pearson, T. J., and Readhead, A. C. S. 1984, *Ann. Rev. Astron. Astrophys.*, **22**, 97.
Rees, M. J. 1966, *Nature*, **211**, 468.
Rees, M. J. 1967, *Monthly Notices Roy. Astron. Soc.*, **135**, 345.
Rogers, A. E. E., and Morrison, P. 1972, *Science*, **175**, 218.
Rossi, B. B. 1971, *Records Am. Acad. Arts Sci.*, 1970–1971, p. 14.
Sanders, R. H. 1974, *Nature*, **248**, 390.
Scheuer, P. A. G. 1984, in *IAU Symposium 110, VLBI and Compact Radio Sources*, ed. R. Fanti, K. Kellermann, and G. Setti (Dordrecht: Reidel), p. 197.
Scheuer, P. A. G., and Readhead, A. C. S. 1979, *Nature*, **277**, 182.
Schilizzi, R. T., and de Bruyn, A. G. 1983, *Nature*, **303**, 26.
Shapiro, I. I. 1967, *Science*, **157**, 806.
Stubbs, P. 1971, *New Scientist*, **50**, 254.
van der Laan, H. 1966, *Nature*, **211**, 1131.
Whitney, A. R., Shapiro, I. I., Rogers, A. E. E., Robertson, D. S., Knight, C. A., Clark, T. A., Goldstein, R. M., Marandino, G. E., and Vandenberg, N. R. 1971, *Science*, **173**, 225.

Summary of Known Superluminal Sources

RICHARD W. PORCAS

Max-Planck-Institut für Radioastronomie, Bonn

1. Introduction

It is a great pleasure to take part in this Workshop on Superluminal Radio Sources, with which we celebrate Marshall Cohen's sixtieth birthday. He himself could easily give this review, so it is with some trepidation that I now give a "summary of known superluminal sources" in his presence.

The phenomenon of superluminal motion is rather younger than Marshall; the first reports of apparent faster-than-light motion in the quasars 3C 279 and 3C 273 were published only 16 years ago, soon after the introduction of the Very Long Baseline Interferometry (VLBI) technique (Whitney et al. 1971; Cohen et al. 1971; see also Harwit (1981) for a historical account). My first introduction was as a final-year undergraduate when I read a popular article in *New Scientist* (Stubbs 1971) in which this new phenomenon was used to question the cosmological nature of the quasar redshifts. Since its discovery, superluminal motion has remained one of the most intriguing topics in radio astronomy, and lies at the very heart of the problem of understanding the nature of quasars and active galactic nuclei.

Two further landmarks may be noted in the short history of this subject. First, the *Nature* review by Marshall and his colleagues (Cohen et al. 1977), in which the properties of the four then-known superluminal sources were summarized. This article presented a consolidation of earlier work, contributed the definition of "superluminal motion", and, by virtue of its authority, established values for the cosmological parameters H_0 and q_0 which (perhaps because of lack of progress in observational cosmology!) have been widely used in the superluminal source literature ever since. The second event was the previous Workshop on Superluminal Sources, held at Jodrell Bank in February 1983 (Porcas 1983). By then the number of known superluminal sources had risen to seven, and the participants spent two days discussing the properties of these sources and the various theories proposed to explain them.

Now, nearly four years later, we have assembled for another workshop devoted to this topic. Monitoring the structural changes in superluminal sources is largely carried out by individuals or institutions who "adopt" sources and who thus can maintain the effort necessary to execute series of VLBI experiments over extended periods. Nearly all of these source "owners" are participants here and will describe individual sources or samples of sources. What I will attempt to do is give a short guide to some basic considerations relating to observations of superluminal sources. It is the tradition in such talks to present a table giving an inventory of known superluminal sources; this I will endeavor to do, although in this active field we can expect such a table to become out of date between the time of writing (November 1986) and publication.

In this Workshop we will hear how observations can be related to theories of superluminal motion, and, in particular, to the popular relativistic jet model. Here, however, I shall adopt a purely phenomenological position, and confine myself to the observed structure and kinematics of the radio emission.

2. What is a Superluminal Source?

Cohen et al. (1977) defined the term "superluminal motion" to describe structural changes in radio sources when the "apparent transverse velocity of expansion is greater than the velocity of light". (A number of other terms such as "super-light" or "super-relativistic" expansion were in use before that time.) A superluminal source is any source in which superluminal motion has been observed. Clearly, if we are not to abandon some very basic notions in physics, finding an explanation of the phenomenon commands our immediate attention and interest.

For quasars and other radio sources at large distances, we use the following formula to derive the apparent velocities:

$$\beta_{\mathrm{app}} = \underbrace{\frac{q_0 z + (q_0 - 1)\left(\sqrt{1 + 2q_0 z} - 1\right)}{H_0 q_0^2 (1+z)^2}}_{\text{cosmological distance in light units}} \times \underbrace{(1+z)}_{\substack{\text{time} \\ \text{dilation}}} \times \underbrace{\mu}_{\substack{\text{proper} \\ \text{motion}}}$$

where β_{app} is the (apparent) faster-than-light transverse velocity, in units of the velocity of light, H_0 and q_0 are the Hubble constant and deceleration parameter, z is the redshift of the object identified with the radio source, and μ is the observed angular motion measured by VLBI.

Here we assume that the redshift can be used in this way to indicate cosmological distance (see Narlikar (1986) for a contrary view) and that Friedmann cosmological models apply (again, see Segal (1986) for an alternative view).

It is noteworthy that whilst enormous progress has been made since 1977 in improving our VLBI measurements of angular motion μ (see below), very little progress has been made towards agreement on the values of H_0 and q_0. The two leading schools still span a factor of two in their estimates of H_0 (50 km s^{-1} Mpc^{-1} for Sandage; 100 km s^{-1} Mpc^{-1} for de Vaucouleurs) as indeed they did in 1977. Even worse, determination of q_0 has apparently "reverted entirely to the theorists" (Maddox 1984). The resulting uncertainty in deriving superluminal velocities is not so great that, by taking extreme values, the phenomenon can be made to go away, but it is one factor which hinders progress towards a satisfactory explanation. By the same token, if a particular explanation becomes compelling for independent reasons, we might be justified in turning the problem around, and using the superluminal phenomenon to directly determine the cosmological parameters.

In this review velocities are evaluated by including a dimensionless parameter h, which is the Hubble constant in units of 100 km s^{-1} Mpc^{-1}; thus the accepted range for h lies between 0.5 and 1.0. If the angular proper motion μ is measured in milliarcseconds per year, then the formula for superluminal velocity becomes

$$\beta_{\mathrm{app}} = 47.4 \times \frac{q_0 z + (q_0 - 1)\left(\sqrt{1 + 2q_0 z} - 1\right)}{h q_0^2 (1 + z)} \times \mu.$$

In Tables 1–3, velocities for superluminal sources are given for two values of q_0; the "traditional" value, $q_0 = 0.05$, which roughly corresponds to that expected from the observed distribution of matter in the universe, and $q_0 = 0.5$, which is the value corresponding to the universe having the "critical" density. The difference becomes significant at high redshifts. It seems we are free, at present, to choose values of h and q_0 on purely psychological grounds: high values for conservatives who wish to "minimize the embarrassment", low values for the more rebellious to "maximize the damage"! At the editors' request, a value of $q_0 = 0.5$ is used for the remainder of this text.

We can be rather more proud of our progress in finding new superluminal sources, and in investigating some of their detailed properties. The initial discovery of superluminal motion was based on model fitting to just a few VLBI measurements of amplitude in the (u, v) plane (Figure 1). μ then represented the relative motion between two model Gaussian components which satisfactorily fit the rather sparse data. Nowadays, we can observe sources with arrays of ten telescopes or more, gaining vastly more coverage of the (u, v) plane, and a large fraction of the phase information as well from the "closure phase" measurements. The values of μ which are obtained today represent the systematic relative motion between two or more brightness peaks in high resolution "hybrid" maps of the source brightness distribution, which nearly a decade of experience has taught us to have confidence in.

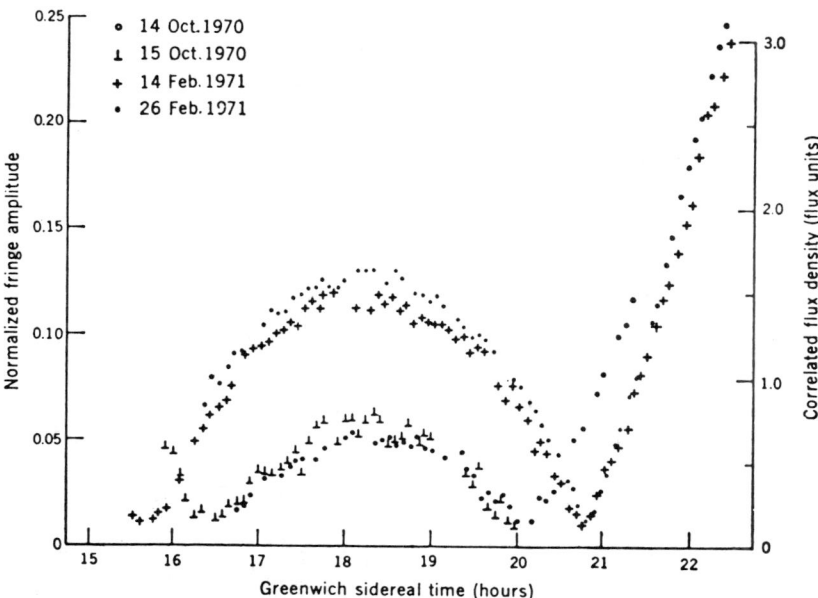

Figure 1 Amplitude of visibility function of 3C 279 (from Whitney et al. 1971).

In this Workshop, we will be able to wallow in an orgy of such data, and therefore I make no excuse in starting the trend with the source I have been monitoring, 3C 179, which can be used by way of example to illustrate a number of features. Some properties of this quasar are given by Porcas (1981) and Shone, Porcas, and Zensus (1985). Figure 2 shows a sequence of seven VLBI hybrid maps of the radio core, each made at 10.7 GHz using four or five of the most sensitive telescopes available. The first four maps were made using the Mark II recording system and with the help of a new algorithm for incorporating antenna-based fringe rate and delay residuals (Alef and Porcas 1986). The last three maps were obtained using the more sensitive Mark III system. In order to establish superluminal motion, it is first necessary to make a correct registration of the maps from the different epochs. As these are hybrid maps, the production of which loses the absolute position information, we must try and identify some feature, present in all the maps, which can be assumed stationary. In common with many compact radio sources, 3C 179 has one compact, flat-spectrum component (the eastern one) visible at all epochs; we assume this is stationary and denote it the "core". A second component is also visible in each map. At the first epoch (1979), it is a brightness maximum just over 1 mas to the west of the core in position angle $-89°$; at subsequent epochs, this separation

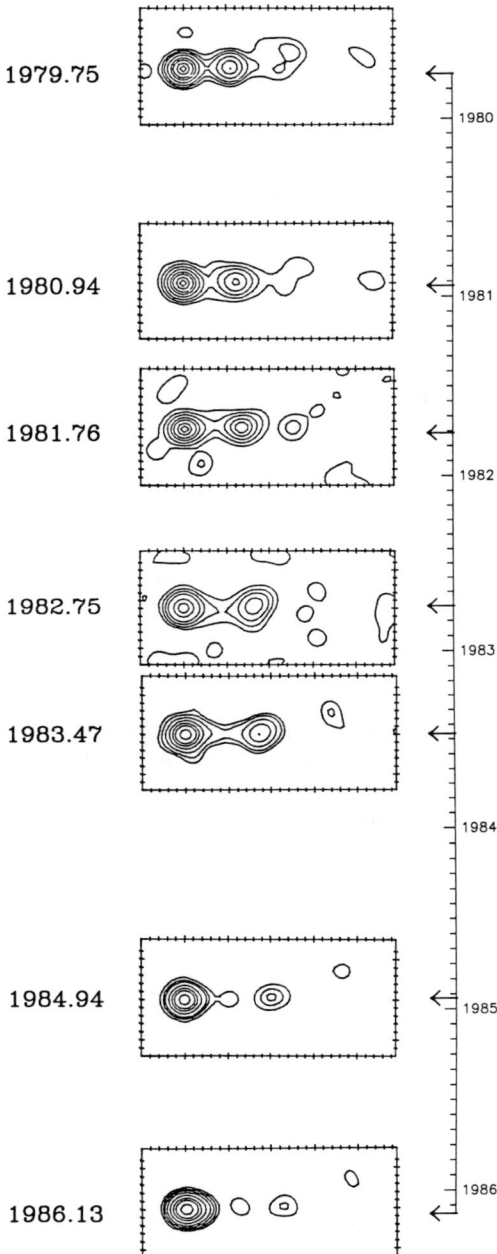

Figure 2 Hybrid maps of 3C 179 made at 10.7 GHz at seven epochs. Tick mark interval is 0.2 mas.

increases, and in the last two maps the brightness peak has clearly faded. The average rate of change of this separation is 0.19 mas yr^{-1}, which corresponds to an apparent transverse velocity of $4.8h^{-1}c$.

This component of 3C 179 demonstrates very clearly the phenomenon of superluminal motion. It is a well-defined component, present at all epochs, and undergoing systematic outward expansion away from the core, at a velocity exceeding c even with the most conservative values of h and q_0. Even if we supposed that both this component and the "core" were separating with opposite velocities from some common "event" midway between them, the minimum velocity would be $2.4h^{-1}c$.

The maps of 3C 179 also show other radio emission, however, even further west of the core. At most epochs, there is a less well-defined component at a brightness level of one or two contours, which also moves systematically away from the core, apparently at a different position angle of $\sim -75°$. The persistence of this feature in seven maps spanning some $6\frac{1}{2}$ yr lends support to the reality of its existence, although it is never brighter than a few percent of the core. If it were the only feature apart from the core, and if only two or three maps existed, it would be hard to identify it as a component with superluminal motion.

3. An Inventory of Superluminal Sources

As can be seen from the example of 3C 179, the classification of a source as superluminal will depend on many factors and preconditions. It must be active at the time of monitoring, and it must be observed with VLBI at least twice, but probably many more times if the moving component is not well defined. The observations must cover a long time span if the angular expansion rate is small compared to the beam width, as with sources at high redshift, for example.

In making up a list of known superluminal sources, a certain degree of subjectivity must inevitably prevail. In many cases, the question is not "is this source superluminal or not?", but rather "is the evidence good enough (yet) to include this source?".

An interesting question is how to establish that a source is not superluminal. Of course, no observations can guarantee that at some future date a source will not eject a component with superluminal motion. However, a component which has a well-defined position with respect to the reference feature, and which has been identified at at least two epochs may turn out to have a velocity less than c. As can be seen, as much effort must be expended to demonstrate that a component is not superluminal as is needed to show that it is.

These remarks are made only to point out the danger of using lists of known superluminal sources to identify a class of source whose properties

Table 1 Superluminal Sources

Source		z	Component	μ mas yr^{-1}	$\beta_{app}h$ $q_0 = 0.05$	$\beta_{app}h$ $q_0 = 0.5$	Ref.	see page
1253−055	3C 279	0.538	?	0.5	10.4	9.2	1	
			B2	0.11	2.3	2.0	2	34
1226+023	3C 273	0.158	C3	0.79	5.5	5.3	3	
			C4	0.99	6.9	6.6	3	
			C5	1.20	8.3	8.0	4	26
			C7a	0.76	5.3	5.1	4	26
0430+052	3C 120	0.033	A	1.35	2.1	2.1	5	48
			B	2.53	3.9	3.9	5	48
			C	2.47	3.8	3.8	5	48
			D	2.66	4.1	4.1	5	48
			E	2.54	3.9	3.9	5	48
1641+399	3C 345	0.595	core	<0.02	<0.5	<0.4	6	12
			C2	0.48	10.8	9.5	7	40
			C3	0.30	6.8	5.9	7	40
			C4	0.07, 0.3	1.6, 6.8	1.4, 5.9	7	40
0723+679	3C 179	0.846[a]	−	0.19	5.7	4.8	8	12
0333+321	NRAO 140	1.258	B	0.15	6.2	4.8	9	280
2200+420	BL Lac	0.070	B	0.76	2.4	2.4	10	60
0923+392	4C 39.25	0.699	a−c	<0.006	<0.15	<0.13	11	67
			b	0.16	4.1	3.5	12	67
1901+319	3C 395	0.635	1−2	<0.07	<1.7	<1.4	13	72
			3	0.64	∼15	∼13	13	72
2251+158	3C 454.3	0.859	1−2	<0.05	<1.5	<1.3	14	55
			4	0.35	10.7	8.8	14	55
1845+797	3C 390.3	0.0569	var.	0.74	1.9	1.9	15	
1928+738		0.302	A1−4	0.6	7.5	7.0	16	83, 94
1642+690		0.751[b]	−	0.34	9.3	7.9	17	94
0850+581		1.322	−	0.12	5.1	3.9	18	94
0212+735		2.367[b]	−	0.09	6.0	3.9	19	83
1150+812		1.25	−	0.13	5.3	4.1	19	83
0906+430	3C 216	0.669	−	0.11	2.7	2.4	20	94
1137+660	3C 263	0.652	−	0.06	1.5	1.3	21	114, 126
1951+498		0.466	−	∼0.07	∼1.3	∼1.2	22	126
1040+123	3C 245	1.029	−	0.11	3.9	3.1	23	114
1721+343	4C 34.47	0.206	−	0.36	3.2	3.1	24	148
0735+178		0.424	NE	0.18	3.0	2.8	25	
2230+114	CTA 102	1.037	−	∼0.65	∼23	∼18	26	206
0851+202	OJ 287	0.306	SW1−2	0.28	3.6	3.3	27	193

[a] Wills and Wills 1976. [b] Lawrence et al. 1986.

References: 1. Cotton et al. 1979. 2. Unwin 1986. 3. Unwin et al. 1985; Biretta et al. 1985. 4. Cohen et al. 1987. 5. Walker 1986. 6. Bartel et al. 1986. 7. Biretta, Moore, and Cohen 1986. 8. Porcas 1986. 9. Marscher and Broderick 1985. 10. Mutel and Phillips 1984. 11. Shaffer et al. 1977. 12. Shaffer and Marscher 1985. 13. Simon et al., this Workshop, page 72; Waak et al. 1985. 14. Pauliny-Toth 1987. 15. Alef et al. 1987. 16. Eckart et al. 1985. 17. Pearson et al. 1986. 18. Barthel et al. 1986. 19. Schalinski et al. 1987. 20. Pearson, Readhead, and Barthel, this Workshop, page 94. 21. Zensus, Hough, and Porcas 1987. 22. Zensus and Porcas, this Workshop, page 126; Zensus and Porcas 1986. 23. Hough and Readhead, this Workshop, page 114. 24. Barthel, this Workshop, page 148. 25. Bååth 1984. 26. Bååth, this Workshop, page 206. 27. Roberts and Wardle, this Workshop, page 193.

Summary of Superluminal Sources

Table 2 Possibly Superluminal Sources

Source		z	Comment	Ref.
0415+379	3C 111	0.049	Superluminal fading of jet	1
2223−052	3C 446	1.404	Superluminal brightening in stable double	2
0235+164		0.851	Superluminal flux variations	3
0538+498	3C 147	0.545	X-ray deficit; internal motions	4
1807+698	3C 371	0.051	Structural variations	5
0711+356		1.62	Structural var. in extended component?	5
1038+528A		0.678	Ejection from core?	6

References: 1. Götz et al. 1987. 2. Brown et al. 1981. 3. Scheuer 1976. 4. Preuss et al. 1984. 5. Readhead, Pearson, and Unwin 1984. 6. Marcaide et al. 1985.

Table 3 Subluminal Sources

Source		z	Component	μ mas yr^{-1}	$\beta_{app}h$ $q_0 = 0.05$	$\beta_{app}h$ $q_0 = 0.5$	Ref.	see page
0316+413	3C 84	0.0172		0.24	0.19	0.19	1	76
1228+127	M 87	0.004		<0.3	<0.06	<0.06	2	
1637+826	NGC 6251	0.023		<0.3	<0.32	<0.32	3	162
2134+004		1.936		<0.01	<0.6	<0.4	4	
2021+614		0.2266[a]		<0.06[b]	<0.6	<0.6	5	94

[a] Bartel et al. 1984. [b] 2σ upper limit.

References: 1. Romney et al. 1984. 2. Schmitt and Reid 1985. 3. Jones 1986. 4. Pauliny-Toth et al. 1984. 5. Readhead, Pearson, and Unwin 1984.

can be compared with sources not in the list. The latter includes the vast majority of sources, for which there simply are insufficient data (e.g., none at all) to classify the source.

With this in mind, I introduce my list of superluminal (and other) sources (Tables 1–3), and briefly describe its contents. Table 1 consists of sources for which superluminal motion of a component either is fairly well established, or is claimed by authors on the basis of (as yet) perhaps insufficient data. Table 2 contains sources where the changes in flux density and structure may not be described as "component motion" or where the data are insufficient yet to clearly classify them. Table 3 contains sources with established "non-superluminal" behavior. References to the most recent publications on each source are given; many sources are also the subject of contributions in this volume, as noted in the tables.

The first seven sources of Table 1 are those known at the time of the last superluminal workshop; there is a very large range in the quality and quantity of data amongst these. 3C 345 is certainly the best studied source. At least three components (C2, C3, C4) have now been followed for a number of years at a range of observing frequencies (Biretta, Moore, and Cohen 1986; Biretta and Cohen, this Workshop, page 40). A number of facts about superluminal motion have been established only for 3C 345, owing to the exceptionally large observing

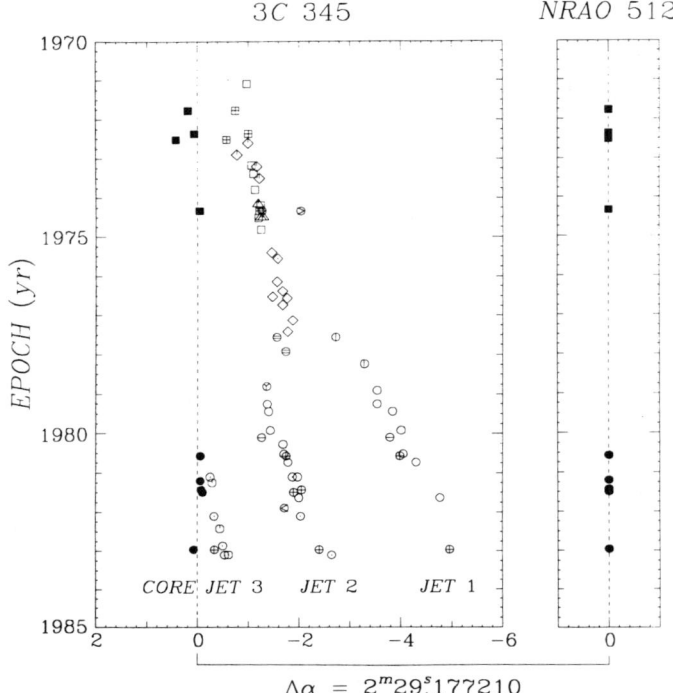

Figure 3 Proper motion of components in 3C 345 (from Bartel et al. 1986).

effort which has been spent on this source. First, the "core" really is stationary ($\beta_{\mathrm{app}} < 0.4h^{-1}$). This has been measured using the VLBI phase referencing technique and referring the position of the core to that of the nearby compact radio source NRAO 512 (Bartel et al. 1986; Figure 3). To first order, then, we are justified in speaking of the core of 3C 345 as stationary and, by extension, the core in other sources as well. However, structural changes in the core at levels below our resolution may cause its centroid to "jitter", especially when a new component emerges. Also, there is evidence that the position of the brightness peak of the "core" is frequency-dependent. This has been shown directly by Marcaide and Shapiro (1984) for the quasar 1038+528A using dual frequency phase referencing. It is also strongly implied by the frequency-dependent separation of component C3 in 3C 345 (Figure 4).

The second important result from the detailed study of 3C 345 is that different components in a given source do not always have the same velocity, and, indeed, any given component may change velocity (both magnitude and direction changes have been reported for C4 in 3C 345,

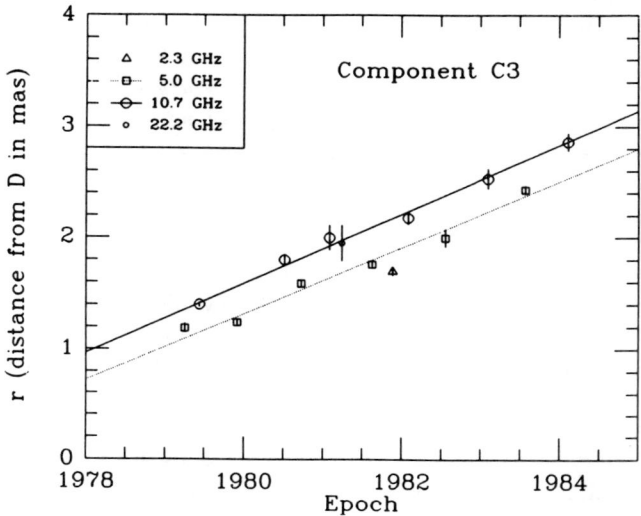

Figure 4 Frequency dependence of the separation of component C3 in 3C 345 (from Biretta, Moore, and Cohen 1986).

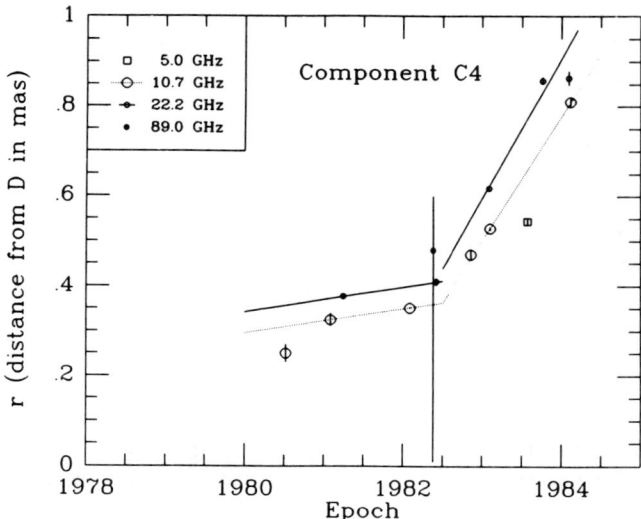

Figure 5 Separation of component C4 in 3C 345, as function of time (from Biretta, Moore, and Cohen 1986).

see Figure 5). Different velocities for successive components have also been reported for two other sources in this group, 3C 120 and 3C 279.

An important question, not yet answered, is whether there is a fixed track in each source which successive components follow. To answer this, we must monitor a source with multiple, clearly defined components (e.g., 3C 345), and wait until a new component reaches the distance from the core formerly occupied by the previous component. Will it then have the same velocity and position angle as its predecessor?

Table 1 contains three sources for which both quasi-stationary and superluminal components have been observed. In 4C 39.25 and 3C 395, the milliarcsecond structure consists of two components whose relative velocity is $\lesssim c$ ($< 0.13h^{-1}c$ for 4C 39.25), whilst a third component moves between them at superluminal velocity. In these cases, it is clearly necessary to determine which component (if any) is the "core" before we can classify the motion as either outward (expansion) or inward (contraction). In 3C 454.3, at least two components remain more or less stationary with respect to the "core" ($< 1.3h^{-1}c$) whilst another component moves away at superluminal velocity. As with many sources in the list, phase referencing VLBI observations which could "anchor" a component in the structure would greatly clarify the source kinematics.

Superluminal motion of components in the core of the galaxy 3C 390.3 has only recently been reported (Alef et al. 1987). In this source, as with 3C 120, it is clear that very frequent monitoring is required to keep track of the different components to ensure an unambiguous identification from epoch to epoch.

Twelve other sources in Table 1 come from VLBI observing programs set up to monitor classes of objects. 1928+738, 1642+690, 0850+581, 0212+735, 1150+812, and 3C 216 come from flat-spectrum source samples; 3C 263, 1951+498, 3C 245, and 1721+343 from monitoring cores of extended doubles; 0735+178 from a sample of BL Lac objects; and CTA 102 from monitoring of low-frequency variable radio sources. A number of other sources monitored in such programs are not included, either because the time-baseline is not yet long enough to detect changes, or because the redshifts are not known (e.g., Barthel et al. 1985; Padrielli et al. 1986).

In Table 2 are listed a number of sources where rapid brightness changes occur in the milliarcsecond structure although no "component motion" can be identified. This is the dominant behavior in the quasar 3C 454.3 (Pauliny-Toth 1987) and it has also recently been reported in the galaxy 3C 111, where an extended jet-like emission region apparently faded between two observing epochs (Götz et al. 1987). For other sources, such as 3C 446 and 0711+356, there is rapid change of component flux densities. For 3C 147 and 3C 371, there are clearly structural changes but it is difficult to identify clear components.

Table 3 lists the only source, 3C 84, in which a component has been seen to move at a speed less than c ($\beta_{app} = 0.19h^{-1}$). In addition, there are upper limits less than c for the two nearby galaxies M 87 and NGC 6251, and two other sources.

4. Summary

It should be stressed again that the list of sources in Tables 1–3 should be viewed more as a tabulation of claims, rather than an attempt to isolate a class of "superluminal sources". In many cases, the lack of astrometric data is critical; the identification of a core, which is assumed to be stationary, can be difficult—4C 39.25, 3C 454.3, 1928+738, and 1721+343 are good examples of this. There is clearly a lot of work for VLBI observers to do in this area. We must also be careful not to force all sources into the mold of those which show classical superluminal motion such as 3C 345 and 3C 179. Sources such as 3C 454.3, 4C 39.25, and 3C 111 may be pointing the way to a better understanding of why some sources seem to be breaking Einstein's speed limit.

I thank Ute Kraus for assistance in producing the maps shown in Figure 2, and Marshall Cohen for discussions concerning the content of the tables.

References

Alef, W., Götz, M. M. A., Preuss, E., and Kellermann, K. I. 1987, *Astron. Astrophys.*, submitted.
Alef, W., and Porcas, R. W. 1986, *Astron. Astrophys.*, **168**, 365.
Bååth, L. B. 1984, in *IAU Symposium 110, VLBI and Compact Radio Sources*, ed. R. Fanti, K. Kellermann, and G. Setti (Dordrecht: Reidel), p. 127.
Bartel, N., Herring, T. A., Ratner, M. I., Shapiro, I. I., and Corey, B. E. 1986, *Nature*, **319**, 733.
Bartel, N., Shapiro, I. I., Huchra, J. P., and Kühr, H. 1984, *Astrophys. J.*, **279**, 112.
Barthel, P. D., Miley, G. K., Schilizzi, R. T., and Preuss, E. 1985, *Astron. Astrophys.*, **151**, 131.
Barthel, P. D., Pearson, T. J., Readhead, A. C. S., and Canzian, B. J. 1986, *Astrophys. J. (Letters)*, **310**, L7.
Biretta, J. A., Cohen, M. H., Hardebeck, H. E., Kaufmann, P., Abraham, Z., Perfetto, A. A., Scalise Jr., E., Schaal, R. E., and Silva, P. M. 1985, *Astrophys. J. (Letters)*, **292**, L5.
Biretta, J. A., Moore, R. L., and Cohen, M. H. 1986, *Astrophys. J.*, **308**, 93.
Brown, R. L., Johnston, K. J., Briggs, F. H., Wolfe, A. M., Neff, S. G., and Walker, R. C. 1981, *Astrophys. Letters*, **21**, 105.
Cohen, M. H., Cannon, W., Purcell, G. H., Shaffer, D. B., Broderick, J. J., Kellermann, K. I., and Jauncey, D. L. 1971, *Astrophys. J.*, **170**, 207.
Cohen, M. H., Kellermann, K. I., Shaffer, D. B., Linfield, R. P., Moffet, A. T., Romney, J. D., Seielstad, G. A., Pauliny-Toth, I. I. K., Preuss, E., Witzel, A., Schilizzi, R. T., and Geldzahler, B. J. 1977, *Nature*, **268**, 405.
Cohen, M. H., Zensus, J. A., Biretta, J. A., Comoretto, G., Kaufmann, P., and Abraham, Z. 1987, *Astrophys. J. (Letters)*, **315**, 489.

Cotton, W. D., Counselman III, C. C., Geller, R. B., Shapiro, I. I., Wittels, J. J., Hinteregger, H. F., Knight, C. A., Rogers, A. E. E., Whitney, A. R., and Clark, T. A. 1979, *Astrophys. J. (Letters)*, **229**, L115.

Eckart, A., Witzel, A., Biermann, P., Pearson, T. J., Readhead, A. C. S., and Johnston, K. J. 1985, *Astrophys. J. (Letters)*, **296**, L23.

Götz, M. M. A., Alef, W., Preuss, E., and Kellermann, K. I. 1987, *Astron. Astrophys.*, in press.

Harwit, M. 1981, *Cosmic Discovery* (New York: Basic Books), p. 140.

Jones, D. L. 1986, *Astrophys. J. (Letters)*, **309**, L5.

Lawrence, C. R., Pearson, T. J., Readhead, A. C. S., and Unwin, S. C. 1986, *Astron. J.*, **91**, 494.

Maddox, J. 1984, *Nature*, **307**, 313.

Marcaide, J. M., and Shapiro, I. I. 1984, *Astrophys. J.*, **276**, 56.

Marcaide, J. M., Shapiro, I. I., Corey, B. E., Cotton, W. D., Gorenstein, M. V., Rogers, A. E. E., Romney, J. D., Schild, R. E., Bååth, L., Bartel, N., Cohen, N. L., Clark, T. A., Preston, R. A., Ratner, M. I., and Whitney, A. R. 1985, *Astron. Astrophys.*, **142**, 71.

Marscher, A. P., and Broderick, J. J. 1985, *Astrophys. J.*, **290**, 735.

Mutel, R. L., and Phillips, R. B. 1984, in *IAU Symposium 110, VLBI and Compact Radio Sources*, ed. R. Fanti, K. Kellermann, and G. Setti (Dordrecht: Reidel), p. 117.

Narlikar, J. V. 1986, in *IAU Symposium 119, Quasars*, ed. G. Swarup and V. K. Kapahi (Dordrecht: Reidel), p. 463.

Padrielli, L., Romney, J. D., Bartel, N., Fanti, R., Ficarra, A., Mantovani, F., Matveyenko, L., Nicolson, G. D., and Weiler, K. W. 1986, *Astron. Astrophys.*, **165**, 53.

Pauliny-Toth, I. I. K. 1987, in *IAU Symposium 121, Observational Evidences of Activity in Galaxies*, ed. E. Khachikian, G. Melnick, and K. Fricke (Dordrecht: Reidel), p. 295.

Pauliny-Toth, I. I. K., Porcas, R. W., Zensus, A., and Kellermann, K. I. 1984, in *IAU Symposium 110, VLBI and Compact Radio Sources*, ed. R. Fanti, K. Kellermann, and G. Setti (Dordrecht: Reidel), p. 149.

Pearson, T. J., Barthel, P. D., Lawrence, C. R., and Readhead, A. C. S. 1986, *Astrophys. J. (Letters)*, **300**, L25.

Porcas, R. W. 1981, *Nature*, **294**, 47.

Porcas, R. 1983, *Nature*, **302**, 753.

Porcas, R. W. 1986, *Mitt. Astron. Ges.*, **65**, 95.

Preuss, E., Alef, W., Whyborn, N., Wilkinson, P. N., and Kellermann, K. I. 1984, in *IAU Symposium 110, VLBI and Compact Radio Sources*, ed. R. Fanti, K. Kellermann, and G. Setti (Dordrecht: Reidel), p. 29.

Readhead, A. C. S., Pearson, T. J., and Unwin, S. C. 1984, in *IAU Symposium 110, VLBI and Compact Radio Sources*, ed. R. Fanti, K. Kellermann, and G. Setti (Dordrecht: Reidel), p. 131.

Romney, J. D., Alef, W., Pauliny-Toth, I. I. K., Preuss, E., and Kellermann, K. I. 1984, in *IAU Symposium 110, VLBI and Compact Radio Sources*, ed. R. Fanti, K. Kellermann, and G. Setti (Dordrecht: Reidel), p. 137.

Schalinski, C. J., Biermann, P., Eckart, A., Johnston, K. J., Krichbaum, T. Ph., and Witzel, A. 1987, in *IAU Symposium 121, Observational Evidences of Activity in Galaxies*, ed. E. Khachikian, G. Melnick, and K. Fricke (Dordrecht: Reidel), p. 287.

Scheuer, P. A. G. 1976, *Monthly Notices Roy. Astron. Soc.*, **177**, 1P.

Schmitt, J. H. M. M., and Reid, J. M. 1985, *Astrophys. J.*, **289**, 120.

Segal, I. E. 1986, in *IAU Symposium 119, Quasars*, ed. G. Swarup and V. K. Kapahi (Dordrecht: Reidel), p. 493.

Shaffer, D. B., Kellermann, K. I., Purcell, G. H., Pauliny-Toth, I. I. K., Preuss, E., Witzel, A., Graham, D., Schilizzi, R. T., Cohen, M. H., Moffet, A. T., Romney, J. D., and Niell, A. E. 1977, *Astrophys. J.*, **218**, 353.

Shaffer, D. B., and Marscher, A. P. 1985, *Bull. Am. Astron. Soc.*, **17**, 609.

Shone, D. L., Porcas, R. W., and Zensus, J. A. 1985, *Nature*, **314**, 603.
Stubbs, P. 1971, *New Scientist*, **50**, 254.
Unwin, S. C. 1986, in *IAU Symposium 119, Quasars*, ed. G. Swarup and V. K. Kapahi (Dordrecht: Reidel), p. 161.
Unwin, S. C., Cohen, M. H., Biretta, J. A., Pearson, T. J., Seielstad, G. A., Walker, R. C., Simon, R. S., and Linfield, R. P. 1985, *Astrophys. J.*, **289**, 109.
Waak, J. A., Spencer, J. H., Johnston, K. J., and Simon, R. S. 1985, *Astron. J.*, **90**, 1989.
Walker, R. C. 1986, *Can. J. Phys.*, **64**, 452.
Whitney, A. R., Shapiro, I. I., Rogers, A. E. E., Robertson, D. S., Knight, C. A., Clark, T. A., Goldstein, R. M., Marandino, G. E., and Vandenberg, N. R. 1971, *Science*, **173**, 225.
Wills, D., and Wills, B. J. 1976, *Astrophys. J. Suppl.*, **31**, 143.
Zensus, J. A., Hough, D. H., and Porcas, R. W. 1987, *Nature*, **325**, 36.
Zensus, J. A., and Porcas, R. W. 1986, in *IAU Symposium 119, Quasars*, ed. G. Swarup and V. K. Kapahi (Dordrecht: Reidel), p. 167.

3C 273: Archetype of Superluminal Sources

J. ANTON ZENSUS
California Institute of Technology

1. Introduction

The radio source 3C 273 has been an enigma to many astronomers ever since it was first identified with a bright quasar ($m_v = 13$) at redshift $z = 0.158$ (Schmidt 1963). Therefore, it comes hardly as a surprise that its nucleus, 3C 273B, has also remained at the focus of attention in the study of superluminal sources following the discovery of changes in its VLBI structure in 1971 (Whitney et al. 1971; Cohen et al. 1971).

It was recognized that these changes could be explained not only by relativistic expansion of a simple double structure, but equally well by intensity variations of stationary components in a more complex (e.g., triple) structure (Kellermann et al. 1974). Although still limited to model-fitting analysis, further observations did reveal that at least three dominant components were present (Rogers et al. 1974; Schilizzi et al. 1975) and that the structure was definitely not collinear (Niell et al. 1975). The source continued to increase in size during the mid-seventies (Cohen et al. 1977 and references therein; Kellermann et al. 1977; Seielstad et al. 1979; Cohen et al. 1979; Pauliny-Toth et al. 1981), but it was not until the advent of hybrid mapping (e.g., Readhead and Wilkinson 1978; Readhead et al. 1978; Readhead et al. 1979) that Pearson et al. (1981) could present "the first direct and unambiguous evidence of superluminal expansion in any radio source" with the "classic" sequence of 10.7 GHz VLBI maps of this source. They demonstrated that the changes in 3C 273B could be explained as the motion of a single knot away from the nucleus in the direction of the well-known optical and radio jet 3C 273A (Greenstein and Schmidt 1964; Hazard, Mackey, and Shimmins 1963; Perley, Fomalont, and Johnston 1980; Röser and Meisenheimer 1986).

Emission from the milliarcsecond jet has now been traced with VLBI resolution at 1.67 GHz out to a distance of ~ 130 mas from the nucleus (R. J. Davis and S. C. Unwin, in preparation; see also Porcas 1986). Maps with arcsecond resolution made at 151 MHz and 408 MHz reveal that the jet is continuous from the core to beyond the limit of the optical jet, with no evidence for a counterjet (Figure 1; Davis, Muxlow, and Conway 1985).

Since 1977, the core of 3C 273 has been monitored with VLBI at 5 GHz and 10.7 GHz; additional observations were made at other frequencies (e.g., at 0.609 GHz by Readhead et al. 1978, at 2.3 GHz by Cohen et al. 1983, and at 22 GHz by Lawrence et al. 1985). The changes between 1977 and 1982 are discussed by Unwin et al. (1985). They find superluminal motion ($v_{\rm app} \sim 6h^{-1}c$) of at least two knots which decay with half-lives of 2–3 yr as they separate from the core along the milliarcsecond jet. The jet curves through $\sim 20°$ within the first 10 mas from the nucleus and is then aligned with the arcsecond jet. First observations at 10.7 GHz with high north-south resolution suggest that this curvature is not monotonic (Biretta et al. 1985).

In this contribution I will summarize some results from further observations with high north-south resolution (Cohen et al. 1987; Zensus, Bååth, and Cohen, in preparation).

2. Results from Recent VLBI Observations

Earlier maps of 3C 273 suffered from the limited north-south resolution caused by the low declination ($\delta = 2°$) of the source. Improvement was achieved by including telescopes in the southern hemisphere in the VLBI networks used in Europe and the U.S. (Itapetinga Radio Observatory in Brazil at 10.7 GHz, and Hartebeesthoek Observatory in South Africa at 5 GHz). Figure 1 shows a hybrid map made from observations at 5 GHz in August 1985 with an array of 16 antennas (Zensus, Bååth, and Cohen, in preparation). A sequence of three hybrid maps made at 10.7 GHz from observations between 1984.1 and 1985.6 is shown in Figure 2 (Cohen et al. 1987). Components of the compact structure are labeled following the convention of Unwin et al. (1985) and Biretta et al. (1985). These observations have the following implications.

(a) Emission from the jet can be seen out to distances from the core of ~ 25 mas at 5 GHz and ~ 15 mas at 10.7 GHz. Even though the various components are not well separated, it is possible to identify certain emission regions in the maps from successive epochs. For this purpose, model fitting of a string of elliptical components was used, which gave an adequate representation of the data when five to seven components were used. The core D remained a prominent feature between 1984.12 and 1985.6 and the maps in Figure 2 were aligned with respect to its position. An extension to the west of D seen in 1984.93 became a new component C8 in the 1985.6 data. Components C5 and C7a, already noted by Unwin et al. (1985) and Biretta et al. (1985), seem to have kept their identity. The identification of components C4 and C3 is possible based on an extrapolation of their superluminal motions (Unwin

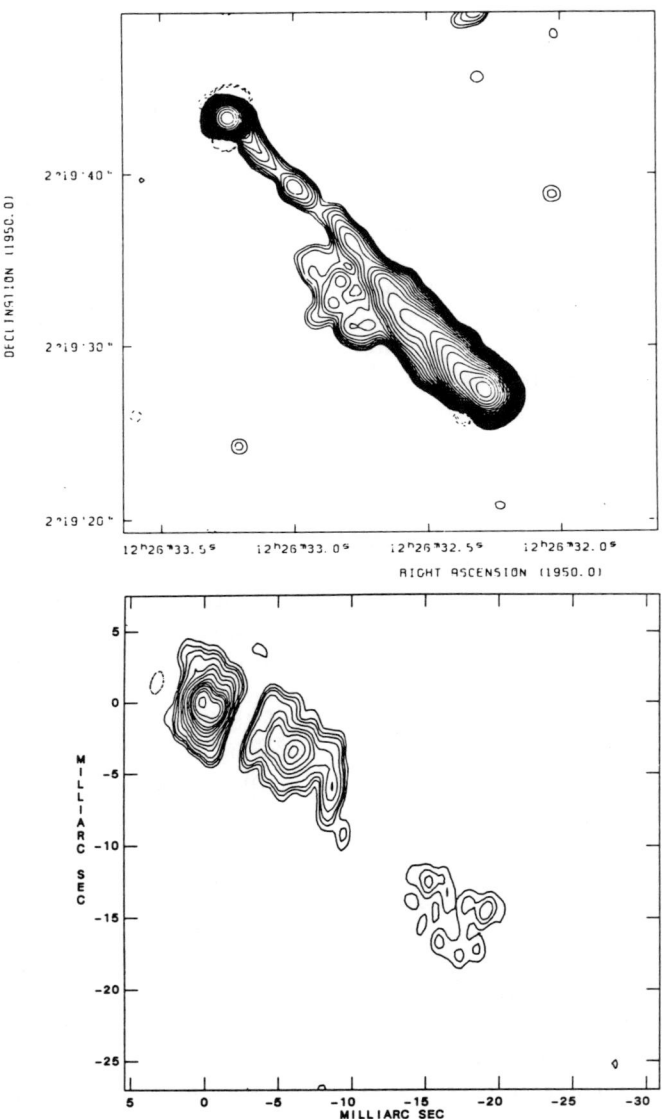

Figure 1 Core-jet structure of the quasar 3C 273. *Top:* MERLIN map at 408 MHz showing the continuous radio jet (Davis, Muxlow, and Conway 1985). *Bottom:* VLBI map of 3C 273B at 5 GHz from a 16-station experiment in August 1985 showing the non-monotonic curvature of the milliarcsecond jet (Zensus, Bååth, and Cohen, in preparation).

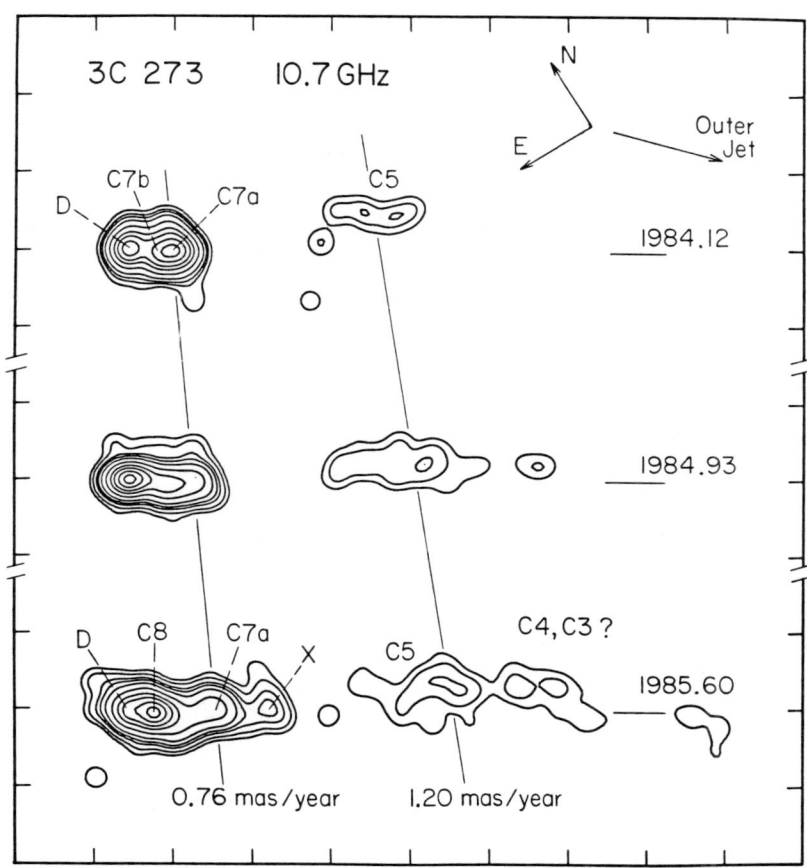

Figure 2 Hybrid maps of 3C 273B at 10.7 GHz (Cohen et al. 1987). The maps are aligned on the eastern component D, rotated counter-clockwise by 32°, and spaced vertically according to epoch. Scale is 2 mas per tick. A circular beam of FWHM=0.6 mas was used. Contour levels are 2, 4, 6, 10, 20, 35, 50, 70, 90% of the peak, with an additional contour at 1% in the maps from 1984.93 and 1985.6. The diagonal lines represent fits to the proper motion of components C5 and C7a.

et al. 1985); for epoch 1985.6 this yields separation of 9.5 ± 1.7 mas and 12.1 ± 0.3 mas from the core, coincident with the location of the region labeled "C4, C3?" in Figure 2. We suspect therefore that the remnants of C2, C3, and C4 are located 6–12 mas from the core at epoch 1985.6. (The 5 GHz map in Figure 1 shows diffuse structure at this separation from the core, too.) The identity of component X is uncertain. Its position is ill-defined and it perhaps is associated with component C6 or C7a at epoch 1984.93.

(b) The sequence of maps at 10.7 GHz reveals that the two new components C5 and C7a are also superluminal. This brings to four the number of superluminal components traced in 3C 273B so far. The lines in Figure 2 correspond to least-squares solutions to the model-fit positions of the corresponding components. The resulting proper motions are $\mu = 1.20 \pm 0.03$ mas yr^{-1} ($v_{app} = 8.0 \pm 0.2c$) for C5 and $\mu = 0.76 \pm 0.05$ mas yr^{-1} ($v_{app} = 5.1 \pm 0.3c$) for C7a. The earlier superluminal components C4 and C3 were moving with $\mu = 0.99 \pm 0.03$ mas yr^{-1} and $\mu = 0.79 \pm 0.03$ mas yr^{-1}, respectively (Unwin et al. 1985). There is furthermore some evidence that structural changes may be present at 5 GHz as far as ~ 25 mas from the core (Zensus, Bååth, and Cohen, in preparation).

(c) An important issue in the discussion of superluminal sources is the question whether the observed motions are ballistic or along fixed tracks (see also the papers at this Workshop by, e.g., Porcas, page 12, and Biretta and Cohen, page 40). Both possibilities have been advocated in the case of 3C 345. Biretta et al. (1985) argued that the curvature of 3C 273B is non-monotonic. Our new observations confirm this, in particular at 5 GHz where the "wiggling" structure is visible out to almost 30 mas with at least five distinct changes in position angle. There is evidence from our observations that some components in 3C 273B move along a fixed track (Cohen et al. 1987), but this is still tentative.

3. Summary

The determination of C5 and C7a as superluminal brings the total number of superluminal components in 3C 273B to four. Motions have been observed at projected distances of 2–8 mas (3.6–14h^{-1} pc) from the core. C5 appears to move significantly faster than the other components. The structure of the jet in 3C273 near the core has a non-monotonic curvature. It remains to be seen whether the components do indeed move along a fixed path which would suggest that the jet is dominated by pressure gradients in the ambient medium.

The study of structural changes in 3C 273 has been pursued under the leadership of Marshall Cohen in collaboration with many colleagues at various institutions. Thanks are due to Marshall Cohen and Tim Pearson for useful comments.

References

Biretta, J. A., Cohen, M. H., Hardebeck, H. E., Kaufmann, P., Abraham, Z., Perfetto, A. A., Scalise Jr., E., Schaal, R. E., and Silva, P. M. 1985, Astrophys. J. (Letters), **292**, L5.
Cohen, M. H., Cannon, W., Purcell, G. H., Shaffer, D. B., Broderick, J. J., Kellermann, K. I., and Jauncey, D. L. 1971, Astrophys. J., **170**, 207.

Cohen, M. H., Kellermann, K. I., Shaffer, D. B., Linfield, R. P., Moffet, A. T.,
   Romney, J. D., Seielstad, G. A., Pauliny-Toth, I. I. K., Preuss, E., Witzel, A.,
   Schilizzi, R. T., and Geldzahler, B. J. 1977, *Nature*, **268**, 405.
Cohen, M. H., Pearson, T. J., Readhead, A. C. S., Seielstad, G. A., Simon, R. S.,
   and Walker, R. C. 1979, *Astrophys. J.*, **231**, 293.
Cohen, M. H., Unwin, S. C., Lind, K. R., Moffet, A. T., Simon, R. S., Wilkinson,
   P. N., Spencer, R. E., Booth, R. S., Nicolson, G. D., Niell, A. E., and Young,
   L. E. 1983, *Astrophys. J.*, **272**, 383.
Cohen, M. H., Zensus, J. A., Biretta, J. A., Comoretto, G., Kaufmann, P., and
   Abraham, Z. 1987, *Astrophys. J. (Letters)*, **315**, 489.
Davis, R. J., Muxlow, T. W. B., and Conway, R. G. 1985, *Nature*, **318**, 343.
Greenstein, J. L., and Schmidt, M. 1964, *Astrophys. J.*, **140**, 1.
Hazard, C., Mackey, M. B., and Shimmins, A. J. 1963, *Nature*, **197**, 1037.
Kellermann, K. I., Clark, B. G., Shaffer, D. B., Cohen, M. H., Jauncey, D. L., Broderick, J. J., and Niell, A. E. 1974, *Astrophys. J. (Letters)*, **189**, L19.
Kellermann, K. I., Shaffer, D. B., Purcell, G. H., Pauliny-Toth, I. I. K., Preuss, E.,
   Witzel, A., Graham, D., Schilizzi, R. T., Cohen, M. H., Moffet, A. T., Romney,
   J. D., and Niell, A. E. 1977, *Astrophys. J.*, **211**, 658.
Lawrence, C. R., Readhead, A. C. S., Linfield, R. P., Payne, D. G., Preston, R. A.,
   Schilizzi, R. T., Porcas, R. W., Booth, R. S., and Burke, B. F. 1985, *Astrophys.
   J.*, **296**, 458.
Niell, A. E., Kellermann, K. I., Clark, B. G., and Shaffer, D. B. 1975, *Astrophys. J.
   (Letters)*, **197**, L109.
Pauliny-Toth, I. I. K., Preuss, E., Witzel, A., Graham, D., Kellermann, K. I., and
   Rönnäng, B. 1981, *Astron. J.*, **86**, 371.
Pearson, T. J., Unwin, S. C., Cohen, M. H., Linfield, R. P., Readhead, A. C. S.,
   Seielstad, G. A., Simon, R. S., and Walker, R. C. 1981, *Nature*, **290**, 365.
Perley, R. A., Fomalont, E. B., and Johnston, K. J. 1980, *Astron. J.*, **85**, 649.
Porcas, R. W. 1986, in *IAU Symposium 119, Quasars*, ed. G. Swarup and V. K. Kapahi
   (Dordrecht: Reidel), p. 131.
Readhead, A. C. S., Cohen, M. H., Pearson, T. J., and Wilkinson, P. N. 1978, *Nature*,
   276, 768.
Readhead, A. C. S., Pearson, T. J., Cohen, M. H., Ewing, M. S., and Moffet, A. T.
   1979, *Astrophys. J.*, **231**, 299.
Readhead, A. C. S., and Wilkinson, P. N. 1978, *Astrophys. J.*, **223**, 25.
Röser, H.-J., and Meisenheimer, K. 1986, *Astron. Astrophys.*, **154**, 15.
Rogers, A. E. E., Hinteregger, H. F., Whitney, A. R., Counselman, C. C., Shapiro,
   I. I., Wittels, J. J., Klemperer, W. K., Warnock, W. W., Clark, T. A., Hutton,
   L. K., Marandino, G. E., Rönnäng, B. O., and Rydbeck, O. E. H. 1974,
   Astrophys. J., **193**, 293.
Schilizzi, R. T., Cohen, M. H., Romney, J. D., Shaffer, D. B., Kellermann, K. I.,
   Swenson Jr., G. W., Yen, J. L., and Rinehart, R. 1975, *Astrophys. J.*, **201**, 263.
Schmidt, M. 1963, *Nature*, **197**, 1040.
Seielstad, G. A., Cohen, M. H., Linfield, R. P., Moffet, A. T., Romney, J. D., Schilizzi,
   R. T., and Shaffer, D. B. 1979, *Astrophys. J.*, **229**, 53.
Unwin, S. C., Cohen, M. H., Biretta, J. A., Pearson, T. J., Seielstad, G. A., Walker,
   R. C., Simon, R. S., and Linfield, R. P. 1985, *Astrophys. J.*, **289**, 109.
Whitney, A. R., Shapiro, I. I., Rogers, A. E. E., Robertson, D. S., Knight, C. A.,
   Clark, T. A., Goldstein, R. M., Marandino, G. E., and Vandenberg, N. R.
   1971, *Science*, **173**, 225.

Observations of 3C 273 at 3 mm Wavelength

ALAN T. MOFFET AND ANTHONY C. S. READHEAD

California Institute of Technology

There have now been five successful millimeter VLBI observing sessions (Readhead et al. 1983; Backer 1984; Backer, this Workshop, page 76). Useful visibility data on 3C 273 were obtained in April–May 1983, April 1984, and March 1985. Fringes were first detected in 1982, but technical factors made the calibration quite uncertain. In 1983 and 1985 careful comparison between 3C 84 and 3C 273 indicates that the correlated flux scale is accurate to ±15% and the relative calibration of individual 700 s integrations is better than 5%.

Rapid changes in the total and correlated flux densities show that the structure was changing significantly between 1983 and 1985, so the two datasets for 1983 and 1985 were analyzed separately (Figure 1). Few observations of 3C 273 were made in 1984, but the visibilities on the Owens Valley–Hat Creek baseline were intermediate between those seen in 1983 and 1985. Model fitting indicates that the brightness distribution was elongated in p.a. 220° at both epochs; models without this feature do not fit the data. This position angle is very close to that of the jet seen at longer wavelengths, so it appears that the elongated structure, and possibly the one-sided jet, persist over a wide range of linear scales within the innermost parsec of the source.

In early 1983, the 3 mm flux density of 3C 273 increased dramatically, perhaps due to the emergence of a new component. We would expect to see this new component in the 1983 VLBI observations. The correlated flux density at this epoch on the longest baseline was 24 Jy, comparable to that on the shorter baselines. The simplest interpretation is that the new component had expanded to 0.7 mas (1 pc) by April 1983, implying a superluminal expansion with $\gamma = 3$ (Backer 1984). The VLBI data from 1984 and 1985 do not show any direct evidence of expansion, but we cannot rule out the possibility of superluminal motion if the source is more complex than a simple double.

These results show that it is possible to map components smaller than a parsec at 3 mm. They also demonstrate the difficulties of VLBI at millimeter wavelengths—it is clear that more effort is required to ensure the technical readiness of each station, however difficult this may be. Such studies will benefit from VLBI observations of comparable

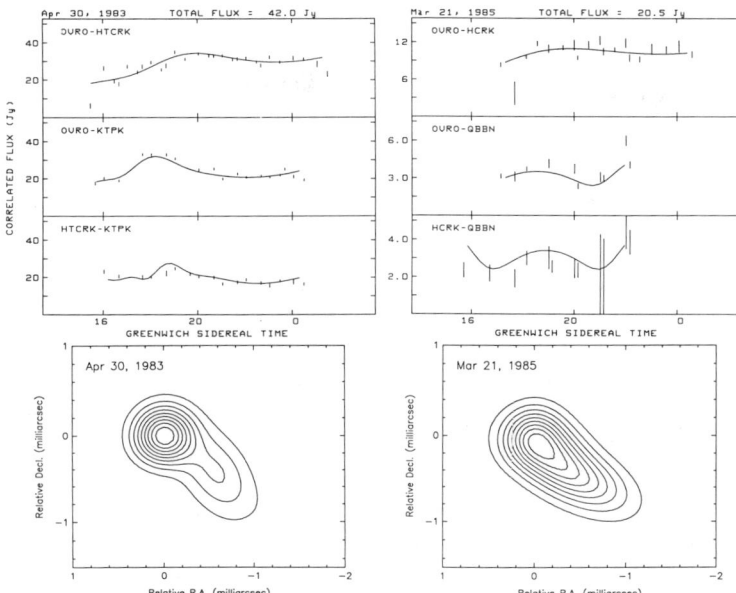

Figure 1 Visibility and models of 3C 273 at 3 mm. *Left:* April 1983 (Hat Creek, Owens Valley, Kitt Peak); resolution 0.6 × 1.2 mas, with fair sampling of the (u, v) plane. *Right:* March 1985 (Hat Creek, Owens Valley, Quabbin); resolution 0.2 × 1.2 mas, with little coverage in v. *Top:* measured visibilities and models fitted by least-squares analysis (*full lines*). *Bottom:* contour maps of the fitted models, convolved with a circular Gaussian of FWHM 0.5 mas. Assuming $H_0 = 100$ km s^{-1} Mpc^{-1} and $q_0 = 0.5$, 1 mas corresponds to 1.8 pc.

resolution at 7 mm and longer wavelengths using orbiting antennas. Without such complementary observations it will be difficult to determine the spectra of individual components and the physical conditions in the emission regions.

The team working on these observations included L. B. Bååth, D. C. Backer, P. F. Goldsmith, H. Hirobayashi, C. R. Masson, J. M. Moran, T. J. Pearson, R. L. Plambeck, C. R. Predmore, B. Rönnäng, A. E. E. Rogers, G. A. Seielstad, J. C. Webber, W. J. Welch, D. P. Woody, and M. C. H. Wright. We gratefully acknowledge assistance from the Jet Propulsion Laboratory, the Smithsonian Astrophysical Observatory, and the National Radio Astronomy Observatory.

References

Backer, D. C. 1984, in *IAU Symposium 110, VLBI and Compact Radio Sources*, ed. R. Fanti, K. Kellermann, and G. Setti (Dordrecht: Reidel), p. 31.
Readhead, A. C. S., Masson, C. R., Moffet, A. T., Pearson, T. J., Seielstad, G. A., Woody, D. P., Plambeck, R. L., Backer, D. C., Welch, W. J., Wright, M. C. H., Rogers, A. E. E., Webber, J. C., Shapiro, I. I., Moran, J. M., Goldsmith, P. F., Predmore, C. R., Bååth, L., and Rönnäng, B. 1983, *Nature*, **303**, 504.

Superluminal Motion in the Quasar 3C 279

STEPHEN C. UNWIN
California Institute of Technology

1. Introduction

The quasar 3C 279 ($z = 0.538$) was one of the first studied with VLBI techniques, and the first in which superluminal motion was recognized (e.g., Moffet et al. 1972). During the early 1970s the motion appeared to be rapid expansion of a simple double, at a speed $v_{app} \simeq 10h^{-1}c$ (Cotton et al. 1979). Observations by Pauliny-Toth et al. (1981) showed the sources becoming more complicated, with indications of a much higher separation speed. It is identified with the brightest peak of a core-jet structure seen at centimeter wavelengths by the VLA (de Pater and Perley 1983). Its optical emission is highly variable on short time scales, and is strongly polarized (Moore and Stockman 1981). Since the mid-1970s, the source structure has become more complicated, and it has received little attention from VLBI observers, apparently because of its unfavorable declination ($-5°$). Even though 3C 279 has been widely regarded as one of the prototypical superluminal sources, published observations since about 1975 are sparse, so my colleagues and I began a systematic program of VLBI monitoring at three frequencies. I present here the results of the program, which has continued for five years now.

2. Observations and Hybrid Maps

Between 1981 and 1985, 13 observations were made, at frequencies of 5, 11, and 22 GHz. All data recordings were made with the 2-MHz Mark II system, and maps were made in the usual way, using the self-calibration "hybrid" method (Cornwell and Wilkinson 1981). We were able to make hybrid maps from every observing run, and typically achieved a map dynamic range of 100 : 1.

The longest and most complete sequence of data is at 10.7 GHz, for which we have a series of five hybrid maps. Contour plots of the maps, convolved with a Gaussian beam equal in size to the dirty beam, are shown in Figure 1. The basic structure is clearly double, with a steady increase in separation of the components. The position angle is aligned to within a few degrees of the kiloparsec-scale (VLA) jet (Figure 2). Maps at the higher frequency (22 GHz) show a very similar morphology,

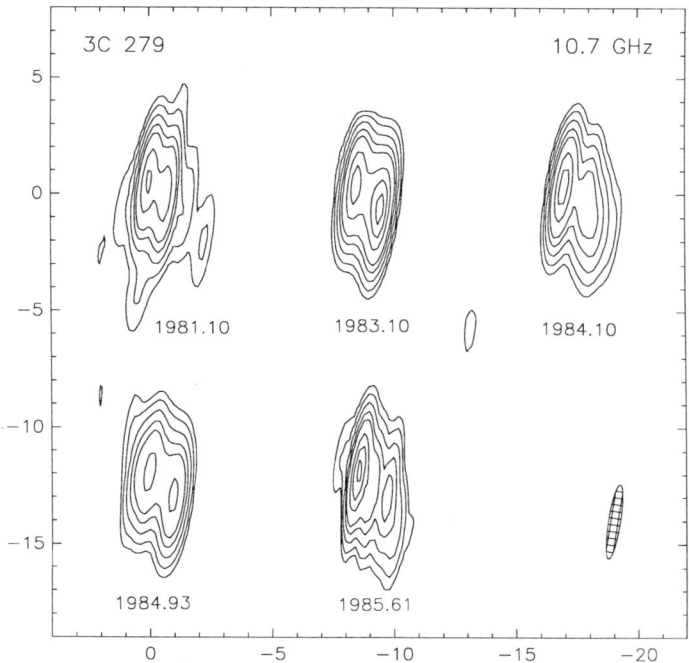

Figure 1 Hybrid maps of 3C 279 at five epochs, at a frequency of 10.7 GHz. The restoring beam (3.2 × 0.5 mas in p.a. −9°) is shown as a shaded ellipse. Contours are drawn at −0.09, 0.09, 0.18, 0.36, 0.65, 1.0, 2.0, 3.0, and 4.0 Jy per beam. The dynamic range of the 1981.10 map is only 50 : 1, much worse than the others.

but with greater resolution. They also show the "core" (northeast) component much stronger than the (southwest) "knot", indicating that it has a higher turnover frequency. We designate the components B1 and B2, respectively.

The four maps at 5 GHz are of lower resolution, and the main components are blended together by the beam. However, the source profile is entirely consistent with the double seen at higher frequencies. In addition, a third component (B3) is seen only at 5 GHz at 4.5 mas from the core, and there is further emission to the southwest on the 1984.25 map, extending out to 12 mas. This observation included baselines to Hartebeesthoek (South Africa), which improved the north-south resolution; the map shows the jet to be very narrow (i.e., unresolved by the beam) perpendicular to its width, as might be expected. A subsequent observation at the same frequency by Pillbratt (1986) shows the jet extending to about 15 mas.

By fitting a source comprising two elliptical Gaussians, we measured component separations, along with error estimates, and hence deduced

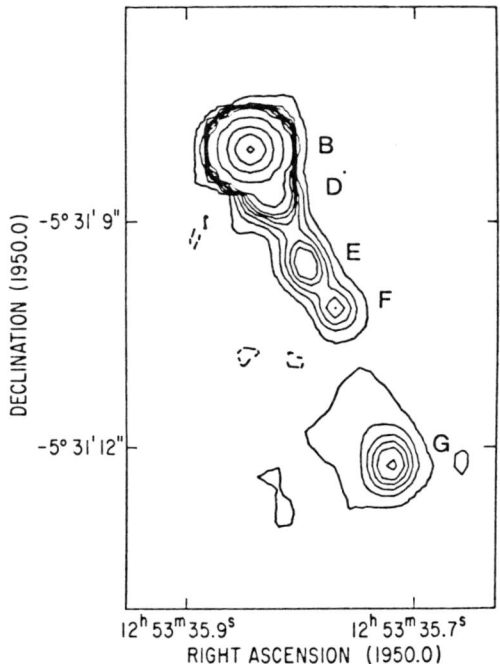

Figure 2 VLA map of 3C 279 at 6 cm with 0″.40 resolution, from de Pater and Perley (1983). A diffuse steep-spectrum component (A) is seen on the northwest side of the core in a lower-resolution map.

the proper motion. Results from the 5-GHz maps were entirely consistent with the higher-resolution data, and were included in the fit, which yields a proper motion $\mu = 0.11 \pm 0.02$ mas yr^{-1}, and an extrapolated zero-separation epoch of 1971.0 ± 1.7. This corresponds to an apparent expansion speed of $v_{\rm app} = 2.1 h^{-1} c$, which is surprisingly slow, being only one quarter of the rate found by Cotton et al. (1979) from early 1970s data. (A very similar rate results from direct measurements off the hybrid maps.) We discuss this apparent discrepancy in Section 4.

3. Component Spectra and Inverse-Compton Calculation

The model-fitting procedure also yielded flux densities for the principal components B1 and B2. For the 5 GHz data, a three-component model fits slightly better, but the third component is weak. We interpolated the measured flux densities to epoch 1983.10 where necessary, since we did not have any simultaneous observations at different frequencies. The results show, as expected, that B1 (the northeast component) is the core, with a turnover at $\simeq 13$ GHz, and that B2 is a "jet" component, or "knot", with a turnover at $\simeq 7$ GHz. B3 is only detected at 5 GHz, so its turnover must be below that frequency.

This basic picture of the spectral-index distribution is the same as seen in other superluminal sources studied at several frequencies (e.g., 3C 345, Unwin et al. 1983; 3C 273, Unwin et al. 1985), and it ties in with the VLA jet as a continuation of the VLBI-scale emission. It follows that the same relativistic twin-jet model (e.g., Königl 1981) proposed to explain many of their properties works also for 3C 279. The lack of a counterjet (on any scale) in 3C 273 is still a problem, but we note that 3C 279 does have a diffuse steep-spectrum "lobe" (component A) on the northeast side of the core on the VLA map by de Pater and Perley (1983), which could represent the end of the (unobserved) counterjet.

Combining the spectral information on individual components with their angular sizes (also obtained by the same model-fitting procedure), we can compute the expected contribution to the X-ray emission from the inverse-Compton process. Comparing this with actual measurements allows a lower limit to δ, the Doppler (blue-shift) factor. The published fluxes show no large variation (Halpern 1982; Zamorani et al. 1981), so the lack of simultaneous X-ray and radio data should not represent a problem. Using a homogeneous sphere model, and the X-ray flux from Zamorani et al. ($S_x = 5.1 \times 10^{-12}$ erg s^{-1} cm^{-2} over the 0.5–4.5 keV energy band), we derived $\delta \gtrsim 1$ for the core (B1), and $\delta \gtrsim 4$ for component B2, for the epoch 1983.10 for which we derived the radio spectra. Because of uncertainties in the measurements and model assumptions, the result for the core is entirely consistent with no relativistic motion ($\delta = 1$). However, the knot B2 must be at least mildly relativistic (Lorentz factor $\gamma \gtrsim 2$), which is consistent with the Lorentz factor required to explain the observed superluminal motion in the relativistic-jet model.

4. Discussion and Conclusions

Perhaps the only really surprising result of the present monitoring program is the slowness of the apparent superluminal expansion, making 3C 279 one of the slowest sources showing the effect. Unless components have been misidentified between epochs, the measurement has a rather small error, and is definitely inconsistent with the earlier results. While the 1970s data were all single-baseline, an examination of the original data suggests that those measurements, though less accurate, are also valid. Therefore, it seems that a change in expansion rate has occurred. This does not, however, imply a change in velocity for individual components—the early 1970s component would have moved out too far, and become too weak, to be detected in our observations, so we are tracking different components in the two series.

This difference may ultimately be resolved with polarization VLBI maps. 3C 279 is strongly polarized, and its polarization behavior is similar to BL Lac, except that the polarized component of 3C 279 varies on much shorter time scales than the total flux density (Aller, Aller, and Hughes 1985), often as short as one month. Hughes, Aller, and Aller (1985; see also Aller, Hughes, and Aller, this Workshop, page 273) propose a model for BL Lac in which the emission comes mainly from jet material compressed by shock waves. By allowing for compression of the magnetic field lines, and for varying opacity effects, the model can explain the rapid polarization changes. The fast expansion of 3C 279 seen in the 1970s could be explained in terms of an expansion event in which the changes in the total flux density and polarized flux density occurred in the same place. If the polarized and unpolarized emission evolved differently during our observing period, one might measure a speed from the (unpolarized) VLBI maps which is much less, perhaps the speed of the steady underlying flow. Polarization VLBI maps may well show much more complicated behavior than unpolarized maps, and provide a crucial test of the shock-wave model.

To summarize our new results in one sentence: mapping observations of 3C 279 have shown that this source is indeed "prototypical" of the class of superluminal objects of which it is the first member, and there appear to have been at least two expansion events, with very different proper motions.

The observations presented here were organized and conducted by the US VLBI Network Users Group. I would like to thank the people who have directly or indirectly contributed to this work, especially Marshall Cohen, John Biretta, Anton Zensus, and Mark Hodges. All the observations involved telescopes of the National Radio Astronomy Observatory, which is operated by Associated Universities Inc., under contract to the NSF. This research was supported by the NSF, via grant AST 85-09822 to the Owens Valley Radio Observatory.

References

Aller, H. D., Aller, M. F., and Hughes, P. A. 1985, *Astrophys. J.*, **298**, 296.
Cornwell, T. J., and Wilkinson, P. N. 1981, *Monthly Notices Roy. Astron. Soc.*, **196**, 1067.
Cotton, W. D., Counselman III, C. C., Geller, R. B., Shapiro, I. I., Wittels, J. J., Hinteregger, H. F., Knight, C. A., Rogers, A. E. E., Whitney, A. R., and Clark, T. A. 1979, *Astrophys. J. (Letters)*, **229**, L115.
de Pater, I., and Perley, R. A. 1983, *Astrophys. J.*, **273**, 64.
Halpern, J. P. 1982, Ph. D. thesis, Harvard University.
Hughes, P. A., Aller, H. D., and Aller, M. F. 1985, *Astrophys. J.*, **298**, 301.
Königl, A. 1981, *Astrophys. J.*, **243**, 700.
Moffet, A. T., Gubbay, J., Robertson, D. S., and Legg, A. J. 1972, in *IAU Symposium 44, External Galaxies and Quasi-Stellar Objects*, ed. D. S. Evans (Dordrecht: Reidel), p. 228.
Moore, R. L., and Stockman, H. S. 1981, *Astrophys. J.*, **243**, 60.

Pauliny-Toth, I. I. K., Preuss, E., Witzel, A., Graham, D., Kellermann, K. I., and Rönnäng, B. 1981, *Astron. J.*, **86**, 371.
Pilbratt, G. 1986, Ph. D. thesis, Chalmers University of Technology, Göteborg.
Unwin, S. C., Cohen, M. H., Biretta, J. A., Pearson, T. J., Seielstad, G. A., Walker, R. C., Simon, R. S., and Linfield, R. P. 1985, *Astrophys. J.*, **289**, 109.
Unwin, S. C., Cohen, M. H., Pearson, T. J., Seielstad, G. A., Simon, R. S., Linfield, R. P., and Walker, R. C. 1983, *Astrophys. J.*, **271**, 536.
Zamorani, G., Henry, J. P., Maccacaro, T., Tananbaum, H., Sołtan, A., Avni, Y., Liebert, J., Stocke, J., Strittmatter, P. A., Weymann, R. J., Smith, M. G., and Condon, J. J. 1981, *Astrophys. J.*, **245**, 357.

Investigations of 3C 345

JOHN A. BIRETTA
Harvard-Smithsonian Center for Astrophysics
MARSHALL H. COHEN
California Institute of Technology

1. Introduction

The radio source 3C 345 is associated with a 16^m quasar ($z = 0.595$) which has many interesting properties. It is a 2 keV X-ray source (Ku, Helfand, and Lucy 1980) and has optical fluxes and polarizations which vary on time scales of weeks (Neugebauer et al. 1979; Pollock et al. 1979; Moore and Stockman 1981). At radio frequencies it is one of the strongest compact sources, and the total flux varies on time scales of years (Bregman et al. 1986). The radio structure consists of a faint halo (Schilizzi and de Bruyn 1983), a $3''$ jet (Browne et al. 1982), and a compact region which dominates the centimeter-wavelength flux.

Superluminal motion in the compact region was first detected in 1975 (Cohen et al. 1976; Wittels et al. 1976) and since then it has been extensively monitored with VLBI. Biretta, Moore, and Cohen (1986) describe results of 2.3–89 GHz VLBI intensity monitoring from 1979.25 to 1984.11; the 10.7 GHz maps from these observations are shown in Figure 1. Component D appears to be stationary on the sky (Bartel et al. 1986) and components C4, C3, and C2 move away from it at superluminal speeds. VLBI polarization observations at 5 GHz have been made by Wardle et al. (1986). Below we will describe the physical properties of the emission regions and the kinematics of the jet. We will assume $H_0 = 100$ km s^{-1} Mpc^{-1} and $q_0 = 0.5$.

2. Synchrotron Self-Compton Model

Because of the high linear and low circular polarization seen in extragalactic radio sources, the dominant emission mechanism is thought to be incoherent synchrotron radiation from relativistic electrons in a magnetic field. We assume this to be the case, and derive the physical conditions needed to account for radio observations. Inverse Compton scattering of radio photons by relativistic electrons in the plasma will produce X-rays, so we will further require that the inverse Compton

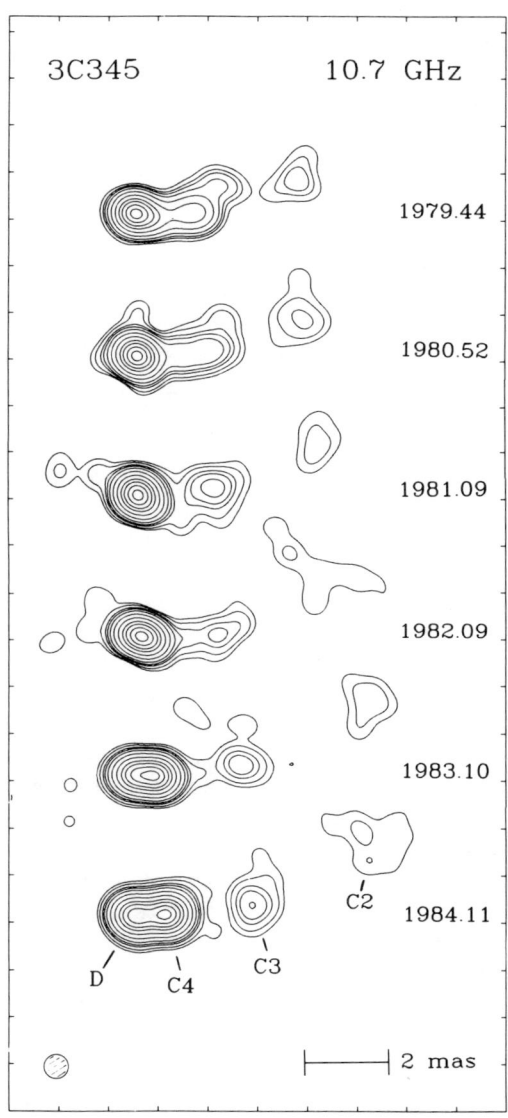

Figure 1 A sequence of maps of 3C 345 at 10.7 GHz (Biretta, Moore, and Cohen 1986). North is at the top; tick marks are 1.2 mas apart. Contours are at 0.5, 1, 2, 3, 5, 10, 20, 35, 50, 70, and 90% of the peak intensity; lower contours have been omitted in some of the maps. The FWHM of the Gaussian restoring beam is 0.6 mas (*hatched circle*).

X-rays from the plasma not exceed the observed X-ray flux. A general discussion of synchrotron models is given by Alan Marscher at this Workshop (page 280).

We will assume that the radio source consists of a conical plasma beam emanating from a "central engine", with a constant bulk relativistic Lorentz factor γ, and at a constant angle θ to the line of sight. The intrinsic half-angle of the cone is ϕ. For simplicity, we model the superluminal knots C4, C3, and C2 (which might be disturbances in the flow) as homogeneous spheres (Gould 1979; Marscher 1977) and model the "core" component D (which would represent the optically thick base of the jet) as an inhomogeneous conical jet (Königl 1981). The "core" has a magnetic field and a relativistic electron density which vary smoothly with distance r along the jet as $B(r) = B(r_0)(r/r_0)^{-m}$ and $N(E,r) = KE^{-p}r^{-n}$, respectively. The spherical components have a well-tangled magnetic field of uniform magnitude, and a power law distribution of relativistic electrons energies $N(E) = KE^{-q}$ throughout. We assume that the spheres move with the same Lorentz factor as the beam (equal fluid and pattern speeds; Lind and Blandford 1985), while for the "core" the conical pattern is fixed relative to the central engine.

Given this model, observed parameters can be used to constrain the physical properties. For the spherical knots (C4, C3, C2), these parameters are the frequency ν_m and flux density S_m of the peak in the spectrum, the optically thin spectral index α_0 ($S_\nu \propto \nu^\alpha$), and the diameter, all of which are directly observable with VLBI. For the conical "core" these are the break frequency ν_{sm}, the observed spectral indices on either side of the break (α_{s1}, α_{s2}), and the flux density $S(\nu_{sm})$ at the break (Königl 1981). Only α_{s1} is directly observable with VLBI, and the other parameters are inferred from comparison of the VLBI spectrum of component D with the total flux spectrum (Bregman et al. 1986). In addition, the observed opening angle of the VLBI jet (Figure 2) can be used to constrain the geometry: $2\phi \csc \theta = 26°$. The proper motion of the superluminal knots gives a lower limit to γ and an upper limit to θ. The total X-ray flux (Ku, Helfand, Lucy 1980) is an upper limit to the inverse Compton radiation from the synchrotron emission regions, and places a lower limit on the relativistic Doppler factor δ.

The above observables are adequate to completely constrain models for the spherical components. However, the inhomogeneous model for the "core" is not completely constrained, and several sets of values for α_0, m, and n are possible. To constrain the model we choose $n = -2$, which conserves the number of relativistic electrons in the jet, and obtain $m = -1$ (i.e., $B \propto r^{-1}$) and $\alpha_0 \equiv (1-p)/2 = -0.6$ for the intrinsic spectral index.

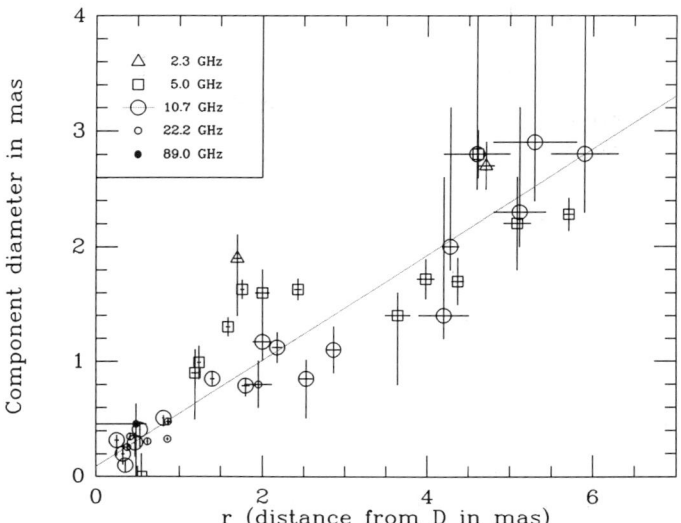

Figure 2 Component diameter as a function of distance from the "core" r for knots C4, C3, and C2. The straight line is a fit to the 11 GHz data, and corresponds to a jet opening angle of 26° (Biretta, Moore, and Cohen 1986).

Our results show that each of the emission regions has a relativistic Doppler factor $\delta = \gamma^{-1}(1 - \beta\cos\theta)^{-1}$ significantly greater than unity, with the "core" having the largest limit of $\delta \geq 10.5$. If δ is in fact near this lower limit, then much of the observed X-radiation must originate from the "core". Unwin et al. (1983) used early data and spherical models for D, C3, and C2, and also found $\delta > 1$. This constitutes evidence for bulk relativistic velocities, and is independent of the observed proper motions.

The observed proper motions and derived Doppler factors give limits on γ and θ, and we obtain $\gamma \geq 7$ and $\theta \leq 5°$. (For simplicity we have ignored the proper motion of C2, which gives $\gamma \geq 10$. Its larger proper motion might be caused by bends or twists in the path of the jet.)

Calculated pressures in the emission regions are significantly greater than those in the narrow-line region (10^{-6}–10^{-9} dyn cm^{-2}). Since the derived relativistic electron energy density is proportional to δ^{-5} and the magnetic fields are proportional to δ, a minimum in the derived pressure occurs at some δ (typically at $\delta \sim 16$). Even these minimum pressures exceed those in the narrow-line region.

We also derive magnetic field strengths in each of the emission regions. Since we determine the magnetic field for three different knots, we may directly examine how the magnetic field varies with distance along the jet. We find that $B \propto r^{-1.0\pm0.5}$ for the three knots, which is in agreement with the relation $B \propto r^{-1}$ obtained for the "core". The energy in the magnetic fields appears to be less than that in the relativistic electrons unless $\delta > 16$.

This model predicts that the distance between the "core" and knots should depend on frequency. This is because at high frequencies the "core" will have optical depth $\tau \sim 1$ at small r, while for low frequencies $\tau \sim 1$ will occur at large r. This effect is observed and is of the correct magnitude. For example, the observed distance between D and C3 is 0.30 ± 0.04 mas larger at 11 GHz than at 5 GHz, whereas the model predicts a difference of 0.27 mas.

Another prediction of the model is the size of the "core" component as a function of frequency. For a given frequency this is approximately equal to $r_{\tau \sim 1} \sin(2\phi)$ where $r_{\tau \sim 1}$ is the radius at which $\tau \sim 1$. The observed values tend to be several times larger than the prediction. For example, at 11 GHz the model predicts a "core" diameter of 0.11 mas whereas 0.30 mas is actually measured. These differences may be due to insufficient resolution, since the predicted sizes are always smaller than the effective antenna beam.

Unwin *et al.* (1985) have applied a similar model to 3C 273. For the inhomogeneous jet "core", they derive α_0 similar to ours, but obtain $m = -2.0$, so that the "core" magnetic field falls off much more rapidly than in 3C 345. Properties of the knots appear to be similar in both sources: they show evidence for $\delta > 1$, particle energies exceed the magnetic field energies unless $\delta \geq 19$, and the emission region pressures exceed those of the external medium. Data in their Table 3 are also consistent with the magnetic field varying as r^{-1} along the knots, as we find for 3C 345.

The above model is useful because of its simple assumptions, but a more complex model may be needed. For example, we model the knots as homogeneous spheres, but their internal structure may be complex. Also, we assume that the jet is straight and can be described by a single γ, but as we will see in the next section, the jet is curved and its components have differing proper motions.

3. Kinematics

The superluminal knots in 3C 345 display interesting kinematics which have not yet been seen in other sources. Component C4 was observed to change position angle from $-135°$ to $-87°$ relative to the "core" as it moved through a distance 0.3 to 0.8 mas from the "core". Components C3, C2, C1, and the 3″ jet continue the northward curvature, and are at position angle $-86°$, $-74°$, $-64°$, and $-31°$, respectively. The misalignment between the milli-arcsecond and arcsecond structure seen in

3C 345 is not uncommon (Readhead et al. 1978), and it seems likely that similar kinematics will be found in other sources. There is evidence that C3 and C2 also change position angle as they move (Biretta, Moore, and Cohen 1986) but the uncertainties are large and further observations are needed. Early evidence that C2 moved along a radial path (Cohen et al. 1983) was based on two-component models which may have been inadequate. Component C4 accelerates as it moves away from the "core" from $v_{app}/c = 1.3$ to 6.5; components C3 and C2 have $v_{app}/c = 6.0$ and 9.5, respectively.

Precession of the central engine's ejection axis has been proposed to explain both the milli-arcsecond-scale jet curvature (Begelman, Blandford, and Rees 1980) and the misalignment of arcsecond jets (Schilizzi and de Bruyn 1983). In such a model individual knots move radially away from the "core", but a group of knots will define a curved pattern on the sky. This mechanism fails in 3C 345 since it cannot explain the changing position angle and acceleration of C4. A model where internal evolution accounts for the position angle changes and precession accounts for the curvature at larger distances would also fail, since the position angle change of C4 is larger than the jet opening angle projected on the sky. Other arguments against precession are given by Biretta et al. (1985).

Bending of the jet by external pressure is another possible way of explaining the observed kinematics. In this model the knots follow a bent track, and pressure equilibrium between the jet and the external medium is required. The fact that the emission region pressures exceed those of the external medium may not be a problem, provided the emission regions are out of pressure balance with an underlying jet (e.g., because of shocks).

If the jet has a non-relativistic equation of state, its bulk velocity will remain approximately constant as it moves through the external medium. A trajectory consistent with the observed kinematics has constant $\gamma \sim 10$ with $\theta \sim 0.5°$ near the "core", and then curves away from the line of sight to $\theta \sim 4°$. However, such a trajectory seems improbable since the initial value of θ is so small. Furthermore, the 4° bend will be difficult to obtain if the Mach number increases as expected for an adiabatically expanding jet.

The alternative case where the jet has a relativistic equation of state is more promising. Such a jet would probably consist of electrons and positrons, and would accelerate as the external pressure decreases. The bending of such jets is discussed by Smith and Norman (1981). A possible trajectory would have $\theta \sim 4°$ and $\gamma \sim 3$ near the "core", and accelerate to $\gamma \sim 10$. This case has the advantages that θ need not be extremely small, and that bends of a few degrees are easy to obtain.

Other evidence for acceleration of the 3C 345 jet is given by Bregman *et al.* (1986), and is based on emission-line strengths. If there is significant acceleration, the "core" emission would require a more complex model than given above in Section 2.

A final possibility is that the jet is twisted or bent by instabilities. The kinematics in 3C 345 can be explained as a helical Kelvin-Helmholtz instability (Hardee 1986), but the jet axis must pass within 0.5° of the line of sight, and again this seems improbable.

4. Summary

We have applied a simple synchrotron self-Compton model to the "core" and superluminal knots. We find that all the emission regions show evidence for $\delta > 1$ and for pressures exceeding those in the external medium. Magnetic fields fall off approximately as r^{-1} both in the "core" and along the knots. The distance between the "core" and knots varies with frequency as predicted by the model, but predictions of the "core" size are too small.

The kinematics of the knots cannot be explained as precession, but may be explained as either bending or instability caused by the external medium, provided the emission regions are out of pressure balance with an underlying jet. A model where the jet has a relativistic equation of state and accelerates is favored, since it does not require the jet to be aligned extremely close to the line of sight. Future observations are needed to better define the internal structure of the knots and their paths on the sky.

References

Bartel, N., Herring, T. A., Ratner, M. I., Shapiro, I. I., and Corey, B. E. 1986, *Nature*, **319**, 733.
Begelman, M. C., Blandford, R. D., and Rees, M. J. 1980, *Nature*, **287**, 307.
Biretta, J. A., Cohen, M. H., Hardebeck, H. E., Kaufmann, P., Abraham, Z., Perfetto, A. A., Scalise Jr., E., Schaal, R. E., and Silva, P. M. 1985, *Astrophys. J. (Letters)*, **292**, L5.
Biretta, J. A., Moore, R. L., and Cohen, M. H. 1986, *Astrophys. J.*, **308**, 93.
Bregman, J. N., Glassgold, A. E., Huggins, P. J., Neugebauer, G., Soifer, B. T., Matthews, K., Elias, J., Webb, J., Pollock, J. T., Pica, A. J., Leacock, R. J., Smith, A. G., Aller, H. D., Aller, M. F., Hodge, P. E., Dent, W. A., Balonek, T. J., Barvainis, R. E., Roellig, T. P. L., Wiśniewski, W. Z., Rieke, G. H., Lebofsky, M. J., Wills, B. J., Wills, D., Ku, W. H.-M., Bregman, J. D., Witteborn, F. C., Lester, D. F., Impey, C. D., and Hackwell, J. A. 1986, *Astrophys. J.*, **301**, 708.
Browne, I. W. A., Clark, R. R., Moore, P. K., Muxlow, T. W. B., Wilkinson, P. N., Cohen, M. H., and Porcas, R. W. 1982, *Nature*, **299**, 788.
Cohen, M. H., Moffet, A. T., Romney, J. D., Schilizzi, R. T., Seielstad, G. A., Kellermann, K. I., Purcell, G. H., Shaffer, D. B., Pauliny-Toth, I. I. K., Preuss, E., Witzel, A., and Rinehart, R. 1976, *Astrophys. J. (Letters)*, **206**, L1.

Cohen, M. H., Unwin, S. C., Pearson, T. J., Seielstad, G. A., Simon, R. S., Linfield, R. P., and Walker, R. C. 1983, *Astrophys. J. (Letters)*, **269**, L1.
Gould, R. J. 1979, *Astron. Astrophys.*, **76**, 306.
Hardee, P. E. 1986, *Can. J. Phys.*, **64**, 484.
Königl, A. 1981, *Astrophys. J.*, **243**, 700.
Ku, W. H.-M., Helfand, D. J., and Lucy, L. B. 1980, *Nature*, **288**, 323.
Lind, K. R., and Blandford, R. D. 1985, *Astrophys. J.*, **295**, 358.
Marscher, A. P. 1977, *Astron. J.*, **82**, 781.
Moore, R. L., and Stockman, H. S. 1981, *Astrophys. J.*, **243**, 60.
Neugebauer, G., Oke, J. B., Becklin, E. E., and Matthews, K. 1979, *Astrophys. J.*, **230**, 79.
Pollock, J. T., Pica, A. J., Smith, A. G., Leacock, R. J., Edwards, P. L., and Scott, R. L. 1979, *Astron. J.*, **84**, 1658.
Readhead, A. C. S., Cohen, M. H., Pearson, T. J., and Wilkinson, P. N. 1978, *Nature*, **276**, 768.
Schilizzi, R. T., and de Bruyn, A. G. 1983, *Nature*, **303**, 26.
Smith, M. D., and Norman, C. A. 1981, *Monthly Notices Roy. Astron. Soc.*, **194**, 771.
Unwin, S. C., Cohen, M. H., Biretta, J. A., Pearson, T. J., Seielstad, G. A., Walker, R. C., Simon, R. S., and Linfield, R. P. 1985, *Astrophys. J.*, **289**, 109.
Unwin, S. C., Cohen, M. H., Pearson, T. J., Seielstad, G. A., Simon, R. S., Linfield, R. P., and Walker, R. C. 1983, *Astrophys. J.*, **271**, 536.
Wardle, J. F. C., Roberts, D. H., Potash, R. I., and Rogers, A. E. E. 1986, *Astrophys. J. (Letters)*, **304**, L1.
Wittels, J. J., Cotton, W. D., Counselman III, C. C., Shapiro, I. I., Hinteregger, H. F., Knight, C. A., Rogers, A. E. E., Whitney, A. R., Clark, T. A., Hutton, L. K., Rönnäng, B. O., Rydbeck, O. E. H., and Neill, A. E. 1976, *Astrophys. J. (Letters)*, **206**, L75.

3C 120

R. CRAIG WALKER AND JOHN M. BENSON
National Radio Astronomy Observatory
STEPHEN C. UNWIN
California Institute of Technology

1. Introduction

3C 120 was one of the four sources found in the early 1970s to have superluminal motions. With a redshift of 0.033, it is still the closest of the known superluminals and shows the highest angular expansion rates. Between 1972.5 and 1974.4, relative motion of two components at a rate of 1.5 mas yr^{-1} was observed (Seielstad et al. 1979). From 1974.4 to 1979, numerous observations were made, but the source structure changed so fast that a convincing story on the motions could not be derived. A second period of clear superluminal motion was seen in 1979 (Walker et al. 1982) at about the same rate as seen earlier. In 1981, we began more frequent observations at a single frequency in an effort to understand the rapid motions. Preliminary results from that monitoring effort are reported in Section 2 (see also Walker 1984; Walker et al. 1984). In addition to the superluminal components in the inner few parsecs, 3C 120 has radio structures on all scales out to hundreds of kiloparsecs (Soboleva et al. 1982; Balick, Heckman, and Crane 1982; Schilizzi and de Bruyn 1983; de Bruyn and Schilizzi 1984). Section 3 summarizes the results from our extensive observations of these large scale structures (Walker, Benson, and Unwin 1987).

The galaxy associated with 3C 120 is usually classified as a Seyfert although its spiral nature is not certain. Baldwin et al. (1980) confirm the presence of stars at the same redshift as the nuclear emission lines, and show that gas throughout the galaxy is ionized by the nuclear emission and has a rather disturbed velocity field. The nuclear region of 3C 120 is a powerful and variable emitter of radiation at all observed frequencies from radio to X-ray. In most respects, 3C 120 resembles objects classified as QSOs. Because it is relatively nearby and because the nuclear region is not so bright that detailed studies of the underlying galaxy are precluded, it is potentially a very important source of information on the QSO phenomenon.

2. The VLBI Observations

Prior to 1980, 3C 120 was observed as part of the superluminal source monitoring program led by Marshall Cohen at Caltech. Eventually it became clear that more frequent observations were needed, so in 1981 we began observing it every four months at 6 cm. The maps from the first eight of these observations are shown in Figure 1. Observations from more recent epochs are still being reduced.

The motions of components since 1978 are shown in Figure 2 along with profiles of the source. The symbols show the positions of the components as determined by Gaussian fits to the profiles. The error bars for the positions of the moving components represent the error in the measurement of the separation from the eastern feature that is assumed to be the core. For the core, the vertical error bars show the width of the feature. When the core width is large, it is likely that a new jet component is about to appear. In such circumstances, the separations to jet components are likely to be underestimated. This effect is especially clear in 1981.92. The solid lines connecting features show the motions of components whose apparent velocities are given in the figure caption. Components more than about 6 mas from the core are very weak and have poorly determined positions, so they are not used in the velocity determinations. Note that between 1981 and 1984 an interpretation in terms of motions *toward* the core is possible. We assume that this is an aliasing problem related to the regular observing intervals. It should be resolved when the newer data are reduced.

It is apparent from Figures 1 and 2 that 3C 120 ejects new components about once per year. The components are brightest as they emerge from the core and decay as they move away. The rate of decay varies between components. After roughly two years, a component becomes too weak to be identified reliably in our maps. The apparent velocity of each component is constant (within the measurement errors) over the range of the observations. However, the velocities of different components can be different—the velocities of the more recent components are nearly a factor of two higher than those of the components seen in 1973 and in 1979. For the years between 1980 and 1984, four new components appeared at regular intervals, all showing about the same apparent velocity. During this period, the total flux density of the source was decaying slowly and fairly smoothly. It seems to have been a time of relatively constant activity. Perhaps the components seen during this period were merely bumps on a nearly steady jet. However, something changed in 1984. At the time that a fifth component of the sequence might have been expected (late 1983), no new component appeared. Then, in mid 1984, the source flux density doubled during the

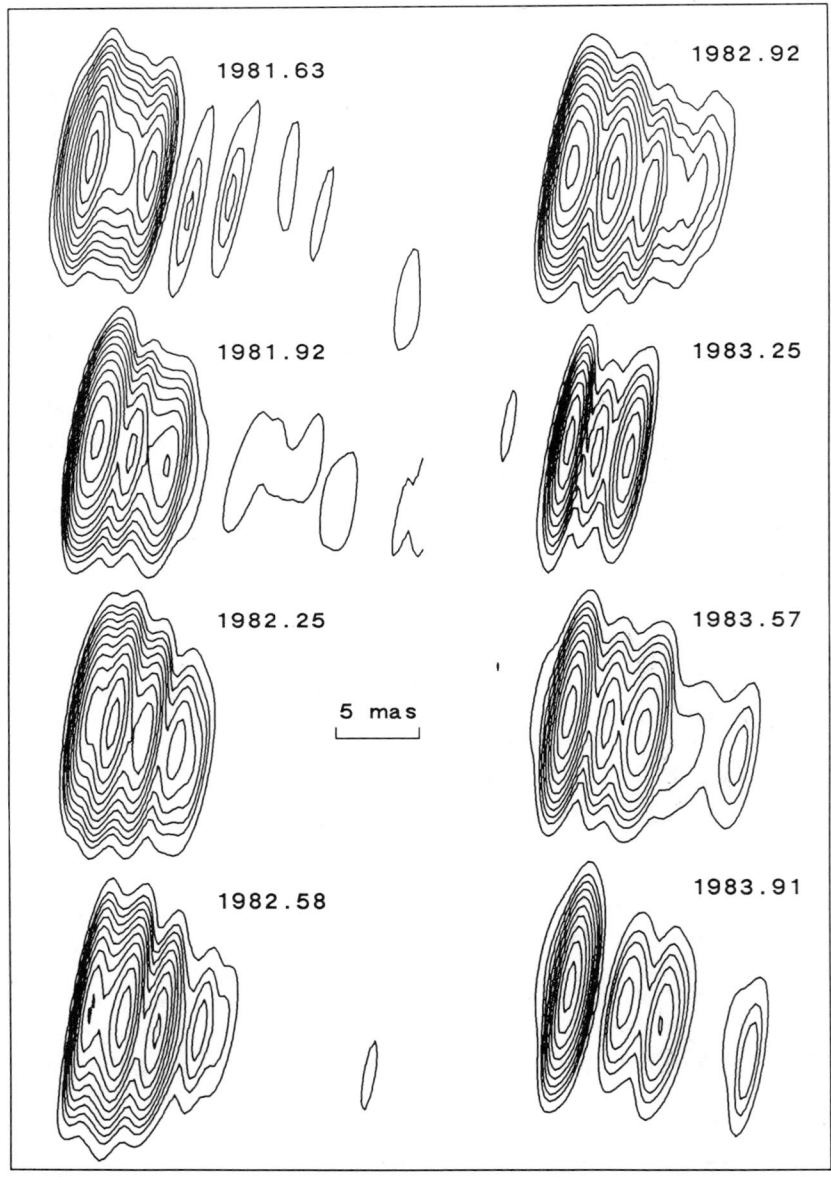

Figure 1 VLBI maps of 3C 120 from the first two years of the regular monitoring program. The contours are logarithmic with levels at −54, −39, −20, 20, 39, 54, 75, 104, 144, 200, 277, 386, 537, 746, and 1036 mJy per beam. All maps have been convolved with an elliptical beam of 1×7 mas elongated along p.a. $-10°$. The highly elongated beam on this low-declination, east-west source provides good measurements of component separations but little information on feature widths or jet straightness.

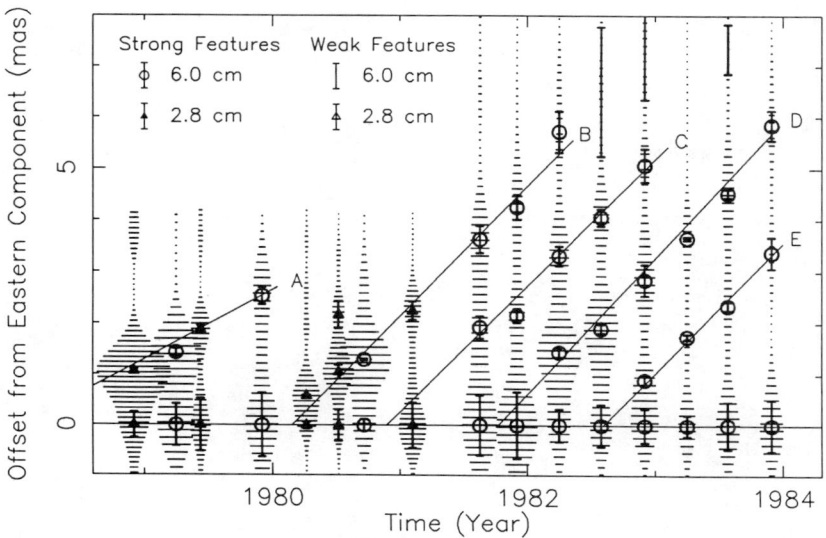

Figure 2 Profiles and positions of features of components in 3C 120 maps since 1978. The horizontal line patterns give profiles of the source taken along the jet axis. The maps used were convolved with the proper beam for each experiment and differ from those shown in Figure 1. The meanings of the symbols and error bars are described in the text. The angular rates of separation from the core for features A, B, C, D, and E are respectively 1.35, 2.53, 2.47, 2.66, and 2.54 mas yr^{-1}. To obtain $v_{\rm app}/c$, multiply the angular rate (mas yr^{-1}) by $1.53h^{-1}$.

largest flare seen in many years. VLBI data for this period exist and should be reduced soon.

3. The Large Scale Structure

To explore the structure in 3C 120 on scales larger than those seen in the VLBI monitoring observations, we have made 18 cm VLBI observations with up to 18 stations and observations with all configurations and most frequencies of the VLA. A summary of the results is given in Figure 3, which shows a sample of images from the full range of scales observed. On the smallest scales are the superluminal components observed with a resolution of about $0.5h^{-1}$ pc. A jet is seen on all scales up to about $100h^{-1}$ kpc. The total extent of the observed source is about 14 arcmin or about $400h^{-1}$ kpc in projection. On the largest scales, the structure is rather bizarre with features extending in many directions. This could be the result of observing along the axis of a fairly typical source with jets and lobes. The superluminal motions suggest that the source is oriented less than 25° from the line of sight so the deprojected size could be very large. However, the relatively large angle to the line of sight, the pronounced bending on large scales, and the likelihood of a

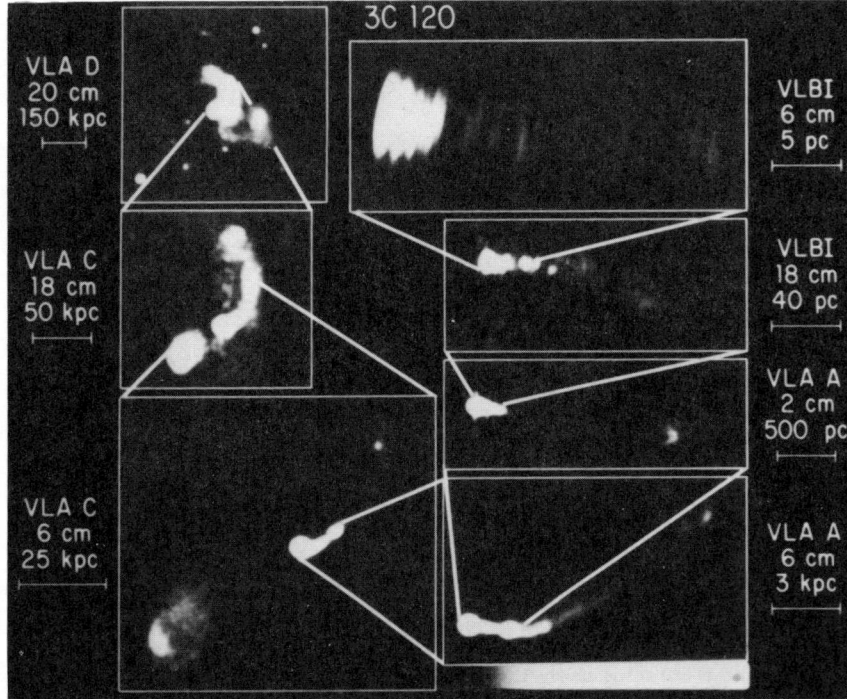

Figure 3 Images of 3C 120 on many scales. The bright components in the image at the upper right are the superluminal features. The core in every image is strongly saturated to allow the very weak, extended features to be seen.

significant transverse size allow the projection factor to be of order two or less. This keeps the source size within the range observed among other extragalactic radio sources.

A principal result of the observations of the large scale structures is the simple evolution of central brightness with width along the jet over 3.5 orders of magnitude in jet width, as shown in Figure 4. A similar simple evolution of emission per unit length with core distance over five orders of magnitude in core distance suggests that the simple behavior shown in Figure 4 extends over an even wider range, including the region in which superluminal motion is seen. However, the width measurements needed for calculations of physical parameters are not reliable over that full range. The solid line in Figure 4 represents a power law with an index of -2.40 ± 0.08, estimated by a least squares fit to the data. With the usual assumptions and caveats (see Walker, Benson, and Unwin 1987), minimum-energy calculations give the following results for the evolution of the physical parameters along the jet ($h = 1$):

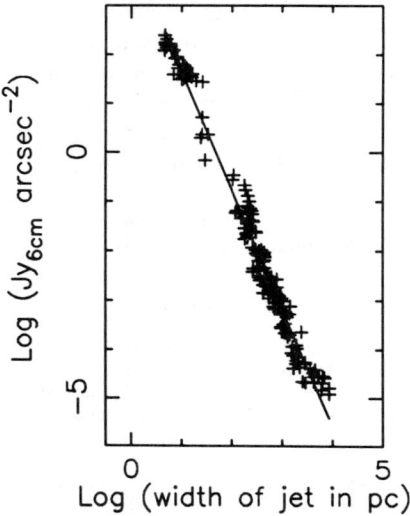

Figure 4 The flux density per square arcsecond (brightness) of the 3C 120 jet vs. the width of the jet in parsecs.

Emissivity: $\epsilon = 7.3 \times 10^{-26} r^{-3.40}$ erg s^{-1} cm^{-3} sr^{-1},
Minimum energy density: $U_{\min} = 1.5 \times 10^{-10}\ r^{-1.95}$ erg cm^{-3},
Magnetic field: $B = 4.1 \times 10^{-5} r^{-0.97}$ G,
Lifetime of observed particles: $t = 17(R_{\text{l.y.}})^{1.2}$ yr,
Pressure: $p/k = 3.7 \times 10^5 r^{-1.95}$ cm^{-3} K.

where r is the FWHM of the jet in arc seconds and $R_{\text{l.y.}}$ is the projected distance from the core in light years.

Based on these calculations, the derived energy density per unit length of the jet is constant, within the errors. The magnetic field is that expected for flux-conserving expansion of a transverse field. However, the observed field between 0.2 arcsec (the highest resolution of VLA observation) and about 1 arcmin from the core is longitudinal. The lifetimes of the particles are longer than the light travel time to the core. However, the particles observed on arcsecond scales could not have survived passage through the high magnetic fields near the core without reacceleration. The minimum pressure is such that the jet is underpressured with respect to expected conditions in the broad and narrow line regions but in balance with typical interstellar and hot coronal conditions. It is possible that the jet is in pressure balance for the full distance, but is out of the minimum energy state near the core. Note that if the jet radiation is Doppler boosted as suggested by the superluminal motions, the energy density and pressure will be overestimated.

The fact that the jet maintains its character over so many orders of magnitude in size is one of the most important results of these observations. The basic parameters of the jet are established on parsec or smaller scales and evolve relatively smoothly while passing from near the central engine to well outside the galaxy. Any models for the evolution and stability of jets must work over this full range.

The National Radio Astronomy Observatory is operated by Associated Universities, Inc., under contract to the National Science Foundation.

References

Baldwin, J. A., Carswell, R. F., Wampler, E. J., Smith, H. E., Burbidge, E. M., and Boksenberg, A. 1980, *Astrophys. J.*, **236**, 388.
Balick, B., Heckman, T. M., and Crane, P. C. 1982, *Astrophys. J.*, **254**, 483.
de Bruyn, A. G., and Schilizzi, R. T. 1984, in *IAU Symposium 110, VLBI and Compact Radio Sources*, ed. R. Fanti, K. Kellermann, and G. Setti (Dordrecht: Reidel), p. 165.
Schilizzi, R. T., and de Bruyn, A. G. 1983, *Nature*, **303**, 26.
Seielstad, G. A., Cohen, M. H., Linfield, R. P., Moffet, A. T., Romney, J. D., Schilizzi, R. T., and Shaffer, D. B. 1979, *Astrophys. J.*, **229**, 53.
Soboleva, N. S., Berlin, A. B., Nizhel'skij, N. A., and Spangenberg, E. E. 1982, *Pis'ma Astron. Zh.*, **8**, 205. English translation: *Soviet Astron. Lett.*, **8**, 108.
Walker, R. C. 1984, in *Physics of Energy Transport in Extragalactic Radio Sources*, NRAO Workshop No. 9, ed. A. H. Bridle and J. A. Eilek (Green Bank: National Radio Astronomy Observatory), p. 20.
Walker, R. C., Benson, J. M., Seielstad, G. A., and Unwin, S. C. 1984, in *IAU Symposium 110, VLBI and Compact Radio Sources*, ed. R. Fanti, K. Kellermann, and G. Setti (Dordrecht: Reidel), p. 121.
Walker, R. C., Benson, J. M., and Unwin, S. C. 1987, *Astrophys. J.*, **316**, 546.
Walker, R. C., Seielstad, G. A., Simon, R. S., Unwin, S. C., Cohen, M. H., Pearson, T. J., and Linfield, R. P. 1982, *Astrophys. J.*, **257**, 56.

Structural Variations in the Quasar 3C 454.3

IVAN I. K. PAULINY-TOTH
Max-Planck-Institut für Radioastronomie, Bonn

The quasar 3C 454.3 ($z = 0.859$) is an optically violent variable and also varies strongly at decimeter and centimeter wavelengths (e.g., Hunstead 1972; Fanti et al. 1979; Kellermann and Pauliny-Toth 1967). Even though the decimeter variations may not be intrinsic (Rickett, Coles, and Bourgois 1984; Rickett 1986), the centimeter variations are large and rapid enough to suggest the presence of bulk relativistic motion. For these reasons, we began VLBI monitoring of the source in mid-1981, when a large flux density outburst reached its peak at 2.8 cm.

The monitoring was carried out with global arrays of five to seven antennas at 2.8 cm, giving an angular resolution of 0.36 mas, with some additional observations at 18, 6, and 1.3 cm. The data were recorded with the Mark II system (Clark 1973) and correlated at the MPIfR. The data analysis involved a new fringe-fitting algorithm (Alef and Porcas 1986), calibration (Cohen et al. 1975), and standard mapping procedures (e.g., Pearson and Readhead 1984) using software developed by J. Romney and W. Alef.

The structure of 3C 454.3 on arcsecond scales (Figure 1) and milliarcsecond scales is of the core-jet type: at 18 cm we find a 20 mas jet at a position angle varying from $-65°$ near the core to $-53°$ at the end of the jet (Pauliny-Toth et al., in preparation). A comparison of several maps near epoch 1981.5 shows that the core region was then self-absorbed below 2.8 cm, while the milliarcsecond jet was transparent.

In Figure 2 a series of 2.8 cm maps of the core region are presented. They demonstrate structural variations within this region which are very different from those observed in "classical" superluminal sources such as 3C 120, 3C 273, and 3C 345 (e.g., Cohen and Unwin 1984).

At the first two epochs, near the peak of the outburst at 2.8 cm, the core region is only slightly resolved. Model-fitting suggests that it consists of an unresolved component $\lesssim 0.2$ mas (0.8 pc; we assume $H_0 = 100$ km s^{-1} Mpc^{-1}, $q_0 = 0.5$), and a larger, concentric component ~ 0.5 mas (2 pc) in size. Between the two epochs, the larger component increased in flux density by about 50%: this represents a superluminal brightening, which requires the exciting signal to have propagated at an apparent rate $\gtrsim 10c$.

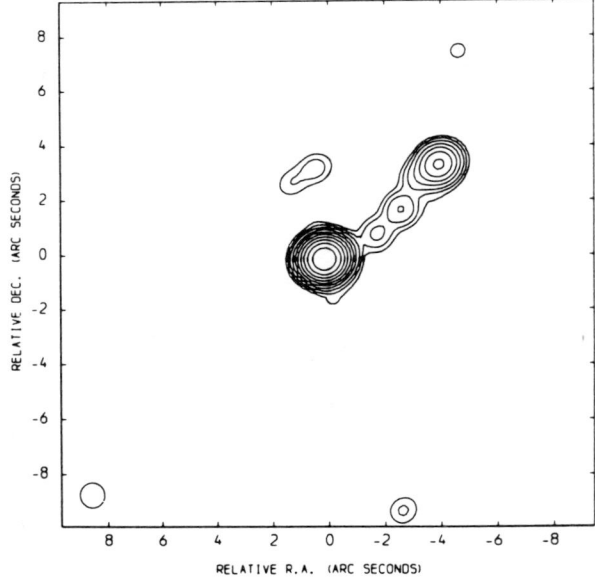

Figure 1 Arcsecond structure of 3C 454.3 at 408 MHz (from Browne et al. 1982).

The map for epoch 1983.8 shows a dramatic change in the structure: the core region became elongated in p.a. $-95°$ and three distinct features became visible (we refer to these features by number, starting from the east), with a separation between feature 1 and feature 3 of 1 mas. A fourth feature lies a further 0.4 mas to the west. If these features were ejected from a region < 0.3 mas in size at epoch 1982.1, in one direction, the apparent velocities required are $\gtrsim 4.6$, 13, and $18c$ respectively.

Following this dramatic increase in size of the core region, however, the separations of features 2 and 3 from 1 remained nearly constant: their apparent velocities were $< 1.4c$. Feature 4, on the other hand, separated from 1 at a rate of 0.35 mas yr^{-1}, or $9c$. A weaker feature even further west seems to have a similar rate of separation. After 1984.9, these weak outer features faded, and the presence of any motion became difficult to establish.

All the features, with the exception of 1, have shown this fading: feature 3 disappeared by 1984.9, and even 2 has gradually become fainter since that date. Feature 1, however, has brightened steadily since 1984.1, when the total flux density at 2.8 cm also began to increase.

The brightening of feature 1 (in our 1.3 cm maps, it is the most compact feature) and the superluminal brightening between the first two epochs suggest (a) that feature 1 is the true core (or closely associated

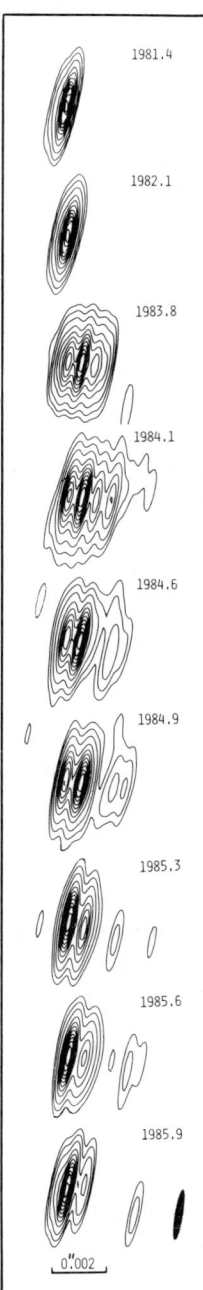

Figure 2 VLBI maps of the core region of 3C 454.3 at 2.8 cm (from Pauliny-Toth 1987). The restoring beam, of FWHM 2 × 0.3 mas, is shown in the lower right corner (*shaded ellipse*). The contour intervals are 2.5, 5, 10, 20, ..., 90% of the maximum brightness temperature, which is 1.9, 1.9, 0.38, 0.33, 0.41, 0.44, 0.63, 0.72, 0.84 × 10^{11} K, in order of epoch. The maps are aligned on the assumption that the easternmost component is the true core, and that this is the only component visible in the first two epochs.

with it) and (b) that the feature observed at the first two epochs also corresponds to this component, so the alignment of the maps in Figure 1 is correct.

The behavior of 3C 454.3 is thus much more complex than that of the classical superluminal sources: a rapid increase in size, with $v_{\rm app} \gtrsim 14c$, is followed by a period during which some features are nearly stationary, while others move with $v_{\rm app} \sim 9c$. The simple relativistic models require bulk motion with Lorentz factor $\gamma \gtrsim 13.7$ and an angle to the line of sight $\theta \lesssim 4.2°$ to explain the initial increase in size. The lack of significant motions for some features thereafter requires modification of the model.

We have considered a number of possible explanations, including a decrease in the initial bulk velocity, and a bend in the trajectory. It is also possible that the true core is invisible, because of opacity or a large inclination of its axis to the line of sight. None of these explain the complex behavior of the source satisfactorily.

The most likely explanation is that the features in our maps represent regions of enhanced emission produced by shocks. Lind and Blandford (1985) have shown that the motion of shock fronts (the brightness peaks) may be smaller than that of the underlying fluid, which determines the Doppler boosting. It appears that in 3C 454.3, the relativistic fluid (moving with $\gamma \gtrsim 14$, $\theta \lesssim 4°$) excited quasi-stationary shocks at progressively larger distances from the true core, thus causing the rapid initial increase in size. At larger distances from the core, a change in the physical conditions gives a moving shock pattern and the classical superluminal behavior of the fainter, more distant features.

Although the classical superluminal sources such as 3C 273 and 3C 345 do not show stationary emission patterns, certain other sources show a behavior similar to that of 3C 454.3. The quasar 4C 39.25, which consisted of two features with a constant separation over several years (Shaffer et al. 1977; Bååth et al. 1980), has more recently shown a new superluminal component (Marcaide et al. 1985; Shaffer et al. 1987; Shaffer, this Workshop, page 67). The central component of the radio galaxy 3C 111 has shown a behavior similar to that of 3C 454.3 after the initial increase in size: after a period during which features in a milliarcsecond jet maintained their separations from the core, all the jet features faded, leaving only the core visible (Götz et al. 1987).

Clearly, simple "unified schemes", which invoke only the alignment of the source axis to the line of sight, one value for γ, and a narrow range of the unbeamed core luminosity (Orr and Browne 1982) cannot adequately explain all the variations in the structures of superluminal radio sources. A comprehensive explanation requires more complex models, which however are unable to make definite predictions of the behavior of individual sources, and are thus less attractive and less useful.

This work was carried out in collaboration with R. W. Porcas (MPIfR), J. A. Zensus (California Institute of Technology), S. Y. Wu (Beijing Observatory), K. I. Kellermann (National Radio Astronomy Observatory), G. Nicolson (National Institute of Telecommunications, Johannesburg), and F. Mantovani (Laboratorio di Radioastronomia, Bologna).

References

Alef, W., and Porcas, R. W. 1986, *Astron. Astrophys.*, **168**, 365.
Bååth, L. B., Cotton, W. D., Counselman, C. C., Shapiro, I. I., Wittels, J. J., Hinterregger, H. F., Knight, C. A., Rogers, A. E. E., Whitney, A. R., Clark, T. A., Hutton, L. K., and Niell, A. E. 1980, *Astron. Astrophys.*, **86**, 364.
Browne, I. W. A., Orr, M. J. L., Davis, R. J., Foley, A., Muxlow, T. W. B., and Thomasson, P. 1982, *Monthly Notices Roy. Astron. Soc.*, **198**, 673.
Clark, B. G. 1973, *Proc. Inst. Elec. Electron. Engrs.*, **61**, 1242.
Cohen, M. H., Moffet, A. T., Romney, J. D., Schilizzi, R. T., Shaffer, D. B., Kellermann, K. I., Purcell, G. H., Grove, G., Swenson Jr., G. W., Yen, J. L., Pauliny-Toth, I. I. K., Preuss, E., Witzel, A., and Graham, D. 1975, *Astrophys. J.*, **201**, 249.
Cohen, M. H., and Unwin, S. C. 1984, in *IAU Symposium 110, VLBI and Compact Radio Sources*, ed. R. Fanti, K. Kellermann, and G. Setti (Dordrecht: Reidel), p. 95.
Fanti, R., Ficarra, A., Mantovani, F., Padrielli, L., and Weiler, K., 1979, *Astron. Astrophys. Suppl.*, **36**, 359.
Götz, M. M. A., Alef, W., Preuss, E., and Kellermann, K. I. 1987, *Astron. Astrophys.*, in press.
Hunstead, R. W. 1972, *Astrophys. Letters*, **12**, 193.
Kellermann, K. I., and Pauliny-Toth, I. I. K. 1967, *Nature*, **213**, 977.
Lind, K. R., and Blandford, R. D. 1985, *Astrophys. J.*, **295**, 358.
Marcaide, J. M., Bartel, N., Gorenstein, M. V., Shapiro, I. I., Corey, B. E., Rogers, A. E. E., Webber, J. C., Clark, T. A., Romney, J. D., and Preston, R. A. 1985, *Nature*, **314**, 424.
Orr, M. J. L., and Browne, I. W. A. 1982, *Monthly Notices Roy. Astron. Soc.*, **200**, 1067.
Pauliny-Toth, I. I. K. 1987, in *IAU Symposium 121, Observational Evidences of Activity in Galaxies*, ed. E. Khachikian, G. Melnick, and K. Fricke (Dordrecht: Reidel), p. 295.
Pearson, T. J., and Readhead, A. C. S. 1984, *Ann. Rev. Astron. Astrophys.*, **22**, 97.
Rickett, B. J. 1986, *Astrophys. J.*, **307**, 564.
Rickett, B. J., Coles, W. A., and Bourgois, G. 1984, *Astron. Astrophys.*, **134**, 390.
Shaffer, D. B., Kellermann, K. I., Purcell, G. H., Pauliny-Toth, I. I. K., Preuss, E., Witzel, A., Graham, D., Schilizzi, R. T., Cohen, M. H., Moffet, A. T., Romney, J. D., and Niell, A. E. 1977, *Astrophys. J.*, **218**, 353.
Shaffer, D. B., Marscher, A. P., Marcaide, J., and Romney, J. D. 1987, *Astrophys. J. (Letters)*, **314**, L1.

Superluminal Motion in BL Lac:
Evidence for Deceleration in Two Events

ROBERT L. MUTEL
Department of Physics and Astronomy, University of Iowa

ROBERT B. PHILLIPS
Haystack Observatory

The compact radio source associated with the elliptical galaxy BL Lacertae (Miller, French, and Hawley 1978) has been studied using intercontinental VLBI arrays at frequencies of 5.0 and 10.7 GHz at approximately three-month intervals since April 1980. Maps made during the first few years established that components were repeatedly ejected from a presumed stationary core component along a position angle of $\sim 190°$ with an apparent transverse speed of $v_{\mathrm{app}} \sim 4c$ (Mutel, Aller, and Phillips 1981; Phillips and Mutel 1982; Mutel and Phillips 1984). This source is a particularly interesting member of the growing list of superluminal objects for the following reasons: (a) it is still the only member of the BL Lac class for which superluminal motion has been documented; (b) the emergence of new superluminal components is clearly associated with total flux and polarization outbursts (Aller, Aller, and Hughes 1985); (c) it is one of the closest superluminal sources, at a distance of $d = 212h^{-1}$ Mpc ($z = 0.0695$), which allows extremely high linear resolution ($0.5h^{-1}$ pc at 10.6 GHz).

We have now completed a systematic analysis of 13 maps made at 10.6 GHz and seven maps made at 5.0 GHz covering the period 1980.4 to 1985.4. In this paper, we will describe a particularly interesting aspect of the last two events, namely an apparent *deceleration* of each of the components as they reach a projected distance of ~ 1.5–$2.0h^{-1}$ pc from the core, accompanied by a large increase in angular size. A more complete description of the maps from all epochs will appear elsewhere (R. B. Phillips, R. L. Mutel, and B. R. Bucciferro, in preparation).

Figure 1 shows a panorama of the maps from all 13 epochs at 10.6 GHz. The maps were all made in the standard manner using between four and seven telescopes of the U.S. VLBI Network and the Mark-II recording scheme. The measured visibilities were edited, calibrated, and mapped using the Caltech VLBI mapping algorithms, including self-calibration of the complex telescope gains (e.g., Pearson and Readhead 1984). The dynamic range (defined as the ratio of peak flux

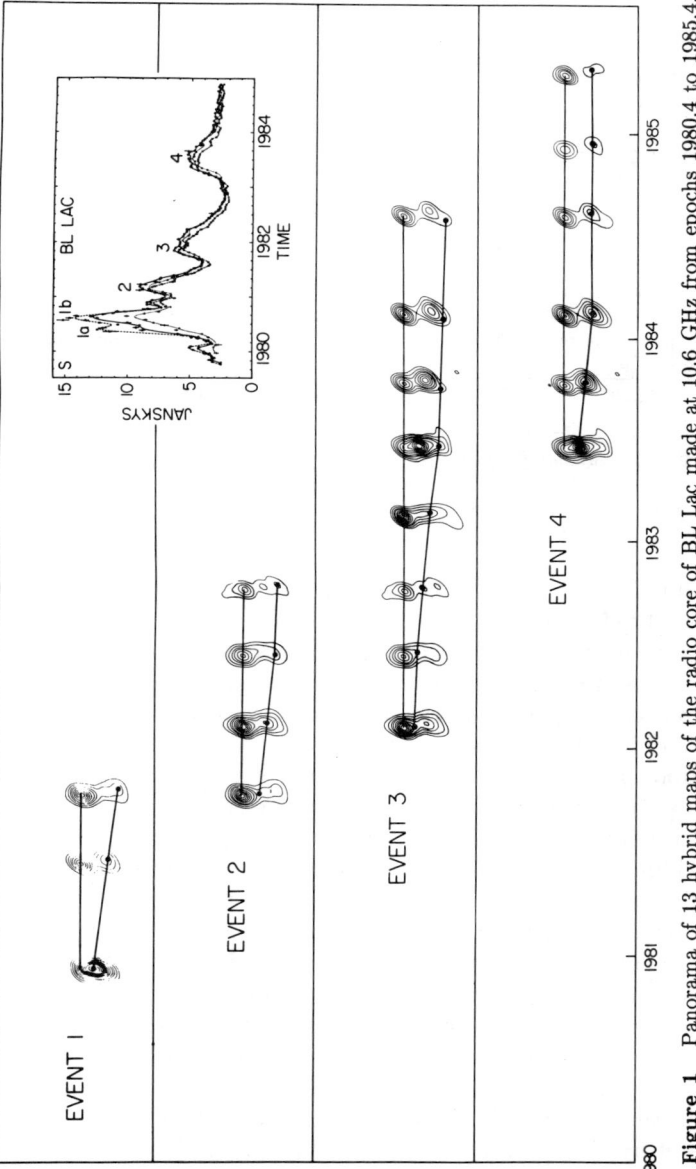

Figure 1 Panorama of 13 hybrid maps of the radio core of BL Lac made at 10.6 GHz from epochs 1980.4 to 1985.4. Each of the four emerging superluminal components can be identified with a flux outburst as shown in the inset at upper right. Contour levels are 100, 200, 300, ..., 2000 mJy per beam for all epochs.

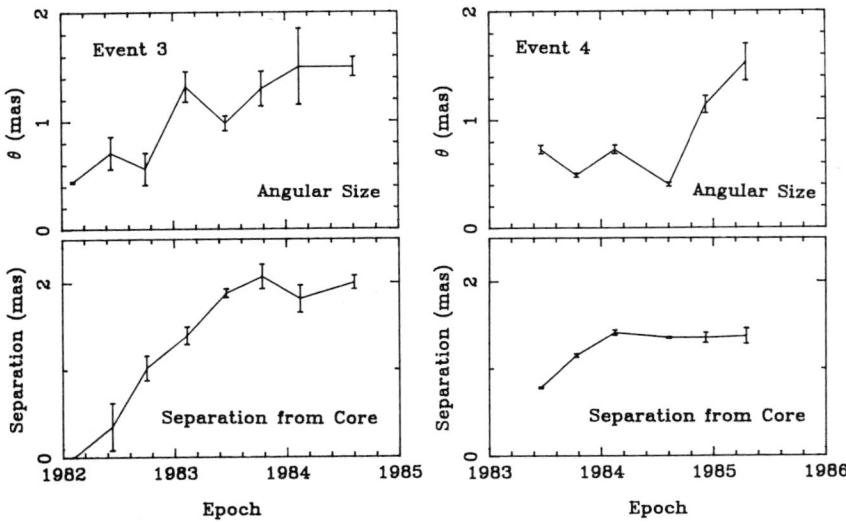

Figure 2 Component sizes and separations from the core for Events 3 and 4 as a function of epoch. All points are derived from Gaussian model components. Error bars are ±1σ uncertainties (see text for discussion of formal errors).

per beam to the lowest "believable" feature) of the final maps varied from 15:1 to 50:1, with an average value of 30:1.

We have identified four distinct events that we will (rather prosaically) refer to as Events 1, 2, 3, and 4. The components from the first two events appear to move nearly uniformly for all epochs in which they are detectable. However, for the latter two events it is clear that the components change their apparent speed after reaching a distance of ~ 1.5 mas from the core. This is the first measurement of *deceleration* in a superluminal source. The only other deviation from uniform speed of individual components in superluminal sources is the *acceleration* of component C4 in the quasar 3C 345 (Biretta, Moore, and Cohen 1986).

In order to quantify the measurement of velocities, fluxes, and sizes of components, we have fit models consisting of elliptical Gaussian components to the calibrated visibility data. Depending on the epoch, either two or three components were necessary for a satisfactory fit. We estimated the standard errors of the parameters of each model component by determining the variation required to produce a $1/N$ fractional increase in the reduced χ^2 of the fit, where N is the effective number of degrees of freedom of the data. Unfortunately, N is not easy to determine, since the visibility points are not independent of each other and the errors do not have Gaussian statistics. To estimate N, we followed the procedure described by Biretta, Moore, and Cohen (1986). The

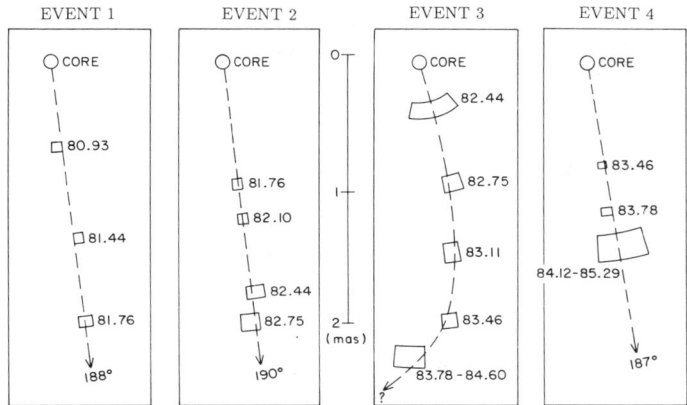

Figure 3 Trajectories of model components for all four events. The boxes represent 1σ formal errors. For Events 3 and 4, the furthest boxes are the combined positions and errors for several epochs, after the apparent deceleration of the components.

resulting errors are internally consistent and well defined, but may be too small if the antenna gains vary more quickly than the assumed time scale of 1 h.

Using these model parameters, we have plotted in Figure 2 the separation and angular size for the components of Events 3 and 4 as a function of epoch. For both events, there is an initial period of approximately uniform motion at a speed $v_{\rm app} = 4.8 \pm 0.2c$ (Event 3) and $3.1 \pm 0.2c$ (Event 4). The components become nearly stationary at projected distances of 1.9 pc (Event 3) and 1.4 pc (Event 4) from the core. They both show large increases in angular size, from less than 0.5 mas to 1.5 mas, as they evolve, particularly during deceleration.

We have also examined the component trajectories for all four events by plotting the radial coordinates (relative to the core) of model components at each epoch (Figure 3). In all four events components were ejected at a very similar position angle (188° ± 2°). This implies that a rather narrow nozzle is attached to the central engine in BL Lac. For Events 1, 2, and 4, the motion appears to be rectilinear, while for Event 3, there is evidence for a curved trajectory. Most components of superluminal sources have rectilinear trajectories. However, extended structure is often highly curved at larger distances from the core (Readhead et al. 1979). In the case of BL Lac, the extended structure is relatively small ($\sim 40h^{-1}$ kpc) and amorphous, with a very weak extension along p.a. −45° (Antonucci 1986).

Canonical Beams and Weak Shocks

Except for the curious kinematics of decelerating components, BL Lac appears to possess many of the features of garden variety superluminal sources: one-sided core-jet morphology, plausible expansion velocities ($3 \lesssim v_{\rm app}/c \lesssim 5$), and components which expand and fade as they move away from the core. As new components emerge, the fractional linear polarization increases markedly. The polarization position angle is nearly parallel to the trajectory (this is especially clear for Events 3 and 4). These data have been interpreted in terms of a weak shock model (Mach number $M \lesssim 2$) in which the ejecta are associated with compressions that propagate from an optically thick to an optically thin region in a relativistic flow (Hughes, Aller, and Aller 1985). The ejecta are driven by a hot "piston" which compresses an initially random ambient magnetic field.

The Doppler factor δ of the superluminal components can be estimated in two ways. First, we may obtain a lower limit by comparing the observed soft X-ray flux to that expected from inverse Compton radiation (Burbidge, Jones, and O'Dell 1974; Cohen 1985). For BL Lac, the observed flux between 0.5 and 4.5 keV was 0.82 μJy in June 1980 (Schwartz and Ku 1983). Unfortunately, this was during a time of very rapid flux variability (Figure 1, *inset*). The closest VLBI maps, made in April and October 1980 at 5 GHz (Mutel, Aller, and Phillips 1981), show a nearly unresolved source, making estimates of component parameters difficult. Nevertheless, if we use the best estimates of component size, flux, and spectral index during that epoch, we obtain $\delta_{\rm min} > 2.1$. A second estimate for δ is available from the shock model of Hughes, Aller, and Aller (1985). They obtained a lower limit $\delta_{\rm min} > 2$ based on the radiation spectrum. By assuming that the magnetic field varies as $B \propto s^{-1.5}$, where s is the channel width, they obtained a component Doppler factor of $\delta_{\rm shock} = 3.5$.

Using this value of δ and the apparent speed $v_{\rm app}$ of ejected components with the canonical beaming geometry, we can easily solve for the Lorentz factor γ and the angle θ between the component trajectories and the observer's line of sight (e.g., Cohen and Unwin 1984). The results are $10° \lesssim \theta \lesssim 30°$ and $3 \lesssim \gamma \lesssim 5$. *Caveat:* The Doppler factor is not well constrained by either the inverse-Compton or the shock model calculation, and could be significantly larger. This has the effect of increasing the Lorentz factor and decreasing the line of sight angle.

Deceleration: Collision with an Interstellar Wall?

BL Lac is the only well studied superluminal source in the BL Lac class. It is also the only example of a superluminal with decelerating components. Hence, it is tempting to associate these properties in a causal way. Specifically, let us assume that the deceleration is *not* a result of some intrinsic property of emission mechanism, but rather is caused by a change in the environment in the region of deceleration. Using $\theta \sim 20°$, the deprojected distance at which deceleration is noticeable is $l \sim 5$ pc. One interpretation of the featureless optical spectra of BL Lac objects is that the region surrounding the central engine has been evacuated, perhaps by accretion onto the central engine (Strittmatter 1985). The optical continuum is beamed radiation from the central engine itself. On the other hand, the central engines of quasars are surrounded by a dense ($n_e \sim 10^{10}$ cm^{-3}) "broad-line" region with a dimension ~ 1 pc. In both quasars and BL Lac objects, there will be a transition to an interstellar medium of the parent galaxy.

The sudden deceleration of the BL Lac radio components may be caused by the change from a hot, tenuous environment to a denser interstellar medium a few parsecs from the central engine. As the radio jet encounters the denser medium, it quickly dissipates. Conversely, radio-loud quasar jets must plow through a much denser broad-line region before encountering the rather less dense ($10^2 \lesssim n_e \lesssim 10^5$ cm^{-3}) narrow-line region (Filippenko 1986). The acceleration of component C4 in the quasar 3C 345 (Biretta, Moore, and Cohen 1986) at a projected distance of $1.5h^{-1}$ pc from the core may be related to a *decrease* in density in the jet environment.

In this simple-minded view, radio jets in BL Lac objects can be seen as rather emasculated versions of luminous quasar jets—they propagate without difficulty in the tenuous environment of the central engine, but are quickly dissipated upon entering the parent galaxy's interstellar medium. Both the relatively low radio luminosity ($L_{\rm BL\,Lac}/L_{\rm quasar} \sim 10^{-2}$) and the lack of significant kiloparsec-scale structure support this view. It will be interesting to see if other BL Lac objects show evidence for decelerating superluminal components.

We are grateful to the U.S. VLB Network for continued support. We thank H. Aller for unpublished data and useful discussions. This research was supported through National Science Foundation grants AST 82-16890 and AST 84-20994 to the University of Iowa.

References

Aller, H. D., Aller, M. F., and Hughes, P. A. 1985, *Astrophys. J.*, **298**, 296.
Antonucci, R. R. J. 1986, *Astrophys. J.*, **304**, 634.
Biretta, J. A., Moore, R. L., and Cohen, M. H. 1986, *Astrophys. J.*, **308**, 93.
Burbidge, G. R., Jones, T. W., and O'Dell, S. L. 1974, *Astrophys. J.*, **193**, 43.
Cohen, M. H. 1985, in *Extragalactic Energetic Sources*, ed. V. K. Kapahi (Bangalore: Indian Academy of Sciences), p. 1.
Cohen, M. H., and Unwin, S. C. 1984, in *IAU Symposium 110, VLBI and Compact Radio Sources*, ed. R. Fanti, K. Kellermann, and G. Setti (Dordrecht: Reidel), p. 95.
Filippenko, A. V. 1986, in *IAU Symposium 119, Quasars*, ed. G. Swarup and V. K. Kapahi (Dordrecht: Reidel), p. 289.

Hughes, P. A., Aller, H. D., and Aller, M. F. 1985, *Astrophys. J.*, **298**, 301.
Miller, J. S., French, H. B., and Hawley, S. A. 1978, in *Pittsburgh Conference on BL Lac Objects*, ed. A. M. Wolfe (Pittsburgh: University of Pittsburgh), p. 176.
Mutel, R. L., Aller, H. D., and Phillips, R. B. 1981 *Nature*, **294**, 236.
Mutel, R. L., and Phillips, R. B. 1984, in *IAU Symposium 110, VLBI and Compact Radio Sources*, ed. R. Fanti, K. Kellermann, and G. Setti (Dordrecht: Reidel), p. 117.
Pearson, T. J., and Readhead, A. C. S. 1984, *Ann. Rev. Astron. Astrophys.*, **22**, 97.
Phillips, R. B., and Mutel, R. L. 1982, *Astrophys. J. (Letters)*, **257**, L19.
Readhead, A. C. S., Pearson, T. J., Cohen, M. H., Ewing, M. S., and Moffet, A. T. 1979, *Astrophys. J.*, **231**, 299.
Schwartz, D. A., and Ku, W. H.-M. 1983, *Astrophys. J.*, **266**, 459.
Strittmatter, P. A. 1985, in *Extragalactic Energetic Sources*, ed. V. K. Kapahi (Bangalore: Indian Academy of Sciences), p. 13.

4C 39.25: Superluminal Motion Between Stationary Components

DAVID B. SHAFFER

Interferometrics Inc.

ALAN P. MARSCHER

Department of Astronomy, Boston University

The first VLBI observations that defined the compact structure of the quasar 4C 39.25 (0923+392; $z = 0.699$, 1 mas = $13h^{-1}$ light years for $H_0 = 100h$ km s^{-1} Mpc^{-1}, $q_0 = 0.5$) were made by DBS in 1972 while he was one of Marshall Cohen's graduate students. (That makes him an "old timer" at this meeting.) Those observations, with only three antennas (NRAO, HRAS, and OVRO) at 2.8 cm, could be fitted very accurately with a two-component model. Repetitions of those early observations showed that 4C 39.25, unlike other now-famous sources that we observed at the same time, did not have rapid changes in structure. Observations at other wavelengths (6, 3.8, and 2 cm) confirmed the simple double structure. The component separation at all frequencies was 2.0 mas.

During the 1970s, DBS observed this source occasionally, to keep an eye on it. Other than slow variations in component strength, not much seemed to change in 4C 39.25. In particular, there was no apparent component motion, to a limit of about $0.25c$. We called this a subluminal source for a long time.

In 1979, DBS joined the VLBI geodesy group at NASA's Goddard Space Flight Center, where his job included worrying about the effect that source structure might have on geodetic observations. Thus, he developed a "CORTEL"-like (Cornwell and Wilkinson 1981) set of mapping programs. For NASA geodesy, 4C 39.25 has been observed many times with various antenna networks at 13 and 3.6 cm, and maps have been made from some of the 3.6 cm data. The first such maps of 4C 39.25, made in 1979, showed an apparent *contraction* in the source! Later maps showed continued apparent contraction. This anomalous behavior (sources are supposed to expand!) led the authors of this paper to start observing 4C 39.25 more vigorously at 2.8 cm, using modern mapping techniques. The 3.6 cm maps kept coming along, too, essentially for free as part of the NASA observations.

Figure 1 shows the changes we have seen in the structure of 4C 39.25 in the last seven years: a new component seems to have been ejected from (the region of) the western component of the double source that had been observed prior to 1979. The angular velocity of this component, since it became truly distinct in the 1982 June 2.8 cm experiment, is 0.16 ± 0.02 mas yr^{-1}. This corresponds to an apparent velocity of $3.5h^{-1}c$ for the central component. The apparent epoch of zero separation is about 1979.5, which coincides with a plateau in the long-term decline of the flux density of the source (Aller et al. 1985), perhaps indicative of a weak outburst.

The structural variations in 4C 39.25 since 1979 are quite unlike those in any other source of which we are aware—there is a stationary outer component or lobe in the superluminal source 3C 395, but it is much farther from the core (Waak et al. 1985; Simon et al., this Workshop, page 72). What can we make of 4C 39.25, with its strange behavior? We will speculate only briefly here; a more complete discussion is given by Shaffer et al. (1987).

Most simply, we are inclined to identify the western component of the source (component c in Figure 1) as the core from which a new component (b) has been ejected. In that case, though, the stationary component a is unlike components observed in other sources such as 3C 120, 3C 273, or 3C 345, where multiple components are seen to move away from the core at comparable speeds. In most superluminal sources, also, the core can usually be identified as the brightest or most compact component in the source (at least at high frequencies). However, in 4C 39.25, as shown in Figure 1 and emphasized in the 1.3 cm map in Figure 2, the western component does not look like these other cores, at least not since 1983. The central, assumed moving, component is the brightest and most compact. This is not a completely unknown situation, since new components have been seen to dominate a source on occasion, but not generally for so long as appears to be the case in 4C 39.25. There is no other potential core feature in the 1.3 cm map, nor does there appear to be a higher-frequency component in the millimeter spectrum.

On the arcsecond scale, 4C 39.25 shows a two-sided jet, with the brightest component on the east (Browne et al. 1982). In almost all other superluminal sources (3C 120 or 3C 273, for example), components move out towards the brightest outer features. If 4C 39.25 is similar, its extended structure is consistent with the western component being the core. The source clearly has had two-sided ejection in the past, but we cannot tell if we are now seeing intrinsically one-sided activity or just the beamed side of a two-sided jet.

We cannot even tell for sure which component or components are

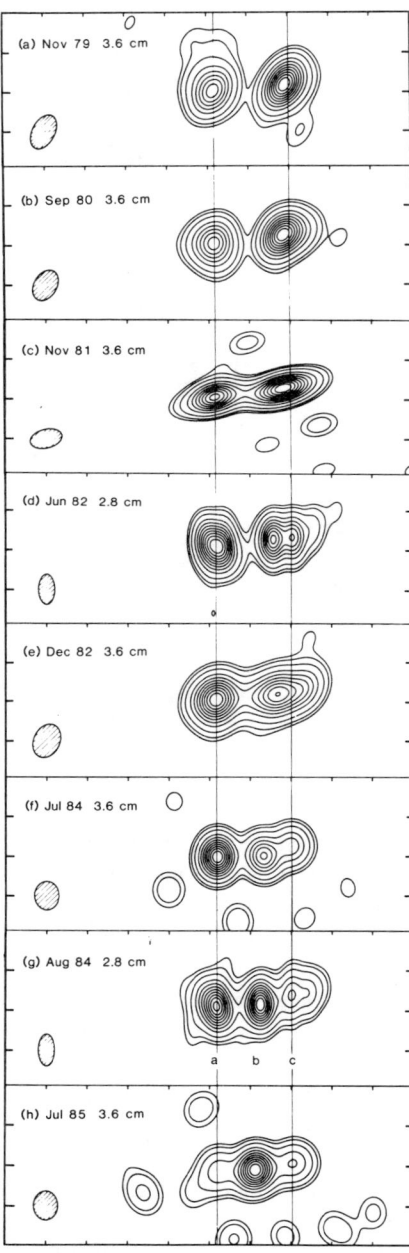

Figure 1 VLBI maps of 4C 39.25, for the given epochs and wavelengths. The restoring beam is shown in the lower left corner of each map. East is to the left and north is up. Tick marks are 1 mas apart. The vertical lines indicate the stationary components a and c. Contour levels are 3, 5, 10, 20, ..., 90% of the peak brightness.

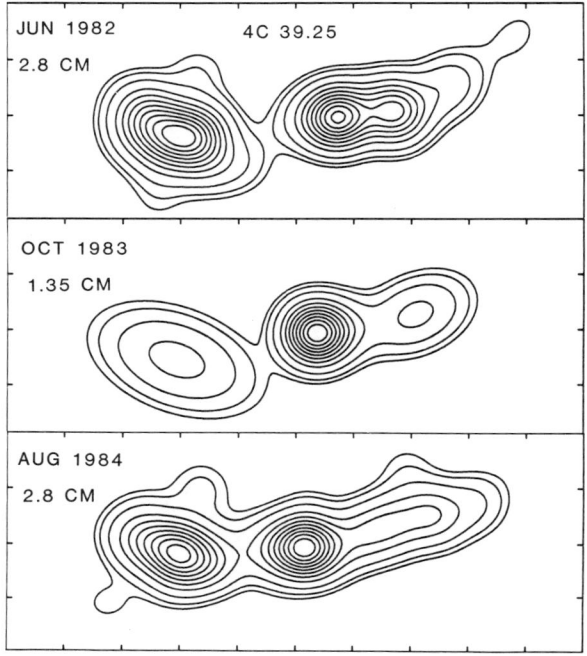

Figure 2 VLBI maps of 4C 39.25 at 2.8 cm (*top and bottom*) and 1.3 cm (*center*). Orientation and contours as in Figure 1. The same restoring beam has been used for all three maps, to emphasize the relative size and brightness of the central component. Tick marks are 0.5 mas apart.

really moving. We do not have any strong limits on the overall motion of the source, except that if its position had changed by more than one or two milliarcseconds over the last five years, we would have detected the motion in the analysis of the NASA geodesy observations. It is possible that the two outer components are moving with a constant velocity, or that all three components are moving but with different velocities.

Both components a and c could be stationary features in a relativistic jet. Depending on its environment, such a jet can alternately expand and contract, showing compact hot spots where the contraction reaches a local maximum. Component b would then be a shock or instability caught up in the flow.

It is possible, although we feel it is unlikely, that components a and c are gravitational lens images of the same compact core. In this case, both "cores" should be seen to eject the same component, although not necessarily at the same time because of path length differences for the two images. There is no evidence for lensing in the arcsecond structure of the source (Browne *et al.* 1982).

Which, if any, of these models is correct may be revealed as we follow the evolution of 4C 39.25. Component b is on a collision course with component a. The actual encounter, due in 1991 at the present pace, may show one of several possible results. If a is a stationary, compact lobe being fed by the jet or caused by an obstruction in the jet flow, b should stop when it gets there, and the region of component a will probably brighten. If a is a contraction in an undulating jet, b should pass through the region at constant velocity, although there may be a brightening as the b material is compressed. If the components are just images, they should show no interaction at all. We are continuing our observations of this fascinating object, waiting to see what happens.

References

Aller, H. D., Aller, M. F., Latimer, G. E., and Hodge, P. E. 1985, *Astrophys. J. Suppl.*, **59**, 513.
Browne, I. W. A., Clark, R. R., Moore, P. K., Muxlow, T. W. B., Wilkinson, P. N., Cohen, M. H., and Porcas, R. W. 1982, *Nature*, **299**, 788.
Cornwell, T. J., and Wilkinson, P. N. 1981, *Monthly Notices Roy. Astron. Soc.*, **196**, 1067.
Shaffer, D. B., Marscher, A. P., Marcaide, J., and Romney, J. D. 1987, *Astrophys. J. (Letters)*, **314**, L1.
Waak, J. A., Spencer, J. H., Johnston, K. J., and Simon, R. S. 1985, *Astron. J.*, **90**, 1989.

Superluminal Motion Towards a Stationary Component in Quasar 3C 395

R. S. SIMON, K. J. JOHNSTON, J. HALL,
J. H. SPENCER, AND J. A. WAAK

E. O. Hulburt Center for Space Research, Naval Research Laboratory

1. Introduction

We have recently confirmed that the quasar 3C 395 (1901+319) is a superluminal radio source, based on a third-epoch VLBI image. The compact structure in 3C 395 is dominated by three components, one of which is moving away from the core towards a stationary component with a velocity of $v_{app} = 0.64 \pm 0.1$ mas yr^{-1}. At the distance of 3C 395 ($z = 0.635$; Phillips and Mutel 1980), this corresponds to an apparent velocity of $\sim 13h^{-1}c$ (assuming $H_0 = 100h$ km s^{-1} Mpc^{-1}, $q_0 = 0.5$).

3C 395 is thus unique among the dozen or so superluminals, in that it has a superluminal component moving between two relatively stationary components (although Shaffer and Marscher report a similar phenomenon in 4C 39.25 at this Workshop, page 67). If the (highly variable) core in 3C 395 were to fade out, the other two components would be seen to be contracting superluminally. It is possible to interpret the structure we observe in terms of a core-jet model with a one-sided radio jet accounting for the milliarcsecond radio emission.

3C 395 is a compact radio quasar with a visual brightness of 17^m (Véron, Véron, and Witzel 1974). It has been observed at 6 cm and 2 cm using the Very Large Array in its "snapshot" mode (Perley, Fomalont, and Johnston 1980; Perley 1982; van Breugel et al. (1984); Pearson, Perley, and Readhead 1985). These observations showed that 3C 395 consists of two components on the arcsecond scale: a bright, unresolved component with a flux density of 1.43 Jy at 6 cm, and a fainter component (0.32 Jy at 6 cm) to the northwest in position angle 307°. The brighter component is variable in flux density and has a rising spectrum between 6 cm and 2 cm.

Early VLBI observations of 3C 395 at 6, 13, and 18 cm (Phillips and Mutel 1980; Phillips and Shaffer 1983; Johnston et al. 1983) all showed that the core, the brighter of the two arcsecond components, was dominated by two subcomponents with a separation of 15.6 mas along position angle 118°. The northwest component (1) was unresolved and had a spectral index of $\alpha = -0.2$ ($S_\nu \propto \nu^\alpha$), while the southwest component (2) was partially resolved and had a steeper spectrum ($\alpha = -0.6$). Component 2 is on the opposite side of the unresolved core from the extended arcsecond component 4, implying that the asymmetry of the small and large scale structures is reversed.

Waak et al. (1985) presented new VLBI observations of 3C 395 at 6 cm which, when combined with a reanalysis of the data of Johnston et al. showed that structural changes occurred between 1979.93 and 1983.26. They tentatively interpreted these changes as superluminal motion of a third component (3) away from component 1 and towards component 2, with an apparent velocity of $14h^{-1}c$.

To test these findings, we repeated the VLBI observations at 6 cm (Simon et al. 1987). These observations confirmed the three-component structure and the superluminal motion of component 3. Figure 1 shows the three 6 cm VLBI images of 3C 395 (Johnston et al. 1983; Waak et al. 1985; Simon et al. 1987) and Table 1 summarizes the component parameters.

2. Discussion

We assume that component 1 is stationary and that it is the active core of 3C 395. This is reasonable because it is essentially unresolved, it has a flatter radio spectrum than component 2, and it has brightened substantially, from 0.6 Jy in 1983 to 1.1 Jy in 1985.

In all three images, component 2 is moderately resolved with a size of 2–3 mas in position angle $\sim 160°$. It is located ~ 15.6 mas southwest of the core (component 1) along position angle 118°. The position of this component relative to the core has changed by *less* than 0.2 mas in six years and its flux density has remained approximately constant.

Component 3 is moving rapidly away from the core towards component 2 at a rate of 0.64 ± 0.1 mas yr^{-1}. This corresponds to an apparent velocity of about $13h^{-1}c$. The flux density of this component was the same in 1983.3 and 1985.4.

A possible interpretation of these observations is that 3C 395 is a core-jet radio source in which a jet, oriented nearly along the line of sight, bends back through the line of sight. In this interpretation, material ejected into the jet at first moves superluminally outwards from the core, as component 3 seems to be doing. As in other superluminal sources such as 3C 345 (e.g., Biretta, Moore, and Cohen 1986), the moving component might fade with time as it moves out, perhaps disappearing completely in the next five years or so. An underlying, continuous jet acts as a channel for the relativistically moving material to the

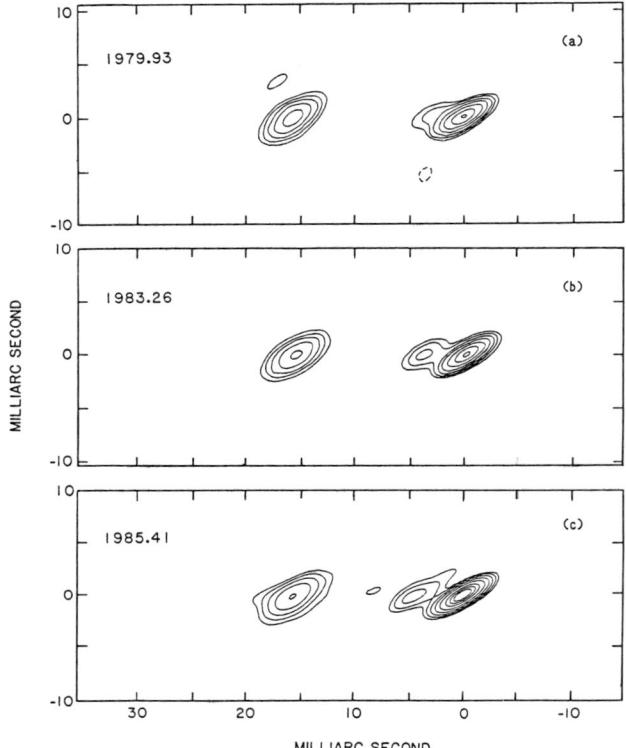

Figure 1 Images at 6 cm of 3C 395 at epochs 1979.9, 1983.3, and 1985.4. The maps have been rotated clockwise by 28° and have been convolved with a beam 3 × 1 mas in size extended along p. a. −30°. Contour levels are −2, 2, 4, 8, 16, 32, 64, and 96% of 415 mJy per beam (1979.9), 576 mJy per beam (1983.3), and −11.5, 11.5, 23, 46, 92, 184, 369, 553, and 737 mJy per beam (1985.4). The 1983.3 and 1985.4 contours are at the same absolute level even though the core has brightened considerably between 1983 and 1985.

region near the arcsecond component. By chance, the channel bends through the line of sight so that a stationary hot spot is observed in the jet. This bright region of the jet is observed as component 2. The apparent change in the jet direction through ∼ 180° (to bend the jet towards the distant arcsecond component) is then easily explained as a projection effect. The stationary component would owe its brightness to a combination of Doppler beaming effects and optical depth back through the jet.

The above picture is extremely speculative, but does suggest a number of questions to be answered by further VLBI observations: (a) is the velocity of the moving component constant? The above would suggest that component 3 will eventually slow as it moves along the jet back to-

Table 1 Components of 3C 395

Component	Offset from Core Distance (mas)	Offset from Core p.a. (deg)	Epoch	Wavelength (cm)	Flux Density (Jy)	Ref.
1 (Assumed core)	...	...	1978.39	18	1.0	1
	...	...	1979.93	6	0.4	2
	...	...	1980.59	13	~0.8	3
	...	...	1983.26	6	0.6	2
	...	...	1985.41	6	1.1	4
2 (Stationary)	15.6 ± 0.2	118.5 ± 0.5	1978.39	18	1.5	1
	15.6	118	1979.93	6	0.4	2
	~15	~120	1980.59	13	~1.6	3
	15.6	118	1983.26	6	0.4	2
	15.1 ± 0.1	118.6 ± 0.5	1984.92	18	0.94	4
	15.7 ± 0.3	118.4 ± 1.1	1985.41	6	0.44	4
3 (Moving)	1.2 ± 0.3	118	1979.93	6	0.09	2
	3.7 ± 0.8	118	1983.26	6	0.12	2
	3.6 ± 0.2	117.5 ± 3.5	1984.92	18	0.30	4
	4.7 ± 0.3	119.7 ± 3.7	1985.41	6	0.09	4
4 (Arcsecond)	~700	−153	1979	6	0.3	5
	...	...	1981.12	6	~0.3	6

References: 1. Phillips and Mutel 1980. 2. Waak *et al.* 1985. 3. Phillips and Shaffer 1980. 4. Simon *et al.* 1987. 5. Perley, Fomalont, and Johnston 1980. 6. Pearson, Perley, and Readhead 1985.

wards the line of sight; (*b*) is component 2 truly stationary? Our present limit to its velocity is $\lesssim c$; (*c*) is the supposed underlying jet detectable if we improve our sensitivity to faint structure on the 10–20 mas scale? We have only hypothezised the existence of a jet, not actually observed it. However, in each of these maps there is 100–600 mJy unaccounted for when we compare the sum of the three VLBI components plus the arcsecond component to the total flux density. We may be observing for the first time the resupply of a stationary hotspot in an extragalactic radio source.

References

Biretta, J. A., Moore, R. L., and Cohen, M. H. 1986, *Astrophys. J.*, **308**, 93.
Johnston, K. J., Spencer, J. H., Witzel, A., and Fomalont, E. B. 1983, *Astrophys. J. (Letters)*, **265**, L43.
Pearson, T. J., Perley, R. A., and Readhead, A. C. S. 1985, *Astron. J.*, **90**, 738.
Perley, R. A. 1982, *Astron. J.*, **87**, 859.
Perley, R. A., Fomalont, E. B., and Johnston, K. J. 1980, *Astron. J.*, **85**, 649.
Phillips, R. B., and Mutel, R. L. 1980, *Astrophys. J.*, **236**, 89.
Phillips, R. B., and Shaffer, D. B. 1983, *Astrophys. J.*, **271**, 32.
Simon, R. S., Hall, J., Johnston, K. J., Spencer, J. H., Waak, J. A., and Mutel, R. L. 1987, *Astrophys. J. (Letters)*, submitted.
van Breugel, W., Miley, G., and Heckman, T. 1984, *Astron. J.*, **89**, 5.
Véron, M. P., Véron, P., and Witzel, A. 1974, *Astron. Astrophys. Suppl.*, **13**, 1.
Waak, J. A., Spencer, J. H., Johnston, K. J., and Simon, R. S. 1985, *Astron. J.*, **90**, 1989.

Subluminal Expansion in NGC 1275

DONALD C. BACKER

Radio Astronomy Laboratory and Astronomy Department
University of California, Berkeley

The nearby, active galaxy NGC 1275 (3C 84, $z = 0.018$) has been the subject of many investigations in the past 20 years. In this paper, I first review the evolution of the microwave flux density and the structure seen in 10.7 GHz VLBI images. Then I summarize recent VLBI observations at 22 GHz and 90 GHz. Rapid changes in the core and the presence of a "hot spot" in the outflowing material are prominent features of these short-wavelength VLBI data. The case for subluminal proper velocities of the outflow is strong.

1. Flux Density Evolution

NGC 1275 was one of the first radio sources to display centimeter-wavelength variability (Dent 1966). However, this source is unusual in that its "light" curve suggests that a single outburst event has occurred over the past 30 years. Figure 1 displays data from Dent et al. (1983) and O'Dea, Dent, and Balonek (1984) at four frequencies. The contribution from the steep-spectrum "3C" component is indicated by dashed lines; this component arises in large-scale structure that is irrelevant to the present discussion. The slow evolution of the outburst is indicated in Figure 1 by solid lines. The outburst emission can be modeled by synchrotron radiation that has become optically thin at successively lower frequencies over the past decades. O'Dea, Dent, and Balonek have shown that the spectrum of the outburst below 2 GHz is steeper than expected for synchrotron self-absorption. They suggest that the emitting region is surrounded by thermal gas which is optically thick below 2 GHz. The data at 8 GHz indicate that the time origin of the outburst is around 1959. Superposed on the outburst flux density are smaller flares: e.g., one centered around 1975 and another in 1981. These flares appear to move to lower frequencies as time progresses. This qualitative decomposition of the light curves will be discussed further following a description of the VLBI results. Gear et al. (1987) discuss the spectrum of NGC 1275 above 300 GHz. They decompose the spectrum into the variable tail of the centimeter-wavelength synchrotron emission, a thermal component peaking at 100 μm, and a stellar component. The integrated radio luminosity emitted in the outburst is around $10^{52}h^{-2}$ erg if a 100-yr lifetime is assumed. The flares are less energetic by two orders of magnitude.

The variations seen at centimeter wavelengths suggest two nuclear components: a large, slowly evolving component associated with the outburst, and a compact component that produces the rapidly varying flares. If the large component expanded at a constant speed equal to c, then its size would now be $9h^{-1}$ pc ($33h^{-1}$ mas). The flares would have only parsec dimensions. These components would be visible in VLBI observations.

Rothschild et al. (1981) proposed that a compact radio source generates the point-like X-ray source in NGC 1275 by the synchrotron self-Compton process. Unpublished Einstein Observatory observations of NGC 1275 suggest that the X-ray point source increased in flux by a factor of two between 1979 and 1980. The giant flare at 90 GHz that peaked in 1980 (Figure 1) is easily associated with an X-ray increase. The brightness temperature of the flare is $10^{10-12}h^2$ K at the peak of this outburst, using a light travel time dimension and a range of possible frequencies for the maximum. An enhancement of the X-ray flux from synchrotron self-Compton radiation is expected.

2. Centimeter VLBI Structure and Variability

Romney et al. (1984) have demonstrated that the nuclear radiation at 10.7 GHz (2.8 cm) arises in a subluminally expanding structure with scale size of 6 mas (Figure 2). The source is elongated in the north-south direction and is more complex than the core-jet structures of superluminal sources. The core, which one identifies with the central engine, is not obvious. In the eight-year interval of observations, the overall dimension has grown by two milliarcseconds. Romney et al. estimate an apparent rate of 0.24 ± 0.06 mas yr^{-1}; this corresponds to an apparent transverse velocity of $0.2h^{-1}c$.

The brightening of the northernmost region of the 3C 84 images in 1981 (Figure 2) was associated by Romney et al. with the millimeter flare that began in 1979 (Figure 1). This suggests that the northernmost region is the site of the central engine, or at least is nearest to a fresh supply of energetic particles. Unwin et al. (1982) reached the same conclusion from a multi-frequency VLBI study; the northernmost region of 3C 84 had the flattest spectral index. The emission southward of the core can then be identified with the effluent that is slowly expanding away from the core. With this description, 3C 84 becomes

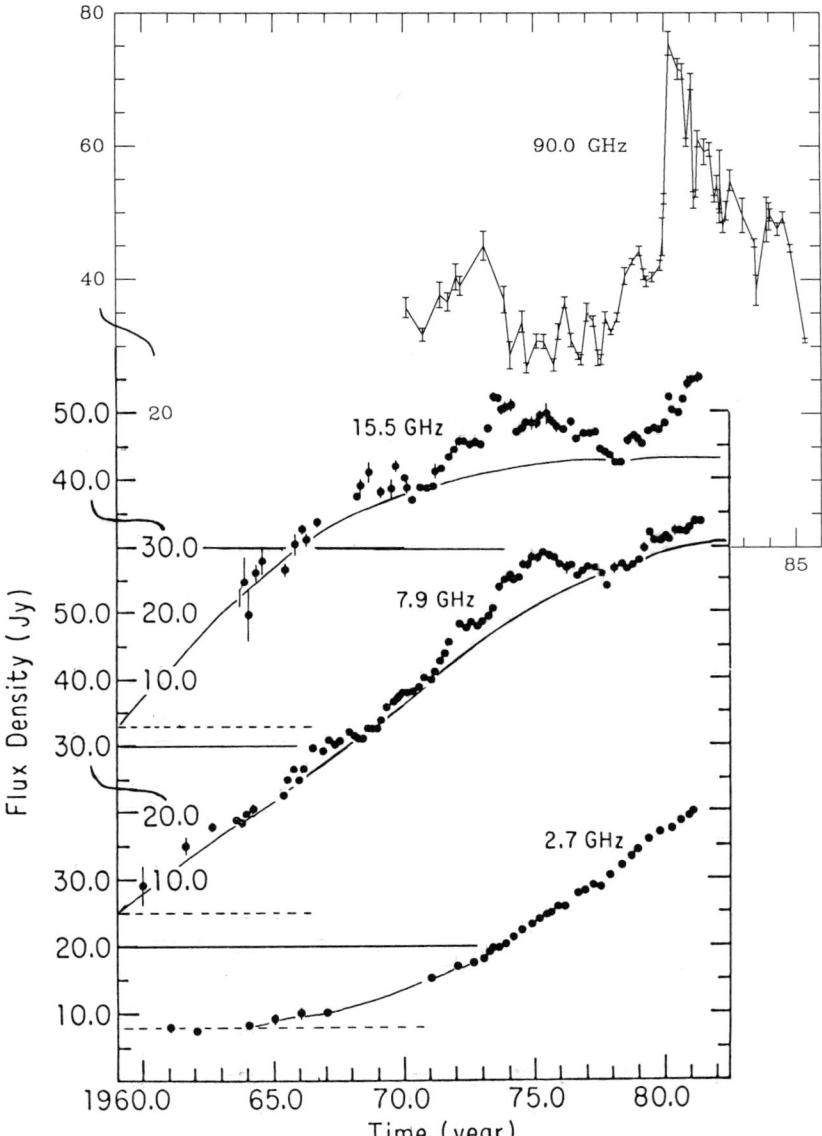

Figure 1 Flux density variations of 3C 84 from O'Dea, Dent, and Balonek (1984) at 2.7, 7.9 and 15.5 GHz, and from Dent *et al.* (1983) at 90.0 GHz. W. M. Kinzel (private communication) has kindly provided an updated table of 90.0 GHz data for this plot. The zeros for each scale are shown as solid horizontal lines. The contribution of the large-scale, steep-spectrum "3C" component is shown by dashed horizontal lines.

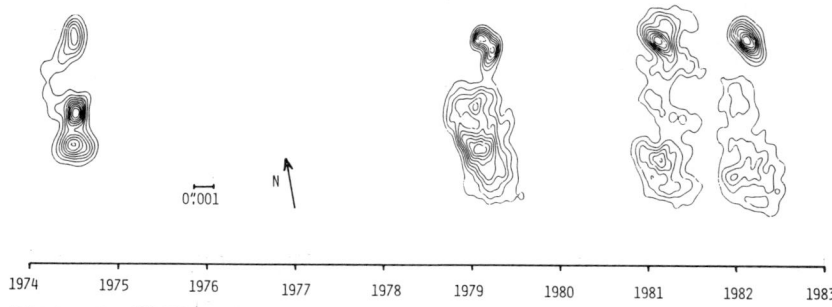

Figure 2 VLBI images of 3C 84 at 2.8 cm (Romney et al. 1984).

morphologically similar to the superluminal core-jet VLBI objects. The northernmost region, or core, can be associated with the rapidly varying flare flux density, while the complex, slowly expanding effluent can be associated with the outburst flux density.

If one accepts the identification of the northernmost region with the central engine in 3C 84, then the asymmetry of the source can be used to place model-dependent limits on kinematics of the source. Romney et al. (1984) summarize two possibilities. (a) If the effluent from the core is intrinsically bidirectional, then the outflow velocity must approach c and must be directed very close to (and away from) the line of sight. This combination would produce the observed slow apparent expansion and Doppler boosting and attenuation. (b) On the other hand, the effluent could be ejected toward the south alone at the observed subluminal velocities. The wide cone angle for the proposed effluent supports the single-sided flow.

3. 1.3 cm and 3.3 mm VLBI

Readhead et al. (1983a) produced the first 1.3 cm VLBI image of 3C 84. They resolved the northern region into two very compact components (A, B). They also identified two large components with lower brightness temperature in what I call the effluent (C, D). Their analysis of the spectra and physical properties of the components supports a core-effluent morphology.

A series of VLBI experiments have been conducted at 90 GHz (wavelength 3.3 mm). These experiments are technically challenging and have required the efforts of scientists and support staff at many institutions. Initial efforts are summarized in several papers: Readhead et al. (1983b), Backer (1984a, 1984b), and Rogers et al. (1984).

The 3.3 mm VLBI observations indicate a core-halo brightness distribution on scales below 1 mas. The halo is evident only on the Hat

Creek–Owens Valley baseline since it has a size of 0.6 mas. It is a persistent feature over the period October 1981 to March 1985, and has a slowly decreasing flux density, from 18 to 12 Jy. This component accounts for part of the decay of the total flux density seen in Figure 1. The most variable component is the core which is unresolved even on the transcontinental baseline from Quabbin to Owens Valley ($10^9 \lambda$). This component also contributes to the time variations.

The millimeter VLBI structure is easily identified with the northernmost region in the centimeter VLBI images since that is the location of the most compact structure. The submilliarcsecond core and its milliarcsecond-scale halo most probably correspond to the component labeled A by Readhead et al. (1983a).

The 3.3 mm VLBI observations are complemented by a series of 1.3 cm VLBI observations that have been conducted by me and Jon Marr in collaboration with others engaged in the millimeter VLBI. Our goal is to assess component spectra and evolution. The 1.3 cm data provide comparable resolution to the shorter baseline millimeter results, more frequent time sampling, and greater sensitivity and dynamic range.

Figure 3 compares the April 1981 image from Readhead et al. (1983a) with the first image from our recent series of observations. Component D in the effluent has moved from about 7 mas to 9 mas when measured with respect to the northernmost component A. Component C in the effluent has faded. Component B in April 1981 has either faded or moved. There is a new component B south of A. One goal of the present series of 1.3 cm VLBI observations is to describe the kinematics of this second component. In observations of October 1985 and February 1986 this secondary component has varied in strength, but has not moved appreciably.

A notable feature of the long-baseline, 1.3 cm VLBI data is a beat between ultracompact components with a separation of about 10 mas in February 1985. A similar feature was noted by Readhead et al. in the April 1981 data with the exception that the separation corresponded to 8 mas. Although the contours chosen for Figure 3 do not emphasize the location of this feature, we conclude from other displays and analysis that there is a milliarcsecond "knot" in the southernmost component. The beat in the long-baseline data is between this knot and the nuclear component in the north. The knot has a flux density of 0.5 Jy, a brightness temperature of 10^9 K, and is entrained in the outflow. If the magnetic field in the knot is near the equipartition field of 1 G, then the particle lifetime is two years, and reacceleration is required. The continued activity in the core suggests that it may supply energy to the knot, which corresponds to the working surface in extragalactic double sources and in galactic objects such as SS 433 and Sco X-1. From model

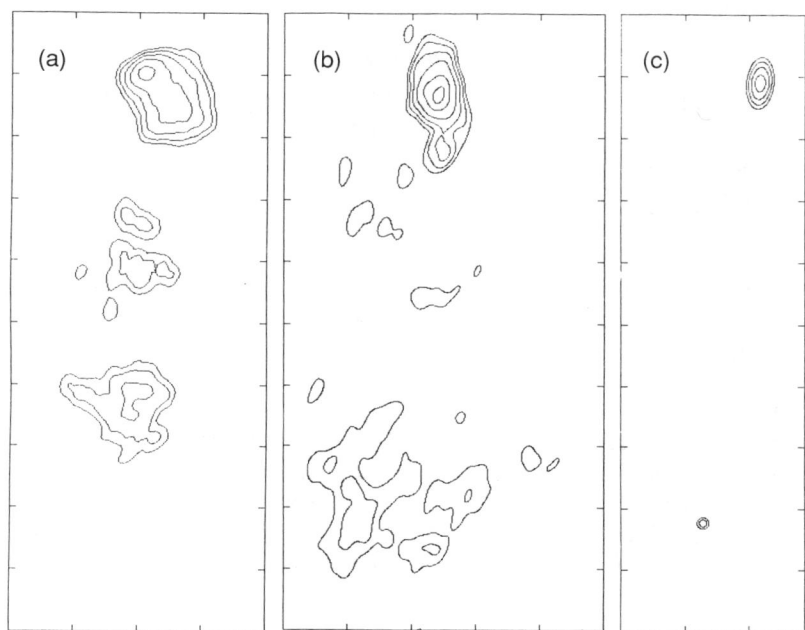

Figure 3 VLBI images of 3C 84 at 1.3 cm and 3.3 mm. The tick marks are 1.4 mas apart. (a) 1.3 cm, April 1981: a hybrid map from the data of Readhead et al. (1983a). (b) 1.3 cm, February 1985: a hybrid map from a seven-station experiment. (c) 3.4 mm image from a model fit to the March 1985 data using antennas at Hat Creek, Owens Valley, and Quabbin.

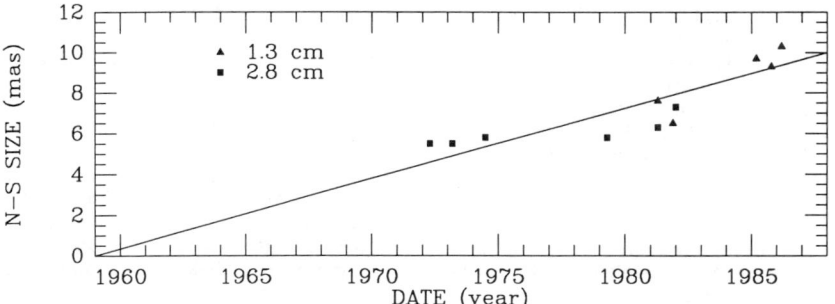

Figure 4 Determinations of the north-south angular extent of 3C 84 from 2.8 cm (10.7 GHz) and 1.3 cm (22.3 GHz) VLBI observations. Data prior to 1983 are from Romney et al. (1984) and Readhead et al. (1983a). Data since 1985 are from unpublished observations by the author.

fits to the VLBI data, we have placed an upper limit of 1.0 Jy on the flux density of this knot at 3.3 mm.

4. Summary

Figure 4 displays estimates of the north-south length of 3C 84 from observations at 2.8 cm and 1.3 cm discussed above. The line on this plot corresponds to a linear expansion at 0.33 mas yr^{-1} since the origin of the centimeter-wavelength outburst in 1959. The observations are consistent with expansion at this average rate for that entire interval. The southernmost emission in the images corresponds to material ejected at the highest transverse velocities in 1959. Emission in the middle of the source arises from particles that were ejected later or at lower transverse velocities. Light travel time effects may also be important since the outflow is not well collimated.

The overall extent of 3C 84 in 1985 is $2.75h^{-1}$ pc. The average expansion rate corresponds to $0.28h^{-1}c$. The proper velocity of the expansion is most likely subluminal and the effluent comes out from one side of the nucleus. The knot in the effluent is also moving at $0.28h^{-1}c$, and has dimensions of roughly one light year ($\times h^{-1}$). The millimeter-wavelength core in the nucleus has a dimension of two light months ($\times h^{-1}$).

References

Backer, D. C. 1984a, in *IAU Symposium 110, VLBI and Compact Radio Sources*, ed. R. Fanti, K. Kellermann, and G. Setti (Dordrecht: Reidel), p. 31.
Backer, D. C. 1984b, in *URSI International Symposium on Millimeter and Submillimeter Wave Radio Astronomy*, Granada, Spain, September 11–14, 1984 (Union Radio-Scientifique Internationale), p. 93.
Dent, W. A. 1966, *Astrophys. J.*, **144**, 843.
Dent, W. A., O'Dea, C. P., Balonek, T. J., Hobbs, R. W., and Howard, R. J. 1983, *Nature*, **306**, 41.
Gear, W. K., Gee, G., Robson, E. I., and Nolt, I. G. 1987, *Monthly Notices Roy. Astron. Soc.*, in press.
O'Dea, C. P., Dent, W. A., and Balonek, T. J. 1984, *Astrophys. J.*, **278**, 89.
Readhead, A. C. S., Hough, D. H., Ewing, M. S., Walker, R. C., and Romney, J. D. 1983a, *Astrophys. J.*, **265**, 107.
Readhead, A. C. S., Masson, C. R., Moffet, A. T., Pearson, T. J., Seielstad, G. A., Woody, D. P., Backer, D. C., Plambeck, R. L., Welch, W. J., Wright, M. C. H., Rogers, A. E. E., Webber, J. C., Shapiro, I. I., Moran, J. M., Goldsmith, P. F., Predmore, C. R., Bååth, L., and Rönnäng, B. 1983b, *Nature*, **303**, 504.
Rogers, A. E. E., Moffet, A. T., Backer, D. C., and Moran, J. M. 1984, *Radio Sci.*, **19**, 1552.
Romney, J. D., Alef, W., Pauliny-Toth, I. I. K., Preuss, E., and Kellermann, K. I. 1984, in *IAU Symposium 110, VLBI and Compact Radio Sources*, ed. R. Fanti, K. Kellermann, and G. Setti (Dordrecht: Reidel), p. 137.
Rothschild, R. E., Baity, W. A., Marscher, A. P., and Wheaton, W. A. 1981, *Astrophys. J. (Letters)*, **243**, L9.
Unwin, S. C., Mutel, R. L., Phillips, R. B., and Linfield, R. P. 1982, *Astrophys. J.*, **256**, 83.

Superluminal Motion and Other Indications of Bulk Relativistic Motion in a Complete Sample of Radio Sources from the S5 Survey

ARNO WITZEL

Max-Planck Institut für Radioastronomie

Dedicated to Alan T. Moffet

1. Introduction

In the late seventies it became obvious that VLBI observations of complete samples of extragalactic radio sources were needed for statistical investigations of source properties and their changes. Especially, the predictions of the various beaming models of apparent superluminal motion deserve to be tested by statistical studies of the kinematics of nuclear components.

At the same time facilities for observations at numerous wavelengths ranging from the radio to the X-ray regime were available or under construction, and it became clear that such a study would be most successful if it included multifrequency observations, also covering a vast range of angular resolution from arcminutes to milliarcseconds.

A suitable sample could be chosen from the S5 survey (Kühr et al. 1981), which in 1979 completed the NRAO-MPIfR 5 GHz strong source surveys. The S5 survey, done entirely with the 100 m telescope at Effelsberg, covers the declination range 70° to 90° and lists 185 objects down to its completeness limit of 250 mJy. About half of the sources exhibit flat radio spectra, and thus it is highly probable that they contain compact components. We selected the 13 sources from the S5 survey which had flux densities in excess of 1 Jy and spectral indices between 6 cm and 11 cm $\alpha \gtrsim 0.5$ ($S_\nu \propto \nu^\alpha$). Comparisons of their properties at all accessible wavelengths with those of larger samples indicate that these 13 objects are representative of all strong flat spectrum radio sources found at 6 cm wavelength (e.g., Eckart 1983; Eckart et al. 1986, 1987).

In this paper, we first summarize the present observational status of the survey. Then we concentrate on the detection of superluminal motion in repeated VLBI observations of individual sources, and on supporting evidence for the existence of bulk relativistic motion in these objects.

Table 1 The S5 Sample

(1) Source	(2) ID[a]	(3) z	(4) m_V[b]	(5) m_R[c]	(6) S_X[d]	(7) > 50	(8) 18	(9) 13	(10) 6	(11) 3.6	(12) 2.8	(13) 1.3 cm
0016+73	Q	1.76	18.0	18.8	0.12		1		1		1	1
0153+74	Q	2.34	16.0	17.5	$\lesssim 1.0$		2		3			
0212+73	BL	2.37[e]	19.0	19.0	0.49		1	6	3	6	1	1
0454+84	BL	–	16.5	20.3	0.095		1		2			1
0615+82	Q	0.71	17.5	18.7	$\lesssim 0.2$		1		2		1	1
0716+71	BL	–	13.2	15.9	0.31	1	2		2			1
0836+71	Q	2.16	16.5	16.7	$\lesssim 1.0$		1		3		1	1
1039+81	Q	1.26	16.5	16.7	$\lesssim 1.0$		1		2			1
1150+81	Q	1.25	18.5	17.6	$\lesssim 0.2$		1		3			1
1749+70	BL	–	16.5	17.2	0.22		1		2			
1803+78	BL	0.68[f]	16.4	15.3	0.24		1	6	3	6		1
1928+73	Q	0.30[e]	15.5	15.7	0.55	2	2		5			2
2007+77	BL	–	16.7	16.8	0.19		1		4			2

[a] Optical identification: quasar (Q) or BL Lac object (BL).
[b] Optical magnitude estimated from the Palomar Sky Survey prints.
[c] R-band magnitude measured in 1985 with the Steward Observatory 90-inch telescope (A. Eckart, private communication).
[d] X-ray flux density at 1 keV (in μJy) or upper limit based on observations with the HEAO-A1, Einstein, and EXOSAT telescopes.
[e] Lawrence et al. (1986).
[f] C. Lawrence (private communication).

Data in columns 2–4 and 6 are taken from Eckart et al. (1986, 1987), unless stated otherwise. Columns 7–13 list the numbers of our VLBI observations at the respective wavelengths.

2. Observational Status

Optical and X-ray properties of the sources and the present observational status of our VLBI observations at wavelengths from 50 cm to 1.3 cm are summarized in Table 1. Optical spectroscopy gave redshifts for nine objects ranging from 0.3 to 2.6. First results on ultraviolet, optical, and infrared properties of these sources are discussed by Eckart (1983).

Radio spectra are available for all 13 sources at wavelengths $\lambda \gtrsim$ 1.2 mm (e.g., Kühr et al. 1981; Chini et al., in preparation). As an example, the overall spectrum for the BL Lac object 1803+78 is shown in Figure 1. All sources were detected at 1.2 mm and the median spectral index between 5 and 250 GHz is $\alpha = -0.26 \pm 0.18$. All 13 objects are variable radio sources. The structures on scales exceeding 250 mas have been mapped with MERLIN at 18 cm (Eckart 1983) and with the VLA at 20 cm and 6 cm wavelengths with dynamic ranges of up to 5 000:1 (e.g., Perley et al. 1980; Ulvestad et al. 1981; Perley 1982; Antonucci et al. 1986).

For the quasar 1928+73, symmetric two-sided emission on the kilo-

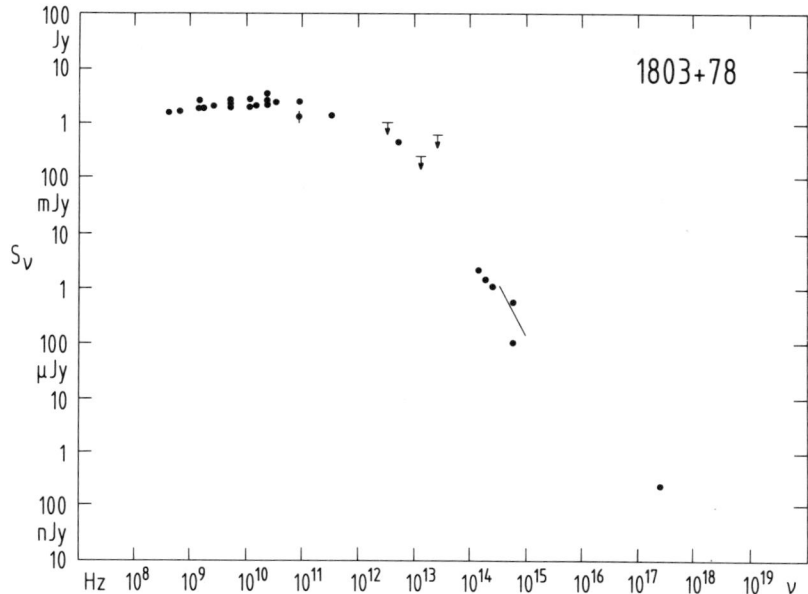

Figure 1 The overall spectrum of the BL Lac object 1803+78.

parsec scale was found for the first time with a dynamic range of 24 000:1 (Johnston *et al.* 1987; Simon *et al.*, this Workshop, page 155). VLA maps of further S5 objects in our sample with similar dynamic ranges are in preparation.

3. The Milliarcsecond Structures

Figures 2 and 3 show recent 6 cm maps of each of the 13 sources in our sample. Except for 0016+73, these maps were made from observations made after 1983. Maps from earlier observations are given by Eckart *et al.* (1987).

The morphological types of the structures in Figure 2 range from almost point-like objects (e.g., 0615+82) to core-jet type sources with long jets containing numerous identifiable knots (e.g., 1928+73). One source, 0153+74, exhibits a special morphology. The bent emission regions joining the two prominent 6 cm components are confirmed by our 18 cm VLBI maps. We note that all sources classified as "core-jet" show significant bending.

Table 2 lists the numbers of sources in the morphological classes "unresolved", "slightly resolved", and "core-jet" at different wavelengths (0153+74 is listed as a core-jet object here).

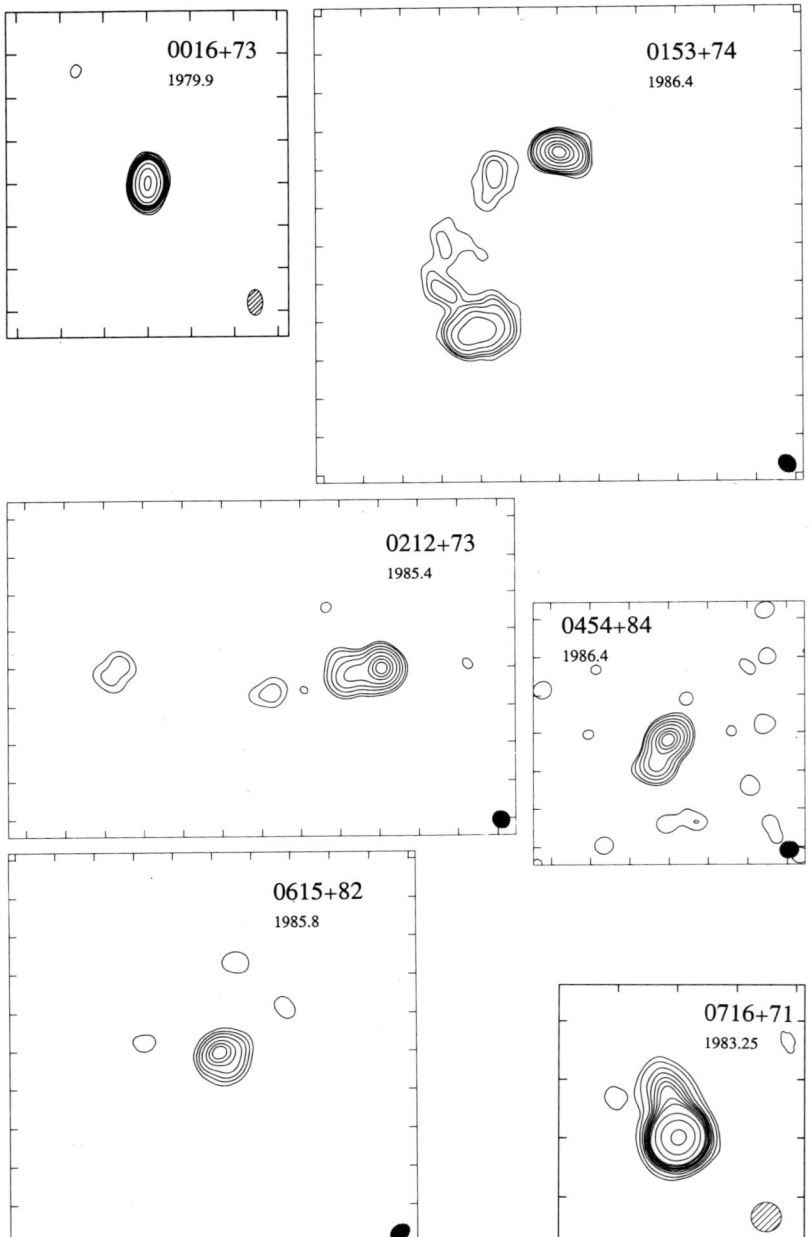

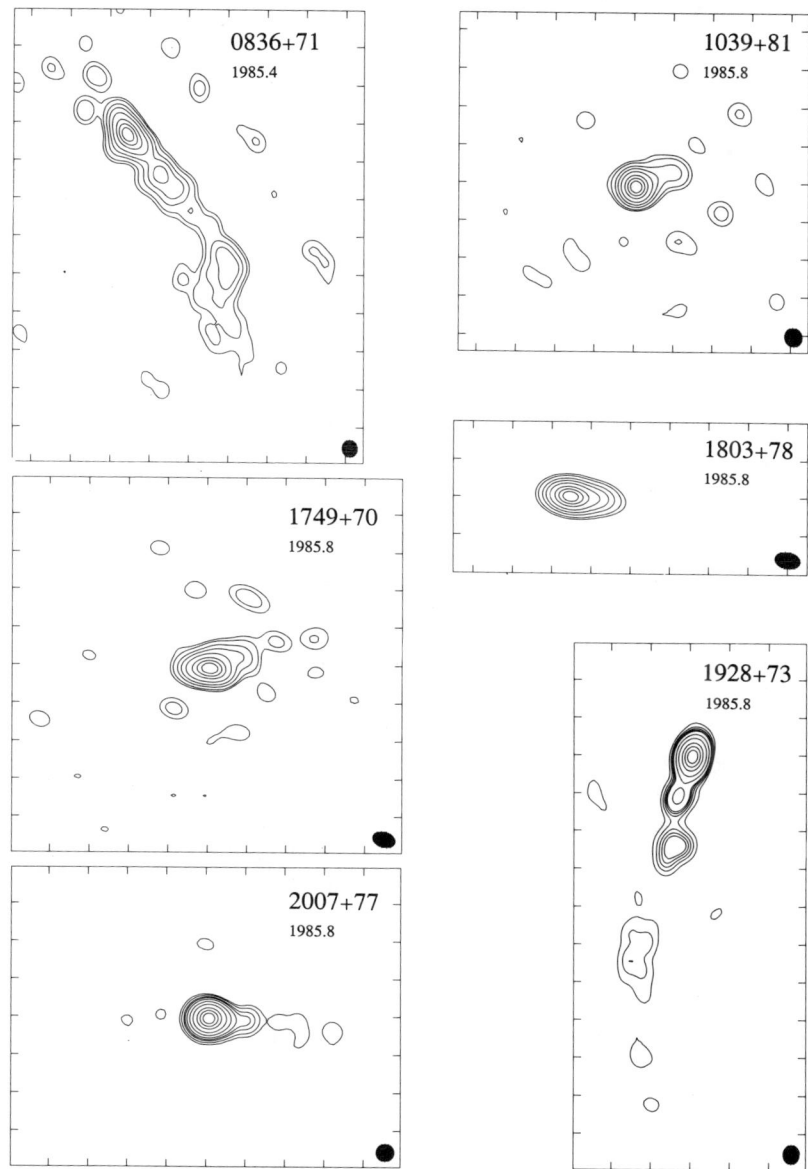

Figure 2 6 cm maps for 12 sources of the complete sample (1150+81 is shown in Figure 3). Contour levels (in percent of the peak flux density) are: for 0016+73, −1, 2, 3, 4, 5, 6, 8, 10, 20, 40, and 80%; for 0716+71, −0.5, 0.5, 1, 2, 3, 4, 5, 6, 8, 10, 20, 40, and 80%; for the remaining sources, 1, 2, 5, 10, 20, 40, 60, and 80%, with additional contours at 3% and 4% for 1928+73 and at 0.5% and 3% for 2007+77. Scale is 2 mas per tick in all maps.

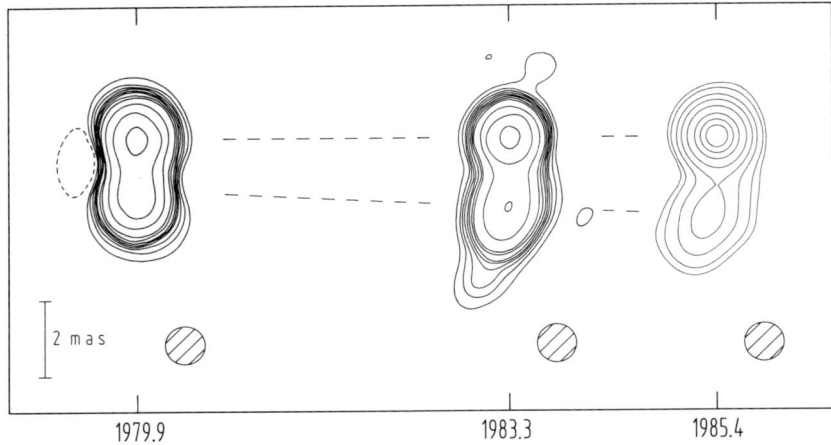

Figure 3 6 cm maps of the quasar 1150+81 ($z = 1.25$) at three observing epochs. The change in separation of the two components is 0.13 mas yr^{-1}, corresponding to $v_{app} = 4.1 h^{-1} c$. Contour levels are $-1, 1, 2, 3, 4, 5, 6, 8, 10, 20, 40$, and 80% of the peak flux density for the maps from 1979.9 and 1983.25, and 1, 2, 5, 10, 20, 40, 60, 80% for the 1985.4 map. The beam is given as a cross-hatched ellipse.

4. Structural Changes and Superluminal Motion

We have obtained repeated observations at 6 cm for all 13 sources except for 0016+73. Several objects were observed repeatedly at other wavelengths (Table 1), but in the following discussion of structural changes we will concentrate on the 6 cm data.

At 6 cm, five sources were observed twice, and the remaining seven sources were observed at least three times. Additional 6 cm maps for three sources are published in the literature (1749+70: Bååth 1984; 0153+74, 0212+73: Pearson and Readhead 1984).

Our observations before 1982 were generally done in a snapshot mode (typically four to five 15-min scans per source over a range of 12 hours) and the dynamic range of these maps is less than 30:1 (Eckart *et al.* 1986). All observations at later epochs yielded maps with dynamic ranges exceeding 100:1 (Eckart *et al.* 1986, 1987; Schalinski *et al.*, in preparation).

Of the 12 sources observed repeatedly, eight have known redshifts which can be used to derive separation velocities. For these objects Table 3a gives the measured rate of change of separation of components and the derived apparent velocities. For the four sources with unknown redshift, Table 3b lists the measured rate of change and the critical redshift z_{crit} beyond which the apparent velocity exceeds the velocity of light.

In the following we describe the sources in some detail.

Table 2 VLBI Structure of Sources in the S5 Sample

	≥ 50 cm	18 cm	6 cm	2.8 cm	1.3 cm
Unresolved	1	3	—	—	—
Slightly resolved	—	2	3	2	—
Core-jet	1	8	10	2	11

Table 3a Measured Expansion Speeds: Sources with Known Redshifts

Source Name	$d\theta/dt$[a] (mas yr^{-1})	v_{app}[b] ($h^{-1}c$)	N[c]
0153+74	<0.03	<1.3	3
0212+73	0.09	3.9	3
0615+82	0.05:	1.1:	2
0836+71	0.15	6.2	3
1039+81	<0.08	<2.5	2
1150+81	0.13	4.1	3
1803+78	<0.03	<0.6	3
1928+73	0.60	7.0	5

Table 3b Measured Expansion Speeds: Sources without Redshifts

Source Name	$d\theta/dt$[a] (mas yr^{-1})	z_{crit}[b]	N[c]
0454+84	0.15	0.16	2
0716+71	0.09	0.28	2
1749+70	0.10	0.25	2
2007+77	0.30	0.07	4

[a] Separation changes $d\theta/dt < 0.2$ mas were not considered significant.
[b] Apparent expansion speeds v_{app} and critical redshifts z_{crit} were derived using $H_0 = 100h$ km s^{-1} Mpc^{-1} and $q_0 = 0.5$.
[c] Number of observing epochs at 6 cm.

1803+78: this source has clearly not exhibited any superluminal motion between 1979.9 and 1985.8. This finding is confirmed by repeated S and X-band observations.

0212+73, 1150+81, and **1928+73:** these sources show apparent superluminal motion, based on observations at three or more epochs. 1928+73 has been shown before to be a superluminal radio source by Eckart et al. (1985); our two 18 cm observations of this quasar confirm the separation velocity of $7.5h^{-1}c$. 0212+73 and 1150+81 have been proposed as candidates for apparent superluminal motion by Eckart et al. (1986). Figure 3 shows a composite picture of the three observing epochs of 1150+81.

0836+71: this quasar shows three "stationary" main components between 1979.9 and 1985.4 ($v_{app} \leq 1.0h^{-1}c$). However, the northern compact component clearly shows a double structure in 1985.4, whereas it was unresolved at the two earlier epochs. The separation velocity of $6.2h^{-1}c$ in Table 3a reflects that change.

1039+81: our two observations of this quasar were closely spaced in time (about 1.5 yr). We determined a separation change of 0.2 mas. Since this corresponds to 1/5 of the interferometer beam, we do not consider this result to be significant and accordingly only derive an upper limit to any separation velocity of $2.5h^{-1}c$. An additional observation of this source in about 1987 will clarify the situation, especially since both our previous measurements have a dynamic range greater than 100:1 and the source does not have a complex morphology.

0153+74: the quoted upper limit of $1.3h^{-1}c$ refers to the two prominent, widely spaced (12 mas) components of this quasar.

0615+82: the case of this very compact source is uncertain. The secondary component apparent in 1985.8 might have been present in 1979.9, but undetected due to inadequate (u,v)-coverage.

0454+84, 0716+71, and **1749+70:** we observed these sources with unknown redshift only twice. However, 0454+84 and 0716+71 have relatively simple structures at 6 cm, and we conclude that both sources are good candidates for motion with $v_{\rm app} > c$, provided their redshifts exceed 0.16 and 0.28, respectively.

1749+70: this BL Lac object was observed three times by Lars Bååth in 1980–81 (Bååth 1984). The separations for these epochs (Bååth, private communication) combined with our measurements suggest 1749+70 as another superluminal candidate (again with the assumption of $z > 0.25$).

2007+77: this BL Lac object is superluminal if its redshift exceeds 0.07. Our four observations of this source also show that the inner core structure became triple in about 1985. This finding is confirmed by our 1.3 cm map.

We conclude that three of the sources with known redshifts have shown superluminal motion and two more are superluminal candidates, whereas in one source, 1803+78, any motion would be clearly subluminal. The cases of 0153+74 and 0615+82 cannot be decided at present.

For the sources without known redshift, $z_{\rm crit}$ ranges from 0.07 to 0.28 (Table 3b), whereas the lowest measured redshift for any source in the complete sample is 0.3. Eckart et al. (1986) presented evidence that the redshifts of the four objects exceed 0.05, and more recent optical observations show that they exceed 0.3 (Kühr, private communication). Thus, even with $H_0 = 100$ km s^{-1} Mpc^{-1} and $q_0 = 0.5$, all four objects are good candidates for apparent superluminal motion.

5. Other Evidence for Bulk Relativistic Motion

In addition to the direct observations of apparent superluminal motion, independent evidence for bulk relativistic motion can be derived from

Table 4 Doppler Factors and Measured Expansion Speeds

(1) Source Name	(2) $\delta_{\min}{}^a$ (SSC)	(3) $\delta_{\min}{}^b$ (Core variation) $h^{2/3}$	(4) $\delta_{\min}{}^{c,d}$ (Total variation) $h^{2/3}$	(5) v_{app}/c^d (VLBI)
0016+73	8	—	2.3	—
0153+74	*	*	1.4	≤1.3
0212+73	2.4	1.4	1.4	3.9
0615+82	1.3	*	*	1.1
0836+71	4.0	1.9	2.1	6.2
1039+81	2.1	1.2	1.6	≤2.5
1150+81	2.2	1.6	1.8	4.1
1803+78	4.4	*	4.0	≤0.6
1928+73	1.7	1.2	1.1	7.0
0454+84	2.1	*	2.3 (3.3)	2.6 (4.2)
0716+71	1.4	*	1.9 (2.7)	1.6 (2.5)
1749+70	1.3	*	* (1.0)	1.7 (2.8)
2007+77	2.8	*	2.3 (3.3)	5.2 (8.3)

The bottom part of this table lists the BL Lac objects without known redshifts.
[a] Doppler factor derived from SSC calculations.
[b] Doppler factor derived from VLBI-core variability at 6 cm.
[c] Doppler factor derived from total flux density variations at 6 cm.
[d] Values in the bottom part are computed for assumed redshift of $z = 0.5$ and $z = 1.0$ (in brackets).

observations of flux density variability and from the comparison of measured and calculated X-ray flux densities ("Synchrotron-Self-Compton" models; e.g., Marscher, this Workshop, page 280). The latter uses the condition that at least part of the X-rays are produced by the inverse Compton effect, whereas the former is based on the assumption that the flux density variations are intrinsic to the source.

Table 4 compares the minimum Doppler factors ($\delta_{\min}$) derived from SSC calculations, from the variations of the total flux density at 5 GHz, and from the variations of the milliarcsecond core components, again at 5 GHz, and lists the measured values for v_{app}/c.

For the BL Lac objects without known redshift, listed in the bottom part of Table 4, the values in columns 4 and 5 are calculated for both $z = 0.5$ and $z = 1.0$.

An asterisk in column 3 denotes sources for which we measured the same core flux density at the VLBI observing epochs. Variations in between cannot be excluded and the absence of measured core flux density variability does not rule out $\delta > 1$. Also, in the case of the variations of the total flux densities the values for $\delta_{\min}$ are lower limits due to the incomplete time coverage of our single-dish measurements. Note that the data used to derive the Doppler factors in column 4 in some cases were taken at epochs other than those of the VLBI observations.

The comparison of measured and predicted X-ray flux densities provides evidence for bulk relativistic motion in all 13 sources, with the exception of 0153+74. This is consistent with the corresponding values for Doppler factors derived from variability arguments and the observed structural changes.

It seems that 0016+73, the only source for which we do not have multi-epoch VLBI measurements, may also be expected to show apparent superluminal motion (Table 4).

Although for 1803+78 our observations did not show any separation change with velocity greater than $0.6h^{-1}c$, the Doppler factors in columns 2 and 4 indicate bulk relativistic motion in this source. Therefore it would not be surprising if this object displayed superluminal motion at other epochs. In fact, it is tempting to compare 1803+78 with the quasar 0836+71, which also showed "stationary" structures for five years but did display an additional moving component close to the core in 1985.4.

6. Concluding Remarks

The majority of the 13 S5 sources show evidence for bulk relativistic motion. In addition, nine of the 12 sources observed repeatedly with VLBI are (at least) good candidates for superluminal motion. Since these objects do not differ significantly from other strong flat-spectrum radio sources found at 6 cm wavelength (Eckart 1983; Eckart et al. 1986), we conclude that bulk relativistic motion is a common phenomenon in core-dominated radio sources. The minimum Doppler factors derived for the 13 S5 sources range up to about 10. Combined with the observed separation velocities, up to $7.5h^{-1}c$, we derive typical required angles to the line of sight of less than $7h°$ and minimum values for the Lorentz factor $\gamma \geq 8h^{-1}$.

Let me finish with one of the most widely used sentences in observational astronomy: "more observations are needed!" In addition to the necessary further monitoring at centimeter wavelengths we feel that VLBI observations at millimeter wavelengths would help in the understanding of the general physics of these sources and, especially, the "early phases" of superluminal motion. Since any detailed investigation of the regions close to the central engine is severely affected by optical depth effects in the centimeter-wavelength regime, millimeter VLBI observations seem to be the only way at present to derive structural information about the innermost cores of active galactic nuclei.

The work reported here would not have been possible without the close cooperation of C. J. Schalinski, T. Krichbaum, C. Hummel, and P. Biermann (MPIfR), K. J. Johnston and R. Simon (Naval Research Laboratory), A. Eckart (MPI für Extraterrestrische Physik), and H. Kühr (MPI für Astronomie).
It is a pleasure to thank A. Eckart, C. Lawrence, and H. Kühr, who made some of their data available in advance of publication, and R. W. Porcas for critically reading the manuscript. Thanks are also due to A. Quirrenbach and R. Wynands for their help with the data reduction.

References

Antonucci, R. R. J., Hickson, P., Olszewski, E. W., and Miller, J. S. 1986, *Astron. J.*, **92**, 1.
Bååth, L. B. 1984, in *IAU Symposium 110, VLBI and Compact Radio Sources*, ed. R. Fanti, K. Kellermann, and G. Setti (Dordrecht: Reidel), p. 127.
Carter, W. E., Robertson, D. S., and MacKay, J. R. 1985, *J. Geophys. Res.*, **90**, 4577.
Eckart, A. 1983, Ph. D. thesis, Westfälische Wilhelms Universität Münster, Westfalen.
Eckart, A., Witzel, A., Biermann, P., Johnston, K. J., Simon, R., Schalinski, C., and Kühr, H. 1986, *Astron. Astrophys.*, **168**, 17.
Eckart, A., Witzel, A., Biermann, P., Johnston, K. J., Simon, R., Schalinski, C., and Kühr, H. 1987, *Astron. Astrophys. Suppl.*, **67**, 121.
Eckart, A., Witzel, A., Biermann, P., Pearson, T. J., Readhead, A. C. S., and Johnston, K. J. 1985, *Astrophys. J. (Letters)*, **296**, L23.
Johnston, K. J., Simon, R. S., Eckart, A., Biermann, P., Schalinski, C., Witzel, A., and Strom, R. G. 1987, *Astrophys. J. (Letters)*, **313**, L85.
Kühr, H., Pauliny-Toth, I. I. K., Witzel, A., and Schmidt, J. 1981, *Astron. J.*, **86**, 854.
Kühr, H., Witzel, A., Pauliny-Toth, I. I. K., and Nauber, U. 1981, *Astron. Astrophys. Suppl.*, **45**, 367.
Lawrence, C. R., Pearson, T. J., Readhead, A. C. S., and Unwin, S. C. 1986, *Astron. J.*, **91**, 494.
Pearson, T. J., and Readhead, A. C. S. 1984, in *IAU Symposium 110, VLBI and Compact Radio Sources*, ed. R. Fanti, K. Kellermann, and G. Setti (Dordrecht: Reidel), p. 15.
Perley, R. A. 1982, *Astron. J.*, **87**, 859.
Perley, R. A., Fomalont, E. B., and Johnston, K. J. 1980, *Astron. J.*, **85**, 649.
Schalinski, C. J., Alef, W., Campbell, J., Schuh, H., and Witzel, A. 1986, in *Die Arbeiten des Sonderforschungsbereiches 78 Satellitengeodäsie der Technischen Universität München 1984 und 1985*, ed. M. Schneider (München: Bayerische Akademie der Wissenschaften), p. 292.
Ulvestad, J., Johnston, K., Perley, R., and Fomalont, E. 1981, *Astron. J.*, **86**, 1010.

The Quest for Superluminal Sources

TIMOTHY J. PEARSON, ANTHONY C. S. READHEAD,
AND PETER D. BARTHEL

California Institute of Technology

1. Introduction

In 1978 we began a systematic study of the milliarcsecond structure of a complete, flux-density limited sample of strong radio sources selected at 5 GHz. While a major goal of this survey was to find new superluminal sources and to determine how common they are, we also wanted to explore the full range of morphologies exhibited by compact radio sources. In addition, we hoped to use this well-defined sample for statistical tests of theories of compact sources, especially the relativistic beaming theories. Our approach is similar to that adopted by Arno Witzel and his colleagues (this Workshop, page 83); our sample is larger, but our observations of each source are less comprehensive.

If the beaming theories are correct, the sources in our flux-density limited sample are not randomly oriented but are preferentially beamed towards us. This bias imposed by our selection criteria must be taken into account when comparing the theories and the observations. A complementary approach to the problem is to try and design a sample that is unbiased in orientation (e.g., Hough and Readhead, this Workshop, page 114; Zensus and Porcas, this Workshop, page 126; Cawthorne *et al.* 1986). This approach, however, has the drawback that it usually excludes from consideration the strongest, most easily observed sources that are likely to be strongly beamed.

There are a number of basic astrophysical questions that we can address by VLBI studies of complete samples, including:

(*a*) Is there only one kind of central engine?
(*b*) On what scale are the nuclear jets first collimated?
(*c*) What is the fluid speed in the nuclear jets?
(*d*) Is there any evidence for precession?
(*e*) Do successive ejecta follow the same path?
(*f*) Is the one-sided appearance of the nuclear jets intrinsic, environmental, or due to Doppler boosting?
(*g*) Are jets recollimated on scales of 0.1–100 pc?

(h) What bends the jets?
(i) Are different morphologies due to the central engine, the environment, or both?
(j) Do extended triple sources and compact, flat-spectrum objects belong in the same class?
(k) How do the *compact double sources* and the *steep-spectrum compact sources* fit in?

2. The Sample

The sample is selected from the 5-GHz S4 and S5 surveys (Pauliny-Toth et al. 1978; Kühr et al. 1981) and is defined by the following criteria: (a) total flux density $S_{5\,\text{GHz}} \geq 1.3$ Jy; (b) declination $\delta > 35°$; (c) galactic latitude $|b| > 10°$. A complete list of the 65 sources in the sample is given by Pearson and Readhead (1984). Most are powerful, high-redshift objects associated with quasars or BL Lac objects. There are a few low-redshift objects (e.g., Markarian 501, 3C 371), but low-luminosity sources such as Seyfert galaxies are very under-represented compared to a volume-limited sample. We have made VLBI images at two or three epochs of most of the 45 compact sources strong enough to detect with the Mark II system (Pearson and Readhead 1981, 1984, and work in preparation). Our complementary optical study of the sources in the sample is described by Lawrence et al. (this Workshop, page 260).

It is possible to distinguish several classes of sources with rather different properties. Certainly, the images are not all the same; but it is difficult to tell which are the salient features of the images that impart important information about the radiation mechanism and the energy source, and which features are merely accidental. We should emphasize that assignment to classes is somewhat arbitrary, and we may be picking the wrong features to do it. At this stage, it is not clear if by grouping the sources we have identified fundamentally distinct classes of sources, or different stages in the evolution of basically similar sources. We have found, though, that having grouped the sources just on the basis of morphology, there are correlations with other observable characteristics, which suggests that there is some physical significance to this classification.

We have divided the sample into five classes, based on the large-scale structures, the radio spectra, and the radio morphologies revealed by the VLBI observations. A few sources are left over because we don't know where to put them.

Extended Triple Sources (21 objects)

These are steep-spectrum, extended objects in which the main emission regions straddle the optical galaxy or quasar. In most of these objects, the central components are too weak to be mapped with Mark II VLBI. One notable exception is 3C 179 (Porcas, this Workshop, page 12).

Compact Steep-Spectrum Sources (7 objects)

The sources in this class (Peacock and Wall 1982; Fanti and Fanti, this Workshop, page 174) usually have an overall angular size $\lesssim 3''$. In addition to their characteristic steep spectra, which usually have a turnover in the range 0.1–1 GHz, these sources are characterized by low polarization and low radio variability. They are rewarding objects to study with VLBI because they have complex structure on a wide range of angular scales. It is found that they basically have a "core-jet" structure, but unlike the flat-spectrum sources, the core is relatively weak and most of the emission comes from the steep-spectrum jet.

Asymmetric (Core-Jet) Sources (14 objects)

The majority of the objects dominated by a compact, flat-spectrum radio component coincident with the optical object fall into this class. Strictly speaking, for an object to be classified as a "core-jet", one component should have a flat or inverted spectrum; thus observations at at least two frequencies are needed. In some cases, we have provisionally assigned sources to this class based on a VLBI image at one frequency only, but experience suggests that morphology is normally a reliable indicator. We find that the objects in this class generally have high levels of polarization (1–5%) and are highly variable in flux density. It is in this class that most of the well-known superluminal sources belong.

Unresolved and Barely Resolved Sources (9 objects)

These sources appear to be more compact versions of the asymmetric core-jet sources. They have similar levels of polarization and variability.

Compact Double Sources (5 objects)

These sources appear to form a distinct class, quite different from other compact, flat-spectrum sources (Phillips and Mutel 1982). Hodges and Mutel (this Workshop, page 168) discuss the defining characteristics of this class. The sources are distinguished by their morphology—there are two well-separated emission regions of comparable brightness—and by their spectra, which are steep (optically thin) at high frequencies, exhibit a maximum in the rage 1–5 GHz, and are inverted at low frequencies.

They all have low polarization (< 0.5%) and show little if any variability. All of the sources in this class in our sample are identified with faint galaxies.

Miscellaneous (9 objects)

There are four compact sources in the sample which were too resolved on transcontinental U.S. baselines to be included in our first-epoch mapping observations. We have subsequently observed these on the European VLBI network, but we are not yet in a position to classify them. We have not classified a further five sources because they do not clearly belong to any of the above classes. In some cases, further observations may clarify the situation—e.g., 4C 39.25 (Shaffer and Marscher, page 67) should perhaps now be assigned to the "core-jet" class—but some of the sources—e.g., NGC 1275 (Backer, page 76) and M82—are apparently unique in this sample.

3. New Superluminal Sources

It is difficult to demonstrate conclusively that a source is superluminal, and it is even more difficult to prove that it is not. The source must have at least two peaks whose separation can be measured, and the peaks must be detected at at least two epochs. It is difficult to quantify the selection effects involved, but it is clear that we cannot show that a source is superluminal if (a) the core is too strong, so we do not have enough dynamic range to see peaks in the jet; (b) the jet is too smooth, and has no distinguishable peaks; or (c) the source is varying too fast, so peaks cannot be correlated from epoch to epoch. In addition, there are instrumental problems. One would like to have similar (u, v) coverage at all epochs, and similar dynamic range. Two images may not be enough: we cannot be sure that we are seeing a coherent motion until we have made four or five images. Then there is the problem of aligning images made at different epochs—which feature is fixed, and which moving? The best solution to this problem is precision astrometry (Bartel et al. 1986).

Thus we can only quote lower limits on the fraction of superluminal sources in the sample. In particular, many of the sources in the "asymmetric" and "barely resolved" classes cannot yet be proven non-superluminal. In only a few sources can we place reliable subluminal upper limits on the internal motions; these are all in the "compact double" class.

When we began this survey, three of the sources in the sample of 65 were known to be superluminal (3C 179, 3C 345, and BL Lac). As a result of the survey, we have found four new superluminal sources

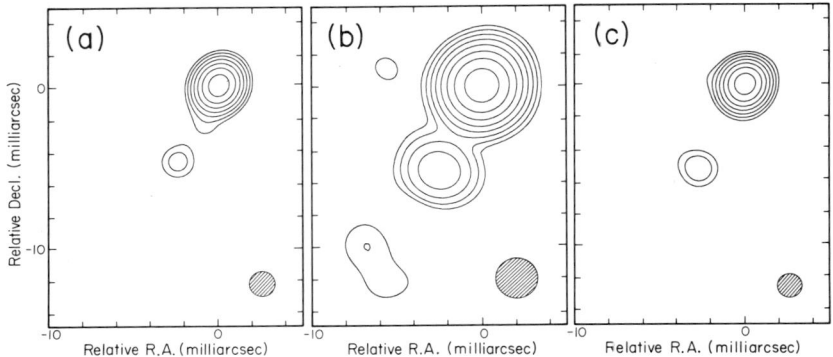

Figure 1 VLBI maps of 0850+581 at 5 GHz. (a) Epoch 1980.53: peak brightness 730 mJy per beam (1.5 mas FWHM); contours at −1, 1, 2, 4, 8, 16, 32, and 64%. (b) Epoch 1984.39: peak brightness 738 mJy per beam (2.5 mas FWHM); contours at −0.5, 0.5, 1, 2, 4, 8, 16, 32, and 64%. (c) Epoch 1985.77: peak brightness 862 mJy per beam (1.5 mas FWHM); contours at −1, 1, 2, 4, 8, 16, 32, and 64% (Barthel et al. 1986).

(0850+581, 3C 216, 1642+690, and 1928+738) which we describe in the following paragraphs. In addition, Witzel and his colleagues (this Workshop, page 83) have shown that at least one more member of the sample (0212+735) is superluminal, for a current total of eight.

0850+581

We have made three maps of 0850+581 at 5 GHz (Figure 1; Barthel et al. 1986) which show an asymmetric structure that can be described as a "core-jet". The secondary component is much weaker than the core, but we can measure a speed of expansion along the jet of 0.12 ± 0.02 mas yr^{-1}, or $v_{app} = (3.9 \pm 0.7)h^{-1}c$ at $z = 1.322$ (assuming $H_0 = 100$ km s^{-1} Mpc^{-1} and $q_0 = 0.5$). On arcsecond scales, the source is triple, and the milliarcsecond jet is aligned with a curved knotty jet running 10″ from the core to the southern lobe. The total angular size is > 15″.2 (Browne, this Workshop, page 129); recent maps by Rusk and Seaquist (1986) show that it is at least 20″.

0906+430 (3C 216)

3C 216 is unlike the other known superluminal sources in that the VLBI observations show a milliarcsecond jet aligned roughly perpendicular to the major axis of the large-scale emission (Pearson, Perley, and Readhead 1985; Browne, this Workshop, page 129); the only comparable source is the quasar 3C 309.1 (Wilkinson et al. 1984a, 1984b, 1986). Both 3C 216 and 3C 309.1 are compact steep-spectrum sources. We have

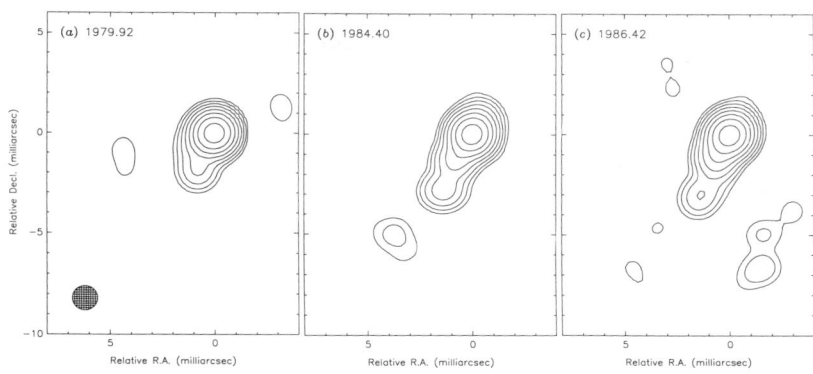

Figure 2 VLBI maps of 0906+430 (3C 216) at 5 GHz. (a) Epoch 1979.92: peak brightness 861 mJy per beam (1.2 mas FWHM). (b) Epoch 1984.40: peak brightness 498 mJy per beam. (c) Epoch 1986.42: peak brightness 494 mJy per beam. Contour levels are 0.5, 1, 2, 4, 8, 16, 32, and 64% of the peak in each image.

5-GHz observations of this source at three epochs (Figure 2). Pearson et al. (1986b) gave a value of $\mu = 0.34$ mas yr^{-1}, measured between the first two epochs; the third-epoch map perhaps suggests that the velocity has decreased, but in view of the difficulty of measuring an accurate separation from the first map, we do not regard the decrease as significant. All three maps are consistent with a uniform expansion at 0.11 mas yr^{-1}, corresponding to $v_{\mathrm{app}} = 2.4h^{-1}c$ at $z = 0.669$.

1642+690

1642+690 has a dominant, compact core with a jet extending 5″ to the south. The two VLBI maps of the core (Figure 3; Pearson et al. 1986a) show that the milliarcsecond structure is similar to that of the superluminal sources 3C 273 and 3C 345. A milliarcsecond jet is closely aligned with the arcsecond jet, and a "knot" in the jet is moving outwards along the jet at 0.34 mas yr^{-1}, an apparent transverse velocity of $v_{\mathrm{app}} = 7.9h^{-1}c$ at $z = 0.751$.

1928+738

The VLBI maps of 1928+738 (Witzel, this Workshop, page 83) show a flat-spectrum "core" and a long "jet" extending 17 mas from the core in position angle 165°. The jet has a spectral index $\alpha \approx -0.5$ and contains at least nine components, five of which were detected at both epochs. The most plausible alignment of the maps from the two epochs indicates that all five of these components have a proper motion relative to the core of ~ 0.6 mas yr^{-1}, or apparent transverse velocities $v_{\mathrm{app}} \approx 7.0h^{-1}c$ (Eckart et al. 1985). The large-scale structure of 1928+738 is shown in

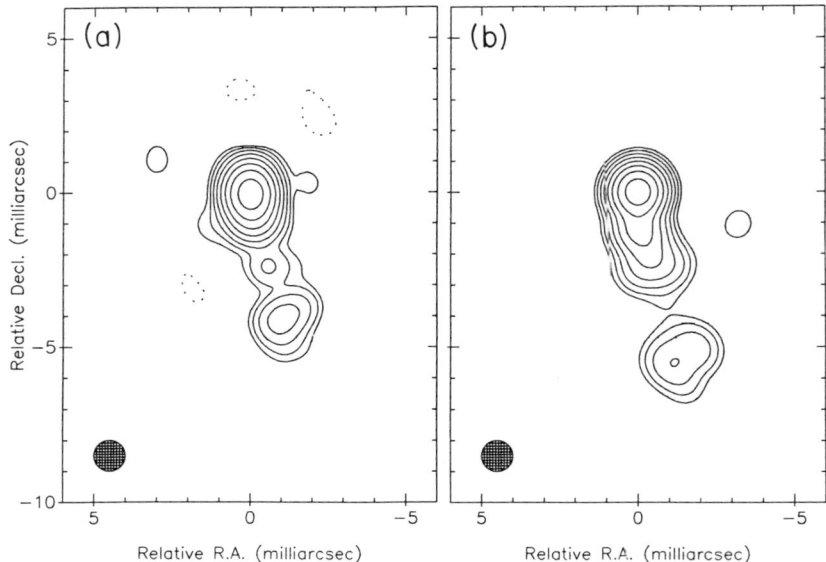

Figure 3 VLBI maps of 1642+690 at 5 GHz. Both maps are presented with a circular Gaussian restoring beam with FWHM 1.0 mas (*hatched circle*) and contour levels −0.5 (*dashed*), 0.5, 1, 2, 4, 8, 16, 32, 64% of the peak. (*a*) Epoch 1980.53: peak brightness 1.32 Jy per beam. (*b*) Epoch 1983.93: peak brightness 0.79 Jy per beam (Pearson *et al.* 1986a).

VLA maps by Simon *et al.* (this Workshop, page 155) and by Rusk and Rusk (1986).

4. Non-Superluminal Sources

We mentioned above that it is very difficult to prove that a source is *not* superluminal. So far, we have only two sources in which the structure is such that we can place subluminal limits on any internal motions. Both are compact double sources (Readhead, Pearson, and Unwin 1984).

0710+439

0710+439 is like the "canonical" compact double sources in that it has a spectrum peaking at about 2 GHz, has low polarization and low variability, and is identified with a faint (20^m, $z = 0.517$) narrow emission-line galaxy. Rather than two components, though, it has three (A, B, and C) lying very close to a straight line. There is no evidence for component motion in the 2.5 yr between our two 5-GHz images. Formally, there is an insignificant subluminal expansion: between components A and B, $\mu = 0.02 \pm 0.02$ mas yr^{-1} ($v_{\text{app}}/c = 0.4 \pm 0.4$); between components A and C, $\mu = 0.05 \pm 0.02$ mas yr^{-1} ($v_{\text{app}}/c = 0.9 \pm 0.4$).

2021+614

2021+614 has been discussed by Bartel et al. (1984). Early observations showed two almost equal components. As it has low polarization and low variability, and is identified with a faint (20^m, $z = 0.2266$) narrow emission-line galaxy, it was natural to place it in the compact double class. Images with higher resolution and higher dynamic range over a range of frequencies show that it is more complicated, however. It consists of at least four components (A, B, C, and D), of which D has a flatter spectrum than the others. The nominal limit on the proper motion between the two bright components, B and D, is certainly subluminal: $\mu = 0.02 \pm 0.02$ mas yr^{-1} ($v_{app}/c = 0.2 \pm 0.2$) (Readhead, Pearson, and Unwin 1984).

5. Discussion

Superluminal motion is common: so far it has been seen in at least eight of the 45 sources that we have mapped. In view of the difficulty of detecting superluminal motion without measurements of high quality at many epochs, we should not be surprised if almost all the sources (except for the compact doubles) eventually turn out to be superluminal.

We have found superluminal motion in compact flat-spectrum sources (e.g., 3C 345), compact steep-spectrum sources (3C 216), and extended triple sources (3C 179, 1928+738). We have not found any superluminal "compact double" sources, and we can place subluminal limits on two (0710+439, 2021+614).

In sources which show superluminal motion, there is almost always a one-sided, large-scale jet, and the superluminal motion is outwards from the core in the approximate direction of the jet. While some superluminal sources (notably 3C 273) are one-sided, many are revealed to have two-sided large-scale structure when examined with sufficient dynamic range. 0850+581, 1642+690, and 1928+738 all have very similar two-sided morphologies. 1928+738 is exceptional in showing some evidence for a counterjet (Simon et al., this Workshop, page 155). The large-scale structures of the superluminal sources are discussed by several contributors to this Workshop; in particular Barthel (page 148) and Simon (page 155) discuss the problems posed by the very large sizes that are derived by naïve deprojection of the superluminal sources.

The detection of superluminal motion in a steep-spectrum compact source (3C 216) is of particular interest. 3C 216 does have a large-scale triple morphology, but its largest angular size ($\sim 2''$) is small compared with sources like 1928+738, and the milliarcsecond jet is roughly perpendicular to the large-scale extension. The place of steep-spectrum compact sources in any "unified scheme" is unclear. The linear sizes of these sources are typically sub-galactic (a few kiloparsecs) and there is in some cases direct evidence for interaction between the radio source and the galactic environment. The feature that most obviously distinguishes the steep-spectrum compact sources from the flat-spectrum sources is the weaknes of the flat-spectrum core. It could be that the core is beamed away from the line of sight (but the jet must be beamed towards us in order to exhibit superluminal motion), but the differences could be more fundamental and related to the environment.

This work was supported in part by the National Science Foundation under grant AST 85-09822 to the Owens Valley Radio Observatory. Once again, we express our gratitude to the participating observatories of the U.S. and European VLBI Networks.

References

Bartel, N., Herring, T. A., Ratner, M. I., Shapiro, I. I., and Corey, B. E. 1986, *Nature*, **319**, 733.
Bartel, N., Shapiro, I. I., Corey, B. E., Marcaide, J. M., Rogers, A. E. E., Whitney, A. R., Cappallo, R. J., Kühr, H., Graham, D. A., and Bååth, L. B. 1984, *Astrophys. J.*, **279**, 116.
Barthel, P. D., Pearson, T. J., Readhead, A. C. S., and Canzian, B. J. 1986, *Astrophys. J. (Letters)*, **310**, L7.
Cawthorne, T. V., Scheuer, P. A. G., Morison, I., and Muxlow, T. W. B. 1986, *Monthly Notices Roy. Astron. Soc.*, **219**, 883; erratum *Monthly Notices Roy. Astron. Soc.*, **222**, 895.
Eckart, A., Witzel, A., Biermann, P., Pearson, T. J., Readhead, A. C. S., and Johnston, K. J. 1985, *Astrophys. J. (Letters)*, **296**, L23.
Kühr, H., Witzel, A., Pauliny-Toth, I. I. K., and Nauber, U. 1981, *Astron. Astrophys. Suppl.*, **45**, 367.
Pauliny-Toth, I. I. K., Witzel, A., Preuss, E., Kühr, H., Kellermann, K. I., Fomalont, E. B., and Davis, M. M. 1978, *Astron. J.*, **83**, 451.
Peacock, J. A., and Wall, J. V. 1982, *Monthly Notices Roy. Astron. Soc.*, **198**, 843.
Pearson, T. J., Barthel, P. D., Lawrence, C. R., and Readhead, A. C. S. 1986a, *Astrophys. J. (Letters)*, **300**, L25.
Pearson, T. J., Barthel, P. D., Readhead, A. C. S., and Lawrence, C. R. 1986b, in *IAU Symposium 119, Quasars*, ed. G. Swarup and V. K. Kapahi (Dordrecht: Reidel), p. 163.
Pearson, T. J., Perley, R. A., and Readhead, A. C. S. 1985, *Astron. J.*, **90**, 738.
Pearson, T. J., and Readhead, A. C. S. 1981, *Astrophys. J.*, **248**, 61.
Pearson, T. J., and Readhead, A. C. S. 1984, in *IAU Symposium 110, VLBI and Compact Radio Sources*, ed. R. Fanti, K. Kellermann, and G. Setti (Dordrecht: Reidel), p. 15.
Phillips, R. B., and Mutel, R. L. 1982, *Astron. Astrophys.*, **106**, 21.
Readhead, A. C. S., Pearson, T. J., and Unwin, S. C. 1984, in *IAU Symposium 110, VLBI and Compact Radio Sources*, ed. R. Fanti, K. Kellermann, and G. Setti (Dordrecht: Reidel), p. 131.
Rusk, R., and Rusk, A. C. M. 1986, *Can. J. Phys.*, **64**, 440.
Rusk, R. E., and Seaquist, E. R. 1986, *Bull. Am. Astron. Soc.*, **18**, 994.
Wilkinson, P. N., Cornwell, T. J., Kus, A. J., Readhead, A. C. S., and Pearson, T. J. 1984a, in *Physics of Energy Transport in Extragalactic Radio Sources*, NRAO Workshop No. 9, ed. A. H. Bridle and J. A. Eilek (Green Bank: National Radio Astronomy Observatory), p. 76.

Wilkinson, P. N., Kus, A. J., Pearson, T. J., Readhead, A. C. S., and Cornwell, T. J. 1986, in *IAU Symposium 119, Quasars*, ed. G. Swarup and V. K. Kapahi (Dordrecht: Reidel), p. 165.
Wilkinson, P. N., Spencer, R. E., Readhead, A. C. S., Pearson, T. J., and Simon, R. S. 1984b, in *IAU Symposium 110, VLBI and Compact Radio Sources*, ed. R. Fanti, K. Kellermann, and G. Setti (Dordrecht: Reidel), p. 25.

Tests of Beaming Models

PETER A. G. SCHEUER

Mullard Radio Astronomy Observatory, Cavendish Laboratory

I propose to comment on five topics. First of all, I want to indicate the current state of the obvious and classical tests of relativistic beaming: 1. The statistics of jet fluxes, and 2. The statistics of jet speeds. Then I shall speak with the zeal of a convert on 3. Why relativistic beaming is true. Finally I want to mention some residual worries under the headings 4. When is a quasar not a quasar? and 5. Curly jets.

1. Statistics of Jet Fluxes

The simplest theory of relativistically beamed jets (Scheuer and Readhead 1979; Blandford and Königl 1979), in which the measured pattern velocity is determined by the Lorentz factor of the radio-emitting matter and the jet's inclination to the line of sight, makes clear predictions about the distribution of the flux densities of the jets. However, bright jets which have been observed at high resolution not only bend from side to side (and therefore have several different inclinations to the line of sight) but appear to have their bright patches immediately after strong shocks. Thus, the pattern speed may be distinctly larger than the material speed. The inevitable conclusion is that the statistics of jet flux densities (or their ratios to flux density in "lobes", presumed slow) cannot be used as unambiguous evidence against relativistic beaming, though they remain potentially useful discriminants between specific physical models. These points have been made at length, and with illustrative models, by Lind and Blandford (1985). Therefore, I shall not consider flux density statistics any further here.

2. Statistics of Jet Speeds

Statistics of measured superluminal velocities could conclusively (?) reject the relativistic beaming hypothesis, provided that we use measurements made on a sample of objects free of bias in orientation, for, regardless of the Lorentz factor, and regardless of the nature of the velocity (i.e., the considerations of Lind and Blandford),

$< 40\%$ have apparent velocity $> 2c$

< 13% have apparent velocity > $4c$
< 3% have apparent velocity > $8c$
< 0.8% have apparent velocity > $16c$

if the relativistic beaming hypothesis is basically true. Note that it is quite unnecessary to observe the whole sample to draw strong conclusions; 10 sources with $v_{app} > 8c$ in a sample of 50 would be plenty to reject any relativistic beaming model contemplated so far.

I want to mention two relevant samples. One is the sample of quasars selected from the Jodrell Bank 966 MHz survey, on which Zensus and Porcas have already done much valuable work (this Workshop, page 126). The other is a subsample of the LRL sample (Laing, Riley, and Longair 1983) with high power in the radio lobes, which is selected at a still lower frequency and does not exclude radio galaxies (Cawthorne et al. 1986). We have not been successful in obtaining observing time for sources from this sample as such, but some of the sources in it are being monitored by Hough and Readhead, and the superluminal velocity of 3C 245, reported at this Workshop (page 114), is a recent product of their work. What strikes me about the results reported so far (see Richard Porcas's admirable summary in these Proceedings, page 12) is that, as the measurements are extended to relatively weaker cores, so the measured velocities are becoming smaller, exactly as one would expect for relativistic beaming and as indicated in the table above.

However, as really testing observations come closer, it is the theoretician's duty to look for ways of escape if the observations should confound the predictions. Here are some queries to justify the question mark in the first sentence of this section:

(a) Is Hubble's constant 50 or 100 km s^{-1} Mpc^{-1}?
(b) Are the samples really free of orientation bias? I shall mention two possible doubts later (Section 4).
(c) Even if the samples are unbiased in orientation, are the measured speeds unbiased (Section 5)?

3. Why Relativistic Beaming is True

Before this summer, I thought that the most important thing to do in the study of superluminal motions was to find out whether relativistic beaming was the correct explanation. I am now convinced that relativistic beaming is basically right; the reason lies in some innocent-looking observations of the polarization of radio lobes by Robert Laing, and, with his permission and connivance, I shall now outline his work. Figure 1 shows a 6 cm VLA map of the quasar 3C 133, made with enough resolution to show the jet very clearly. Figure 2 shows the polarization of the source at 6 cm and 20 cm wavelength, at the resolution of the

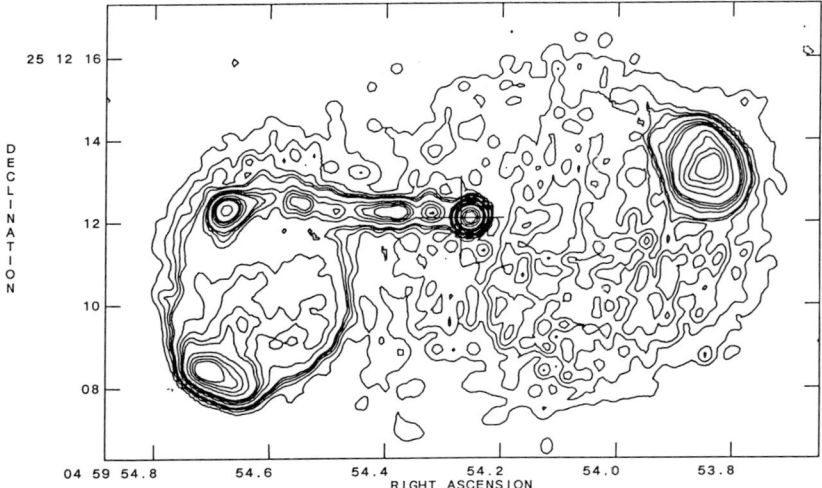

Figure 1 VLA map of 3C 133 at 6 cm, showing the jet and the compact hotspot on which it terminates (R. Laing, private communication).

20 cm map. The 20 cm map shows strong depolarization on the side which does *not* contain the jet, but hardly any depolarization on the side that contains the jet. This is not remarkable in itself; what is remarkable is that the same behavior is seen in every source that Laing has examined, or for which he has found adequate polarization maps in the literature. More precisely: of the nine sources in his data, seven show distinctly greater depolarization on the side opposite the jet; the two others show equal or no depolarization. Cygnus A is, of course, a classic example: the Sf component, on the side opposite the jet and the compact knot B, has long been notorious for its enormous rotation measure.

Now most people would have predicted that any physical difference between the two sides would be in the direction of more matter (and hence more depolarization) on the side in which a jet is visible; to account for Laing's data we should have to suppose that the jet sweeps away plasma, not only from its immediate neighborhood but preferentially from the lobe on its side. Taken together with direct evidence indicating that depolarization arises mainly from plasma outside the source (e.g., Perley, Bridle, and Willis 1984), Laing's evidence points very strongly indeed to the interpretation that the jet is on the side closer to us (and that is why we see it prominently), and the radiation from the lobe on the same side therefore has a shorter path through the intergalactic plasma surrounding the source.

What I find particularly convincing about the evidence from de-

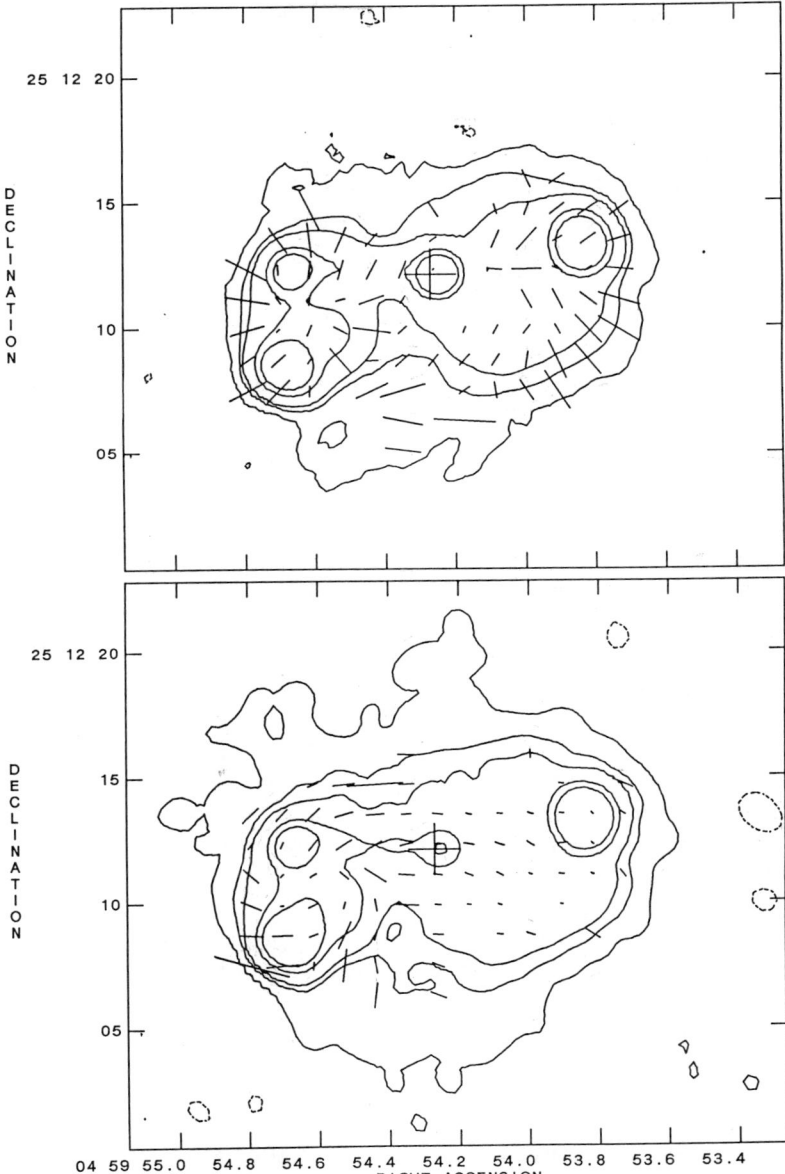

Figure 2 Polarization maps of 3C 133 at 6 cm (*top*) and 20 cm (*bottom*). In each map, the lines show the direction of the electric vector and have lengths proportional to fractional polarization. Both maps have been convolved to the same angular resolution (R. Laing, private communication).

polarization is that it shows relativistic beaming in the large-scale jets, quite independently of any arguments about superluminal motion. It confirms directly the more circuitous argument from the fact that large-scale jets are always continuations of VLBI jets, never appearing to arise from the opposite side of the nucleus. It also makes a coherent story with Laing's observations of "compact hot spots", which he finds in many powerful sources, but only on one side, and always on the same side as any jet (VLBI or large-scale). Evidently, the speed of the jet material is still a substantial fraction of the speed of light after it has entered the "compact hot spot", presumably through an oblique shock.

Thus, I now believe that the reality of relativistic beaming is established, from the nucleus at least to the "compact hot spot", and we can go on to explore the observations of radio sources in this light. There are many implications: in the next section I want to look at two implications relevant to the statistics discussed earlier.

4. When is a Quasar not a Quasar?

What I am really concerned with in this section is whether a sample of quasars selected by flux density in its large-scale radio structures is truly unbiased in orientation.

Folklore has it that as many as 50% of quasars have large-scale jets of a power not totally dissimilar to that of the lobes (though when I looked at maps of as many of the LRL sample as I could find, the proportion seemed distinctly lower). At first this suggested to me that the speeds of the large-scale jets must be $0.5c$ or $0.7c$ rather than $0.99c$, i.e., that there is some deceleration from the superluminal VLBI jets. But this is not necessarily true. First of all, a jet can look very prominent on a good map even when its flux is only a few percent of the total flux. We may also be misled in other ways:

(a) We now have to contend with the fact that some of the large-scale structure is relativistically beamed. Looking at maps suggests that large-scale jets are likely to represent a small but not insignificant infiltration of supposedly unbiased samples.

(b) Quasars may not expose their active nucleus when seen edge-on; thus we are not necessarily choosing an unbiased sample unless we include objects independently of their optical classification. This is an old idea (at least six years old), which gets some support from the well-known (but hard-to-quantify) fact that quasars are, in general, much more unsymmetrical and bent than radio galaxies; Robert Laing's discovery that large-scale jets really are relativistic should encourage us to look at it again.

The number of "hidden" quasars cannot be very large; at most, the number is comparable with the space density of known quasars. It is

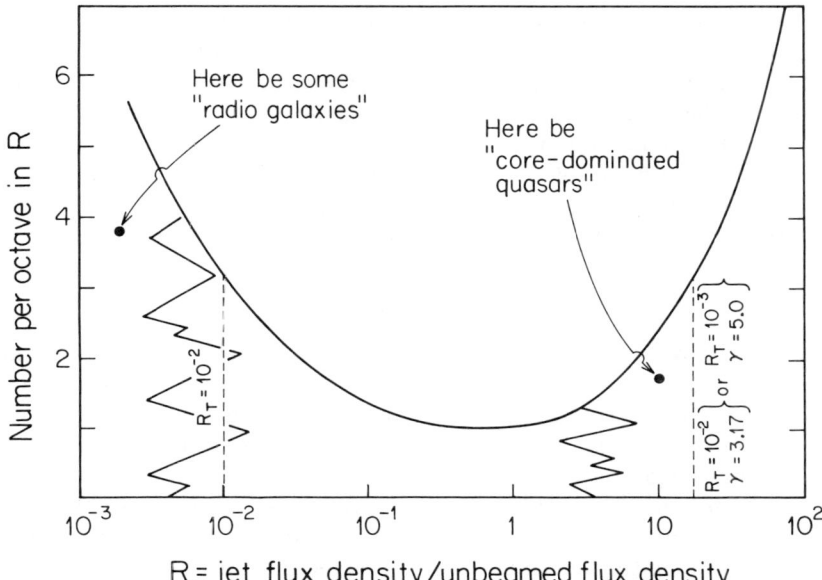

Figure 3 Illustrating possible behavior of statistics for a sample of sources selected by total flux in extended structure. Note that equal areas in this diagram represent equal numbers of sources. In compiling this diagram one needs to make some assumption about the variation of numbers dN of sources with flux density S, and I have assumed $dN \propto S^{-2} dS$ as perhaps appropriate for a sample of very powerful sources, extending to considerable redshifts.

clearly impossible to hide the narrow-line emission region, which is very large. There are some radio galaxies which have astonishingly strong narrow lines, as strong as any quasar's, but they are very rare. Also, in the sample of sources of high radio lobe power (Cawthorne et al. 1986) the total numbers of quasars and radio galaxies in the sample are similar.

Figure 3 shows what the statistics of a sample selected by total flux density in large scale structure might look like. It is a purely illustrative example, designed to show what might happen as a result of suspicions (a) and (b) above, and that the resulting numbers are not totally unreasonable. Clearly, the vertical and horizontal axes are adjustable by arbitrary factors (for sample size and power of jet seen transversely, respectively); in addition, there must be some scatter in the intrinsic power of the jet, so a more realistic diagram would consist of many such figures displaced horizontally and superposed. At low values of $R =$ (jet flux/unbeamed flux) the predicted numbers per octave of R follow the familiar $R^{-1/(2+\alpha)}$ law, but jets seen closer to the line of sight boost the total flux significantly, and hence increase the numbers in a flux-limited sample. Consider, for example, the case: transverse $R = 0.001$, $\gamma = 5$. There are then comparable numbers of sources with R below 0.01 (large-scale jet very weak or not seen) and sources with $0.01 < R < 10$. Many of the former would be classified as radio galaxies. Sources with $R > 10$ (or 1) will be core-dominated sources; the borderline depends the ratio of the intrinsic fluxes of the jet on large and VLBI scales. Increasing γ will increase the fraction of core-dominated sources, but leave the rest of the statistics virtually unchanged.

5. Curly Jets

One of the obvious worries about the interpretation of superluminal motion concerns the bending of many of the superluminal jets. On the one hand, the interpretation of their relatively large bending angles as projection effects has from the start provided circumstantial evidence for a small angle between the jet and the line of sight. On the other hand, large apparent bending angles imply variations comparable with γ^{-1} in the angle to the line of sight too, and hence large brightness variations along the length of the jet. The location of bright spots along the jet might therefore be determined predominantly by small changes in jet direction rather than by intrinsic radio power, and we wonder why these spots move at all. Is it that the intrinsic radio power varies smoothly along a jet whose channel is massaged by a relativistic wind outside it? Or are the bending angles generally too gentle to cause drastic changes in observed brightness? In order to get some feeling for these effects, I did a few simulations of wavy relativistic jets; it is very easy to do, and has of course been done before for a number of specific models—in particular, models of precessing jets. What I learned was that a degree of waviness that regularly produces apparent bends of the order of 90° does indeed produce very large (100 : 1) variations in brightness. However, the simulations showed considerable lengths of bright, tolerably straight jet, and essentially invisible portions at quite different angles. This raises the possibility that in some sources we may be measuring the motion of real irregularities, but only along a portion of the jet particularly close to the line of sight.

These are of course only a few very simple-minded simulations, which can only have heuristic value. (That is why I print no diagrams of them.) But even these simulations of shock-less jets with only a single sine-wave on them show some qualitative resemblances to what we see in sources such as 3C 345 and 3C 309.1. What they suggest to me is that we must be on our guard just in case measurements of superluminal velocity are biased towards high values within a single source.

References

Blandford, R. D., and Königl, A. 1979, *Astrophys. J.*, **232**, 34.
Cawthorne, T. V., Scheuer, P. A. G., Morison, I., and Muxlow, T. W. B. 1986, *Monthly Notices Roy. Astron. Soc.*, **219**, 883; erratum *Monthly Notices Roy. Astron. Soc.*, **222**, 895.
Laing, R. A., Riley, J. M., and Longair, M. S. 1983, *Monthly Notices Roy. Astron. Soc.*, **204**, 151.
Lind, K. R., and Blandford, R. D. 1985, *Astrophys. J.*, **295**, 358.
Perley, R. A., Bridle, A. H., and Willis, A. G. 1984, *Astrophys. J. Suppl.*, **54**, 291.
Scheuer, P. A. G., and Readhead, A. C. S. 1979, *Nature*, **277**, 182.

Discussion

Peter Scheuer's talk was interrupted by a discussion of whether or not quasars could be radio galaxies seen "end-on". The editors asked the principal disputants to summarize the main points of the argument; these summaries are reproduced here.

Ian Browne: I have four arguments against the suggestion that quasars could be radio galaxies seen end-on.

1. Narrow Line Luminosities

Though it seems possible to absorb or scatter emission from the core and broad-line region, it is much more difficult to imagine affecting the narrow-line emission in the same way since it comes from much further out from the nucleus. If radio galaxies and quasars were intrinsically the same, those with similar radio luminosities should have similar narrow-line luminosities. As a rough check of this prediction, I have taken the fifteen 3C and 4C radio galaxies with [O III] $\lambda\lambda 4959, 5007$ data published by Yee and Oke (1978) and Wilkinson, Hine, and Sargent (1981) and which have radio luminosities $> 3 \times 10^{43}$ erg s^{-1}. These radio galaxies should be representative of all radio galaxies in these surveys. I have matched each radio galaxy with a steep-spectrum quasar with a similar extended radio luminosity for which an [O III] luminosity is also available. On average, the [O III] luminosities of the radio galaxies are about a factor of five smaller than those of the quasars. The formal probability that the pairs are drawn from a single parent population is $\leq 0.5\%$.

2. Properties of the Underlying Galaxy

One distinctive feature of radio galaxies is their small spread in absolute magnitudes. A similar small spread would be expected in the absolute magnitudes of the galaxies underlying radio-loud quasars. Examination of the data of Hutchings, Crampton, and Campbell (1984) on 19 radio-loud quasars shows that there is a range of ~ 4 in absolute magnitude with some objects much less luminous than the typical radio galaxy. If these absolute magnitudes are to be trusted, this argues strongly against the quasar–galaxy unification hypothesis.

3. Cosmological Evolution

If radio galaxies and quasars belong to the same parent population, the space density of steep-spectrum radio galaxies and steep-spectrum quasars should evolve in the same way. (Flat-spectrum objects are influenced by beaming and therefore should be excluded.) This does not appear to be so. If we take the revised 3C survey (Laing, Riley, and Longair 1983) for which optical identifications and redshifts are virtually complete, the ratio of radio galaxies to quasars steadily decreases with increasing redshift. In the redshift range $0.3 < z < 0.7$, the ratio is about 2 : 1, decreasing to about 1 : 1 for $z > 1.5$. This means that only a fraction of radio galaxies can be quasars seen in the plane of the sky.

4. Infrared Emission

If a significant proportion of the infrared emission from quasars is from dust, this radiation is likely to be isotropic and be visible in both radio galaxies and quasars. (Dust is one way of obscuring the cores and broad-line region. If this is the case, reradiation in the infrared is a requirement of the model). However, radio galaxies are much weaker in the near infrared than quasars. Thus, the model requires that the nuclear emission is scattered rather than absorbed and that very little of the infrared radiation seen in quasars is thermal in origin. Some people (e.g., Wills 1987) argue strongly that features in the near-infrared spectra of quasars are produced by dust emission.

I also have one argument in support of radio galaxy–quasar unification. Yee (1980) shows that there is a strong correlation between nonthermal optical emission and H β luminosity over a wide range of objects including Seyfert 1 galaxies, radio galaxies, and quasars. Narrow-line radio galaxies and steep-spectrum radio quasars fit well onto this correlation. When the correlation of [O III] luminosity with non-thermal emission is examined, the narrow-line radio galaxies are found to have significantly higher [O III] luminosities than the broad-line objects for a similar core power. This can be understood easily if the optical core and the broad lines (H β in this case) are being obscured in the narrow-line objects, i.e., exactly what the unification hypothesis would predict.

Peter Scheuer: I agree with most of Ian Browne's comments, but they are addressed to a theory very different from the question I raised in Section 4 of my talk. His comments refer to a sort of grand unified theory in which most or all radio galaxies are quasars that are not seen pole-on. What I suggested we should reconsider is the possibility that some quasars masquerade as "radio galaxies" through being seen edge-on to the supposed accretion disc; that involves a space density of "radio galaxies" at most comparable with the space density of known quasars, and thus only a minute fraction of the space density of all radio galaxies. Their properties need not be typical of radio galaxies; indeed, they would be expected to be exceptional in various respects. The optical continuum of these "radio galaxies" could, for all I know, be hidden from us by electron scattering, and might ultimately reappear as kinetic energy. The broad-line emission comes from further out, and could not plausibly be electron-scattered; it would have to be absorbed in dust and reradiated in the infrared. Thus (some or all) quasars would reradiate a substantial part of their broad-line luminosities in the infrared. More strikingly, there would be a few "radio galaxies" with infrared luminosities similar to the broad-line luminosities of quasars, and these would also tend to have very strong narrow lines, as strong as those of quasars. All this seems consistent with Ian Browne's evidence and other observations so far; I do not know whether very strong narrow lines in radio galaxies correlate with strong infrared emission.

References

Hutchings, J. B., Crampton, D., and Campbell, B. 1984, *Astrophys. J.*, **280**, 41.
Laing, R. A., Riley, J. M., and Longair, M. S. 1983, *Monthly Notices Roy. Astron. Soc.*, **204**, 151.
Wilkinson, A., Hine, R. G., and Sargent, W. L. W. 1981, *Monthly Notices Roy. Astron. Soc.*, **196**, 669.
Wills, B. 1987, *Astrophys. J.*, submitted.
Yee, H. K. C. 1980, *Astrophys. J.*, **241**, 894.
Yee, H. K. C., and Oke, J. B. 1978, *Astrophys. J.*, **226**, 753.

Relativistic Beaming and the Nuclei of Double-Lobed Quasars

DAVID H. HOUGH
Jet Propulsion Laboratory

ANTHONY C. S. READHEAD
California Institute of Technology

1. Introduction

If the simple relativistic beaming model for extragalactic radio sources (Scheuer and Readhead 1979; Blandford and Königl 1979) is at all correct, Doppler boosting may bias us toward choosing those objects pointing nearly at us in flux-density limited samples. In order to minimize any such effects in statistical tests of physical theories of these objects, we have attempted to define a complete sample of sources with random orientations. The extended lobes of steep-spectrum, double-lobed sources are presumably not strongly beamed, and the nuclei of double-lobed quasars are on average stronger (and thus more accessible to VLBI) than those in radio galaxies. Therefore we began in 1980 our study of the complete, flux-density limited sample of 26 double-lobed quasars from the 3CR catalog (Laing, Riley, and Longair 1983), with the stipulations that $S_{178 \text{ MHz}} > 10$ Jy *after* subtracting out the nuclear flux density, $\delta > 10°$, and $|b| > 10°$. It is, of course, possible that the exclusion of galaxies introduces an orientation bias into our sample (e.g., Scheuer, this Workshop, page 104). In all there are 16 quasars in this sample for which the flux density of the central component is greater than 30 mJy at 5 GHz, and which can therefore be mapped by VLBI on the U.S. and European VLBI networks. A further six sources could be mapped at 8 GHz using the 64 m telescopes of the Deep Space Network and the VLA, which means that 85% of the sample can be mapped with existing VLBI arrays. Our work on this sample parallels a similar program by Zensus and Porcas (1986; this Workshop, page 126).

2. Sample Statistics

One of our main objectives in this study is to measure the distribution of superluminal velocities for sources in our sample, since this is one of the most powerful discriminants between rival theories. However, the simple relativistic beaming theory predicts the dependence of a number of observable features on the angle to the line of sight, θ, in addition to the superluminal velocity, and we shall discuss these first. The notation which we use is: R = ratio of the core flux density to the sum of the outer components at 5 GHz (emitted), L = projected linear size in kiloparsecs, and C = curvature (supplement of the angle between vectors connecting the core with the outer components). The predicted dependences are as follows:

(a) the jet-to-counterjet brightness ratio is

$$\left(\frac{1+\beta\cos\theta}{1-\beta\cos\theta}\right)^a,$$

where $a = 2 - \alpha$ or $3 - \alpha$, depending on the model (Scheuer and Readhead 1979; Blandford and Königl 1979), α is the spectral index of the core ($S \propto \nu^\alpha$), and βc is the bulk velocity of the radiating material;

(b) $R \propto \gamma^{-a}[(1-\beta\cos\theta)^{-a} + (1+\beta\cos\theta)^{-a}]$;
(c) $L \propto \sin\theta$;
(d) for cases where $0 < \theta < C_{\text{int}}$, where C_{int} is the intrinsic misalignment, $C \to 180°$ as $\theta \to 0°$ (Readhead et al. 1983).

Thus, on the beaming theory we expect a correlation between R and C, and anticorrelations between R and L and between L and C. Unfortunately, interaction with the surrounding medium is likely to increase R and C and reduce L (Readhead et al. 1978), thus producing correlations in the same sense as the beaming theory. Therefore tests based on these correlations can demonstrate consistency with these two models but are unlikely to discriminate between them. Our objective here, therefore, is to see whether the statistics are consistent with the beaming theory.

We begin by examining the dependence of L on redshift z, since we want to reduce as far as possible any effects due to cosmological evolution. The values of L and z for all 26 objects are shown in Figure 1a. There is a strong anticorrelation between L and z, and we have therefore divided the sample into two, at $z = 1.3$, for the purposes of our statistical studies.

The dependence of R on L is shown in Figure 1b for the 18 objects at $z < 1.3$. The following points should be noted: (a) There is a correlation

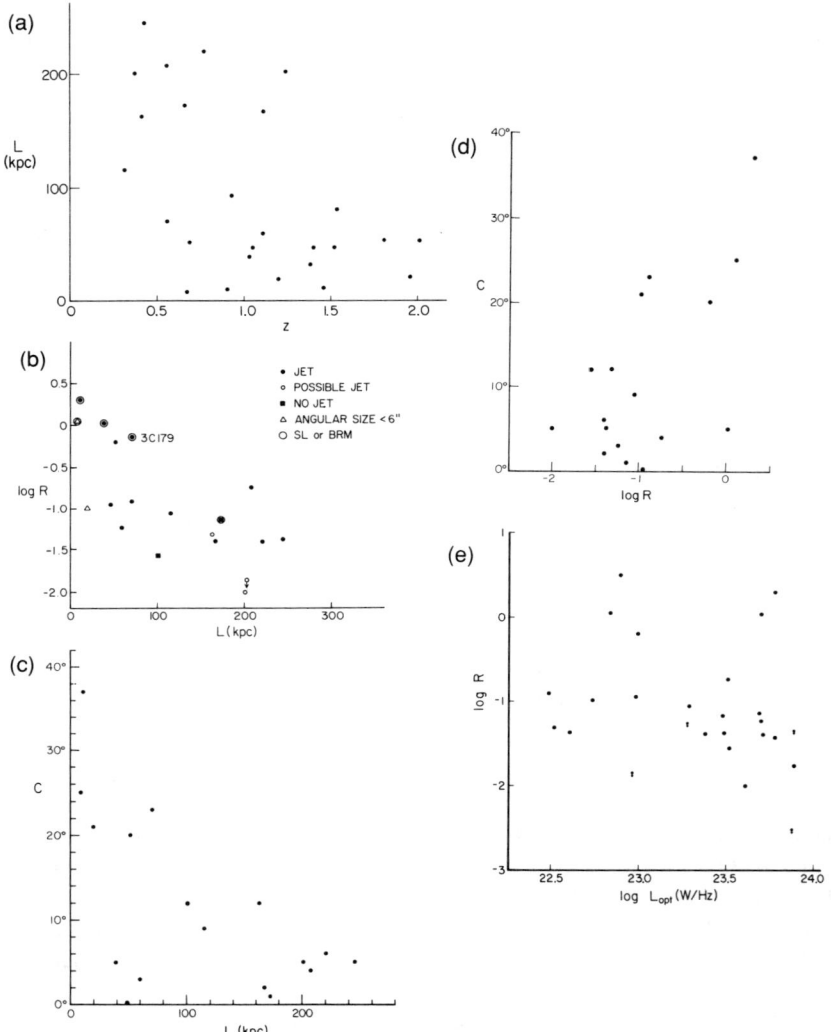

Figure 1 Correlations: (a) Projected linear size L versus redshift z for all 26 quasars. (b) Relative nuclear strength R versus projected linear size L for all 18 quasars at $z < 1.3$. (c) Curvature C versus projected linear size L for objects at $z < 1.3$. (d) Curvature C versus relative nuclear strength R for objects at $z < 1.3$. (e) Relative nuclear strength R versus optical luminosity L_{opt} for all 26 quasars.

between R and the presence of a large-scale jet; (b) thus far superluminal motion or bulk relativistic motion has been detected or inferred only in objects with moderate to high R—this is almost certainly a selection effect; and (c) 3C 179 has been included merely for comparison—it is not a member of the complete sample (Porcas 1984).

The dependences of C on L and of C on R are shown in Figures 1c and 1d, respectively, for the objects with $z < 1.3$. All of these correlations are in the same sense, although not as marked, as those obtained when the flat-spectrum compact objects are included. (Note that C was not defined for one object, 3C 68.1, whose radio core was undetected and plotted as an upper limit in Figure 1b.)

A question that is worth exploring is the possibility that the optical emission in these objects is beamed. In Figure 1e we have plotted the values of R and optical luminosity L_{opt} for the objects in this sample, and we find that there is no evidence for any relationship between these two quantities.

Additional statistical tests (Hough 1986; Hough and Readhead, in preparation) show that the observed properties of this 178 MHz sample are consistent with these objects being drawn from an orientation unbiased sample, and with the correlations being due to relativistic beaming under the following conditions:

(a) The deprojected overall sizes, L_{90}, have a roughly uniform distribution ranging from 0 to 250 kpc ($H_0 = 100$ km s^{-1} Mpc^{-1} and $q_0 = 0.5$);
(b) the scaling factor for R for an object in the plane of the sky is $R_{90} = 0.01$–0.04;
(c) the Lorentz factor $\gamma \sim 5$;
(d) $C_{int} < 10°$.

3. VLBI Maps

We are engaged in a VLBI survey of the nuclei in the complete sample, beginning with those objects whose nuclear flux densities are $S_n > 75$ mJy at 5 GHz. Observations are being made with the Mark III system (Rogers et al. 1983) at a frequency of 10.7 GHz to obtain the best possible combination of sensitivity and resolution. The arrays include telescopes from among those at Effelsberg, Medicina, Haystack, Green Bank, Fort Davis, and Owens Valley.

The six sources in our initial VLBI subsample are listed in Table 1, along with z, the 5 GHz nuclear flux density S_n, R, L, and C.

In each case, the milliarcsecond structure is resolved but very compact. The use of convolving beams from 70% down to 50% of the conventional beam size was necessary to reveal underlying structures in our

Table 1 Six Sources in VLBI Subsample

Source	IAU Name	z	S_n (mJy)	R	L (kpc)	C (deg)
3C 207	0838+133	0.684	588	0.62	52	20
3C 212	0855+143	1.049	148	0.11	47	0
3C 245	1040+123	1.029	910	1.1	39	5
3C 249.1	1100+772	0.311	78	0.087	120	9
3C 263	1137+660	0.656	130	0.071	173	1
3C 334	1618+177	0.555	165	0.18	208	4

hybrid maps. We have resorted to careful intercomparisons of results from model fitting, hybrid mapping, and maximum entropy mapping to draw what we believe are the most definitive conclusions possible about the source structures. Each object shows a double or extended parsec-scale structure. The VLBI source axes are aligned to within < 10° of large-scale jets and compact "hot spots" in every case. For a complete collection of maps, see Hough (1986).

We can tentatively classify these objects as standard "core-jets", if the weaker (and sometimes discernibly more extended) component is identified with the jet. Thus the basic structure appears similar to that often found in cores of strong, flat-spectrum, core-dominated quasars. It is then very interesting to note that in the five objects where a one-sided large-scale jet is seen, both the small-scale and large-scale jets lie on the same side of the compact core (the VLBI jet points toward the more compact outer hot spot in 3C 263, where no large-scale jet has been seen *yet*). There are now over 20 objects in which this is known to be the case. No counter-examples have been found thus far. This suggests that the small-scale and large-scale asymmetries are linked, and that relativistic beaming may explain the one-sided morphology on both scales.

The most encouraging result from these first-epoch maps is that half of the objects have a discrete double VLBI structure which we can monitor for variations. These objects are 3C 245, 3C 263, and 3C 334. Therefore we can optimistically push ahead with a direct test of beaming by measuring any component motions in all the amenable objects.

4. Two New Superluminals

To date, we have observed the central components of two objects at multiple epochs. A collaboration with J. A. Zensus and R. W. Porcas involving three epochs on 3C 263 has revealed superluminal expansion of the VLBI double with an apparent velocity of $v_{\rm app} = (1.3 \pm 0.4)c$, using the measured proper motion $\mu = 0.06 \pm 0.02$ mas yr^{-1} (Figure 2; Zensus, Hough, and Porcas 1987). The weaker of the two components has faded during the expansion and appears slightly elongated along the source

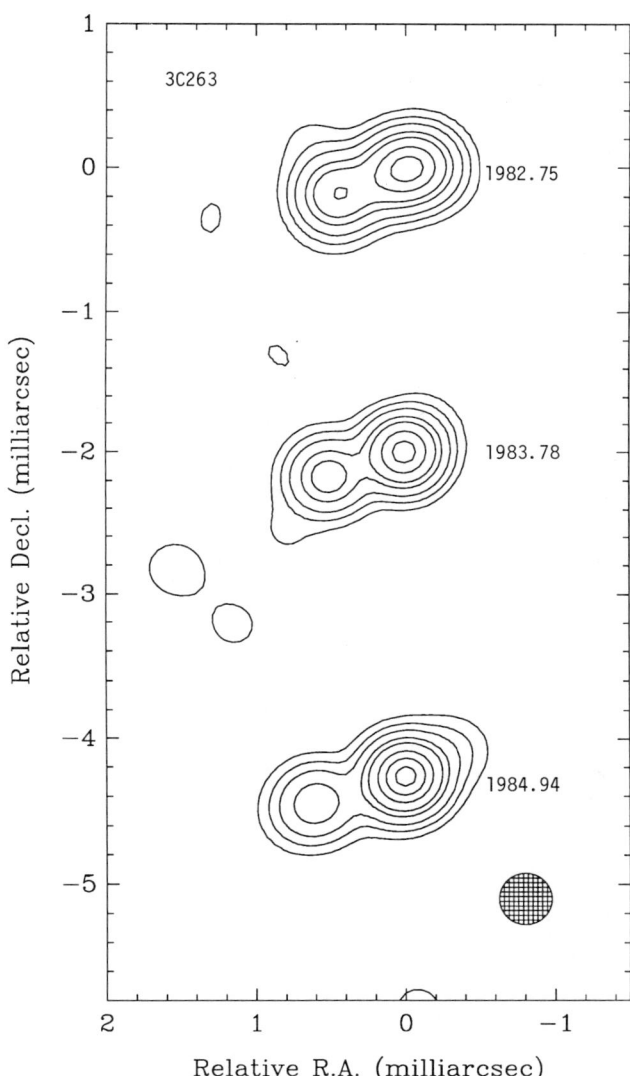

Figure 2 Hybrid maps of the nucleus of 3C 263 at epochs 1982.75, 1983.78, and 1984.94 (Zensus, Hough, and Porcas 1987). Vertical spacings correspond to the time intervals between epochs. Contours are −2, 2, 5, 10, 20, 30, 50, 70, 90 mJy per beam. The circular restoring beam of FWHM= 0.35 mas (70% of the conventional beam size) is shown cross hatched.

axis at the final epoch. This indicates that we may be seeing a typical core-jet object. The nucleus of 3C 263 is the weakest (~ 150 mJy) and "slowest" ($\sim 1.3c$) superluminal source thus far recorded.

3C 245 underwent a strong outburst in total flux density sometime in 1982. We observed an apparent broadening of the core component in 1983, since it could be represented by a circular Gaussian model < 0.3 mas in size in February (i.e., barely resolved) but required a finite size of 0.33 ± 0.03 mas in October. It was finally possible to distinguish two distinct components in 1984 December and 1986 March. These have separated with proper motion $\mu = 0.11 \pm 0.04$ mas yr^{-1}, yielding a superluminal $v_{app} = (3.1 \pm 1.1)c$ (Figure 3). The innermost core component has brightened and broadened at the last epoch, indicating that another component may be emerging.

It is difficult to draw a conclusion regarding the validity of the beaming model based on observations of only two objects. It is interesting to note, however, that 3C 263 (see Browne, this Workshop, page 129) has a large projected linear size, contains a nucleus of medium strength, and displays a "slow" superluminal velocity; 3C 245 (Figure 4) has a small projected linear size, has the strongest central component in the sample, and exhibits "moderate" superluminal motion. A simple explanation of all these facts is that the objects are intrinsically similar, but appear different because the simple beaming mechanism is at work and the axis of 3C 245 points nearer to our line of sight than does that of 3C 263.

If we consider the case of 3C 179 (Porcas 1984), a "classical double" with $v_{app} = 4.5c$, we find that it is roughly similar to 3C 245 in terms of its core strength and overall size. Since it seems unlikely that 3C 179 and 3C 245 would both be pointing at fairly small angles to the line of sight *and* have unusually large Lorentz factors (this is an amplification of the original "Porcas argument" for "Exhibit A", 3C 179, alone; Porcas 1981), we conclude that $\gamma_{min} \sim \beta_{app} \sim 3$–$5$ must be typical for these objects. Thus 3C 245 provides "Exhibit B" to support Porcas's reasoning that γ cannot be typically ≤ 2 as suggested by Scheuer and Readhead (1979). If, for example, $\gamma = 5$ in each of the three objects, we find that the apparent motions can be accounted for if 3C 179, 3C 245, and 3C 263 are oriented at the comfortably probable angles to the line of sight of 18°, 32°, and 70°, respectively. However, we will have to await substantial filling out of the apparent velocity histogram before deciding the fate of the beaming model as applied to this sample.

5. Apparent Velocity and Nuclear Strength

In our parochial view prior to this Workshop, there existed only twelve known superluminal sources (Table 2). Since there are probably now

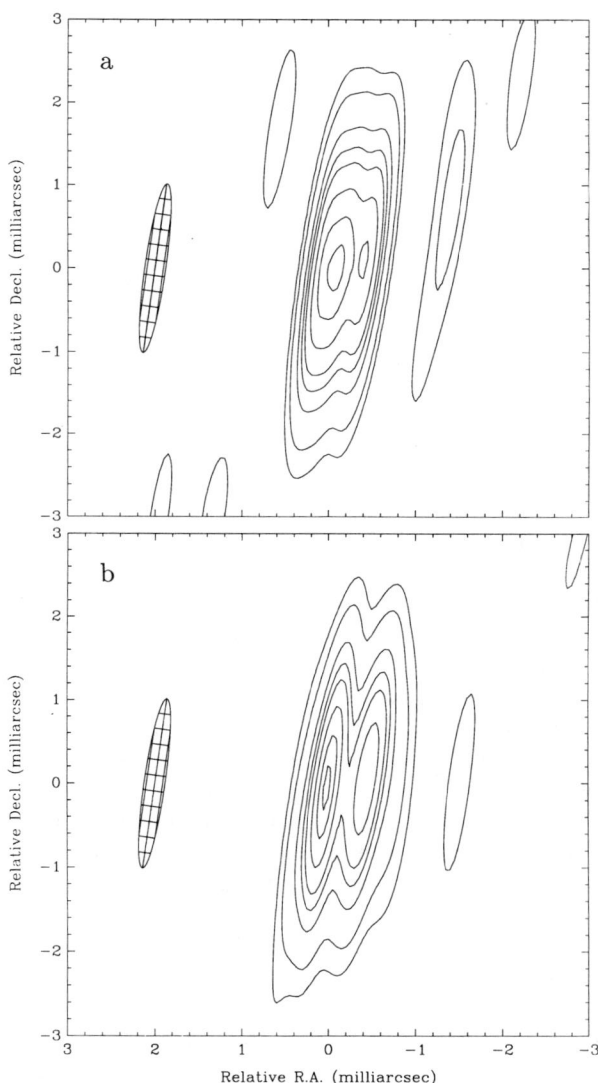

Figure 3 Hybrid maps of the nucleus of 3C 245 at epochs 1984.95 (a) and 1986.18 (b). Contours are 2, 5, 15, 25, 35, 55, 78, 95% of the peak, which is 162 mJy per beam for (a) and 253 mJy per beam for (b). The elliptical restoring beam of 2.04 × 0.24 mas at p.a. 172° (50% of conventional beam size) is shown cross hatched.

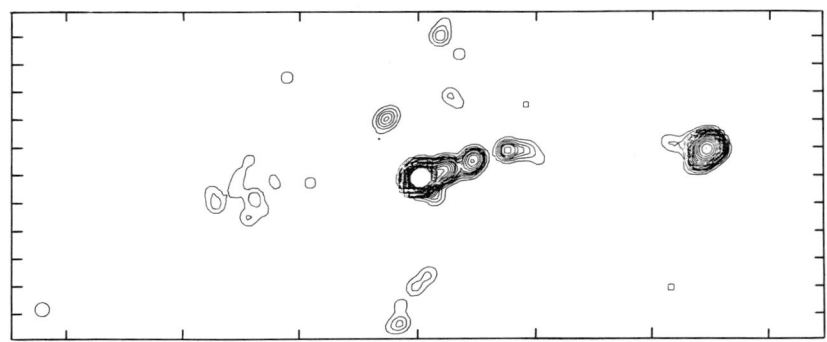

Figure 4 Large-scale structure of 3C 245, mapped with MERLIN at 1.6 GHz (Foley 1982). Vertical ticks are spaced 0″.5 apart, horizontal ticks 2″.0.

at least another half-dozen well-documented examples, the reader is warned that the following conclusions must stand the test of including these. Still, there are some interesting preliminary results worth mentioning.

The two new superluminals reported here, along with 3C 179, can all be termed *lobe-dominated* quasars, in that no more than half the 5 GHz flux density emanates from the nucleus (i.e., $R \leq 1$). The other nine objects, seven of them classified as quasars, are all very much *core-dominated* sources. Taking the largest observed velocity in each case, the average β_{app} for the lobe-dominated quasars (3.0 ± 1.6) is about half that of the core-dominated quasars (5.9 ± 2.0) or of all the core-dominated sources (5.3±2.3). A linear-correlation coefficient test for the ten quasars (excluding BL Lac and 3C 120) yields only a 7% probability that the observed correlation of β_{app} with $\log R$ (Figure 5) could have arisen by chance. It is tempting to attribute this difference to orientation effects on intrinsically similar, beamed nuclear jets. The different selection criteria for observing the two types of object would tend to lead to the core-dominated sources pointing closer to the line of sight and, on average, showing larger apparent motions than the lobe-dominated sources.

What about BL Lac? It has by far the most dominant core of all the objects, yet it is undercut in its apparent velocity only by 3C 263. We can account for this in a unified picture, if BL Lac happens to be oriented *extremely* near the line of sight (e.g., Blandford and Königl 1979). This could then be a case where we observe an object inside the maximum of the β_{app}–θ curve and thus measure a more modest velocity. The Doppler flux boosting would be the greatest of all in this case, since the boosting factor increases monotonically with decreasing angle to the line of sight. Theoretical curves of β_{app} versus $\log R$ for the simple twin-

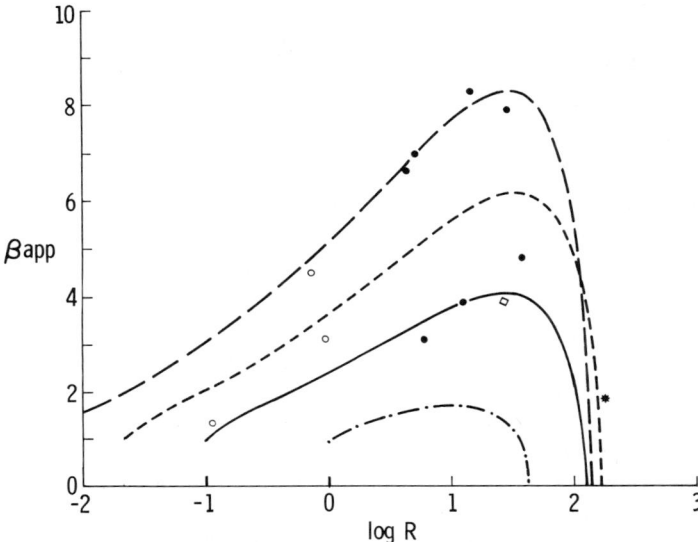

Figure 5 Largest apparent velocity, β_{app}, versus relative core strength, R, for twelve superluminal sources: •, core-dominated quasars; ○, lobe-dominated quasars; ◊, 3C 120; *, BL Lac. Theoretical curves for simple twin-jet beaming model are: —·— for $\gamma = 2.0$, $R_{90} = 1$; —— for $\gamma = 4.2$, $R_{90} = 0.1$; - - - for $\gamma = 6.3$, $R_{90} = 0.022$; — — for $\gamma = 8.4$, $R_{90} = 0.005$. Curves were computed using $\beta_{app} = \beta \sin\theta/(1-\beta\cos\theta)$ and $R = \frac{1}{2}R_{90}[(1-\beta\cos\theta)^{-2.2} + (1+\beta\cos\theta)^{-2.2}]$.

Table 2 A Dozen Superluminal Sources

Source	Optical ID	z	μ	β_{app}	R	References
3C 263	Q	0.656	0.06	1.3	0.11	4, 11
BL Lac	BL	0.070	0.57	1.8	170	2, 6, 10
3C 279	Q	0.538	0.17	3.1	6.2	2, 10
3C 245	Q	1.029	0.11	3.1	0.99	4, this paper
3C 120	G	0.033	2.57	3.9	26	2, 10
0850+581	Q	1.322	0.12	3.9	13	1
3C 179	Q	0.846	0.18	4.5	0.72	2, 9, 10
NRAO 140	Q	1.258	0.15	4.8	38	5, 10
3C 273	Q	0.158	0.99	6.6	4.3	2, 10
1928+738	Q	0.302	0.6	7.0	5.2	1, 3
1642+690	Q	0.75	0.34	7.9	28	7, 8
3C 345	Q	0.595	0.42	8.3	15	2, 10

Note: R calculated assuming a K-correction $(1+z)^{\Delta\alpha}$, with an assumed difference $\Delta\alpha = -0.5$ in the spectral indices of the extended and core emission. No correction to mean or peak R for strongly variable cores.

References: 1. Barthel et al. 1986. 2. Cohen and Unwin 1984. 3. Eckart et al. 1985. 4. Hough 1986. 5. Marscher and Broderick 1985. 6. Mutel and Phillips 1984. 7. Pearson et al. 1986. 8. Perley, Fomalont, and Johnston 1980; 9. Porcas 1984. 10. Schilizzi and de Bruyn 1983. 11. Zensus, Hough, and Porcas 1987.

jet beaming model are shown in Figure 5, assuming various values of γ and R_{90}. While one must allow considerable latitude in fitting the parameter R_{90}, a range of γ from 4 to 8 is all that is required to bracket the observed points and account for BL Lac in a "unified scheme".

6. Discussion

The earliest clear correlations to emerge from the comparison of the morphologies of central and outer components were those between overall size, curvature, and relative strength of the central component (Readhead et al. 1978). It is an unfortunate fact that both Doppler boosting and interaction with the environment produce correlations in the same sense, so that it is very difficult to discriminate between these two possibilities. Thus far, the evidence that has accrued has bolstered the original correlations, but has not provided any clear discriminants, although (relatively modest) superluminal motion and relativistic bulk motion in central components are circumstantial evidence favoring the beaming hypothesis. This deadlock may now have been broken by the discovery of the correlation between jet sidedness and depolarization by Robert Laing described by Scheuer at this Workshop (page 104), but it is possible to construct models of large-scale jets in which such a correlation can be produced intrinsically. There is one piece of evidence which seems to go against the beaming hypothesis in its simplest form, namely the fact that some of the deprojected sizes of the superluminal sources appear to be too large (Barthel, this Workshop, page 148); but, at the least, it is clear that there are now two independent facts which are predicted *directly* on the beaming hypothesis, namely relativistic bulk motion and depolarization, which are not predicted on either the intrinsic or the interaction hypothesis.

We thank A. R. Foley for permission to reproduce his map of 3C 245 shown in Figure 4. D.H.H. is a NRC–NASA Research Associate. This work was supported in part by the National Science Foundation under grant AST 85-09822 to the Owens Valley Radio Observatory.

References

Barthel, P. D., Pearson, T. J., Readhead, A. C. S., and Canzian, B. J. 1986, *Astrophys. J. (Letters)*, **310**, L7.
Blandford, R. D., and Königl, A. 1979, *Astrophys. J.*, **232**, 34.
Cohen, M. H., and Unwin, S. C. 1984, in *IAU Symposium 110, VLBI and Compact Radio Sources*, ed. R. Fanti, K. Kellermann, and G. Setti (Dordrecht: Reidel), p. 95.
Eckart, A., Witzel, A., Biermann, P., Pearson, T. J., Readhead, A. C. S., and Johnston, K. J. 1985, *Astrophys. J. (Letters)*, **296**, L23.
Foley, A. R. 1982, Ph. D. thesis, The Victoria University of Manchester.
Hough, D. H. 1986, Ph. D. thesis, California Institute of Technology.
Laing, R. A., Riley, J. M., and Longair, M. S. 1983, *Monthly Notices Roy. Astron. Soc.*, **204**, 151.

Marscher, A. P., and Broderick, J. J. 1985, *Astrophys. J.*, **290**, 735.
Mutel, R. L., and Phillips, R. B. 1984, in *IAU Symposium 110, VLBI and Compact Radio Sources*, ed. R. Fanti, K. Kellermann, and G. Setti (Dordrecht: Reidel), p. 117.
Pearson, T. J., Barthel, P. D., Lawrence, C. R., and Readhead, A. C. S. 1986, *Astrophys. J. (Letters)*, **300**, L25.
Perley, R. A., Fomalont, E. B., and Johnston, K. J. 1980, *Astron. J.*, **85**, 649.
Porcas, R. W. 1981, *Nature*, **294**, 47.
Porcas, R. W. 1984, in *IAU Symposium 110, VLBI and Compact Radio Sources*, ed. R. Fanti, K. Kellermann, and G. Setti (Dordrecht: Reidel), p. 157.
Readhead, A. C. S., Cohen, M. H., Pearson, T. J., and Wilkinson, P. N. 1978, *Nature*, **276**, 768.
Readhead, A. C. S., Hough, D. H., Ewing, M. S., Walker, R. C., and Romney, J. D. 1983, *Astrophys. J.*, **265**, 107.
Rogers, A. E. E., Cappallo, R. J., Hinteregger, H. F., Levine, J. I., Nesman, E. F., Webber, J. C., Whitney, A. R., Clark, T. A., Ma, C., Ryan, J., Corey, B. E., Counselman, C. C., Herring, T. A., Shapiro, I. I., Knight, C. A., Shaffer, D. B., Vandenberg, N. R., Lacasse, R., Mauzy, R., Rayhrer, B., Schupler, B. R., and Pigg, J. C. 1983, *Science*, **219**, 51.
Scheuer, P. A. G., and Readhead, A. C. S. 1979, *Nature*, **277**, 182.
Schilizzi, R. T., and de Bruyn, A. G. 1983, *Nature*, **303**, 26.
Zensus, J. A., Hough, D. H., and Porcas, R. W. 1987, *Nature*, **325**, 36.
Zensus, J. A., and Porcas, R. W. 1986, in *IAU Symposium 119, Quasars*, ed. G. Swarup and V. K. Kapahi (Dordrecht: Reidel), p. 167.

Superluminal Motion in a Randomly Oriented Quasar Sample

J. ANTON ZENSUS
California Institute of Technology
RICHARD W. PORCAS
Max-Planck Institut für Radioastronomie

1. Introduction

The search for superluminal motion in weak cores of double radio sources has long been recognized as a crucial test of the simple beaming models of superluminal sources (Scheuer and Readhead 1979; Blandford and Königl 1979; Orr and Browne 1982). Invoking a narrow relativistic jet oriented close to the line of sight to the observer, these models imply that the superluminal effect should be rare in a complete sample of sources selected without orientation bias. Peter Scheuer has discussed the relevance of such jet-speed statistics at this Workshop (page 104) and pointed out "ways of escape", in case the observations fail to reproduce the expected (desired?) results.

Despite these worries, there is relatively little known about the distribution or even the abundance of superluminal motion in samples not dominated by strong sources. In particular, the evidence for slower motion in weaker objects (e.g., Hough and Readhead, this Workshop, page 114) is still inferred from a small number of objects. In this paper, we give a status report on the measurement of motions in a sample of quasars with extended double structure.

2. The 966 MHz Sample

Since 1982 we have been engaged in the study of motions in sources from a complete sample of 30 of the 68 quasars from the Jodrell Bank 966 MHz survey. The sample was selected by these criteria: $S_{966 \text{ MHz}} >$ 0.7 Jy during the survey; $m_b < 19$ on the Palomar Sky Survey prints; and extended structure $> 10''$. The sources were mapped with arc-second resolution at 2.695 GHz and 8.085 GHz by Owen, Porcas, and Neff (1978) and at 5 GHz with the VLA by Owen and Puschell (1984). 29 objects show double structure, symmetric with respect to the quasar, and 24 contain compact, flat-spectrum, central components at 5 GHz.

It is probable that the remaining objects also contain such cores below the detection limits of Owen and Puschell.

Originally selected at 966 MHz, where the flux density of the sources is dominated by the steep-spectrum lobes, this sample is not biased towards strong central components. The weak (20–450 mJy) cores can therefore be assumed to be randomly oriented.

Since repeated VLBI observations of such weak objects require the use of the Mark III recording system with its still limited availability, we have concentrated on a subsample of five sources: 3C 179, 3C 263, 3C 268.4, 1732+655, and 1951+498 (see Zensus and Porcas 1984 for a list of source properties). Observations at a frequency of 10.7 GHz with 4–5 antennas in Europe and the U.S. were chosen to achieve the best available resolution and sensitivity.

3. Results

Morphology

Hybrid maps are available for all five sources (Zensus and Porcas 1986), except 1732+655. (A model fit to one-epoch data from this source shows a compact core, possibly double, extended roughly along the overall source axis.) The cores have simple structures similar to those commonly found in core-dominated sources (e.g., at this Workshop, Pearson, Readhead, and Barthel, page 94; Witzel, page 83). They can typically be described by two-component models accounting for large fractions of the total core flux densities. The quasar 3C 179 can be considered a prototype for this class of object. Despite the lack of spectral information for the cores, there is circumstantial evidence that their structures can be classified as core-jets (Zensus 1984). An "orientation memory" is evident between features on parsec and kiloparsec scales. Bending of the structure appears to occur not within the compact regions themselves but rather through interaction of the jets with the external medium.

Motion

Three of the five quasars (3C 179, 3C 263, 1951+498) show structural changes which can be attributed to superluminal expansion. (Preliminary results from second-epoch data for 3C 268.4 also indicate structural changes.) This result suggests that superluminal motion is a frequent phenomenon in this type of object.

3C 179: ($z = 0.846$) The superluminal motion in this source is discussed by Porcas at this Workshop (page 12). The best fit to the monitoring data is an increase of the separation between the two compact components of $\mu = 0.19$ mas yr^{-1} corresponding to $v_{\mathrm{app}} = 4.8 h^{-1} c$.

3C 263: ($z = 0.652$) Next to 3C 179, this is the best studied source in our sample. We found superluminal motion at $v_{app} = 1.3 \pm 0.4 h^{-1} c$ (Zensus, Hough, and Porcas 1987; see Hough and Readhead, this Workshop, page 114). This quasar core is the weakest ($S_{10.7\ GHz} \sim 150$ mJy) superluminal source found so far.

1951+498: ($z = 0.846$) We have observed this quasar at two epochs (1982.8 and 1985.8). Hybrid maps of the core at conventional resolution show an extension in p.a. $\sim 90°$ at both epochs. Model fitting and super-resolving hybrid maps suggest that the structure consists mainly of two components. Their separation changed by ~ 0.2 mas, giving $\mu \sim 0.07$ mas yr^{-1} or $v_{app} \sim 1.3 h^{-1} c$. This result requires confirmation.

References

Blandford, R. D., and Königl, A. 1979, *Astrophys. J.*, **232**, 34.
Orr, M. J. L., and Browne, I. W. A. 1982, *Monthly Notices Roy. Astron. Soc.*, **200**, 1067.
Owen, F. N., Porcas, R. W., and Neff, S. G. 1978, *Astron. J.*, **83**, 1009.
Owen, F. N., and Puschell, J. J. 1984, *Astron. J.*, **89**, 932.
Porcas, R. W. 1981, *Nature*, **294**, 47.
Scheuer, P. A. G., and Readhead, A. C. S. 1979, *Nature*, **277**, 182.
Zensus, J. A. 1984, Ph. D. thesis, Westfälische Wilhelms-Universität Münster.
Zensus, J. A., Hough, D. H., and Porcas, R. W. 1987, *Nature*, **325**, 36.
Zensus, J. A., and Porcas, R. W. 1984, in *IAU Symposium 110, VLBI and Compact Radio Sources*, ed. R. Fanti, K. Kellermann, and G. Setti (Dordrecht: Reidel), p. 163.
Zensus, J. A., and Porcas, R. W. 1986, in *IAU Symposium 119, Quasars*, ed. G. Swarup and V. K. Kapahi (Dordrecht: Reidel), p. 167.

Extended Structure
of Superluminal Radio Sources

IAN W. A. BROWNE

Nuffield Radio Astronomy Laboratories, Jodrell Bank

1. Introduction

Why should we be interested in the extended structure of superluminal sources? The extended structure associated with superluminal cores is often very weak and, even when detected, not very pretty. And, of course, the really exciting action is going on tens or hundreds of kiloparsecs away right down in the core.

The first and obvious reason for interest is that the extended radio structure can be used to put superluminal sources into context. We want to know if objects in which superluminal motion has been detected are special in any other way, or if they are just like the vast majority of other radio sources. In fact the bulk of recent evidence points to the conclusion that superluminal motion is ubiquitous. Motion has been detected in high-luminosity core-dominated quasars, in BL Lac objects, in lobe-dominated quasars, and in a relatively low-luminosity radio galaxy.

Another important reason for studying the extended radio structures of superluminal objects and comparing them with structures of other sources not known to be superluminal is the hope that in this way we might be able to constrain popular relativistic beaming models of what happens in the cores of active radio objects. Predictions of such simple models are easy to make and compare with observations. The results of such comparisons may or may not be conclusive, however.

In this paper, I will first review the observations of known superluminal radio sources and present some simple statistics (e.g., how many are doubles, how many have any extended structure, how big is this structure, does the strength or the extent of the structure correlate with anything else?). Then I will try to tie things together and discuss how consistent the present observations are with simple beaming models.

2. Observations of the Extended Structure

In Table 1 are listed 17 sources in which superluminal motion has been observed or at least claimed. In some, the motion is established beyond doubt and a value (or values) for the apparent velocity β_{app} can be assigned unambiguously, in others it is clear that changes occur which imply superluminal-like effects but unique values of β_{app} are not so obvious (e.g., 3C 454.3), while for some sources β_{app} has been assigned but confirmatory observations are necessary. Most of the subsequent discussion will be based on the 15 quasars, all of which have values of β_{app} available (though they may not all be correct).

Also given in Table 1 are parameters of the extended radio structures. Luminosities have been calculated at a frequency of 5 GHz in the object's frame. R is the ratio of the core to the extended radio luminosity at 5 GHz. Where possible, angular sizes have been measured between the peaks of radio components rather than to the edges of the detected emission since the latter can be a very dynamic-range-sensitive parameter. It should, however, be borne in mind that better observations might reveal new components with larger angular separations. Most of the observations have surface brightness sensitivities of ~ 1 mJy arcsec^{-2}, so it is important that when comparing the structures of superluminal sources with those of radio sources in general, the comparison observations should have similar sensitivities.

For some of the sources, the best available maps are unpublished. In Figure 1, I show some of these maps, mostly from Murphy, Browne, and Perley (in preparation).

Notes on Individual Sources

NRAO 140: A very boring source! The VLA L-band map by Murphy, Browne, and Perley (Figure 1a) shows one secondary component 7″.5 from the core. The map produced by Schilizzi and de Bruyn (1983) shows another component to the north-west of the core but this is not confirmed by the better VLA map.

3C 120: A wealth of extended structure on size scales ranging up to $\sim 700''$ is revealed by the VLA maps (Walker 1986; Walker, Benson, and Unwin 1987; Walker, Benson, and Unwin, this Workshop, page 48).

3C 179: The first classical double to be shown to be a superluminal. The best available map is that of Shone, Porcas, and Zensus (1985) made by combining VLA and MERLIN L-band data.

0850+581: A recently discovered superluminal. The best maps are MERLIN L-band (Figure 1a) and VLA C-band maps (Shone *et al.*, in preparation; Barthel *et al.* 1986). Though the core position angle and

Table 1 Superluminal Sources

(1) Source	(2) z	(3) θ	(4) l	(5) $\log L_e$	(6) $\log L_c$	(7) $\log R$	(8) β_{app}	(9) PA_c	(10) PA_e	(11) ΔPA	(12) Refs.
0333+321 NRAO 140	1.258	7.5	32	25.9	27.5	1.6	4.8	127	149	22	1, 3, 12
0430+052 3C 120	0.033	700	317	23.6	24.4	0.8	2–4	−110	−99	11	4, 6, 11
0723+679 3C 179	0.846	15.5	64	26.9	26.3	−0.6	4.8	92	94	2	11, 13
0850+581	1.322	15.1	65	26.7	27.1	0.4	3.9	153	140	13	2, 8, 14
0906+430 3C 216	0.669	3	12	26.6	26.6	0.0	2.4	151	40	111	5, 15, 34
0923+392 4C 39.25	0.699	4	16	26.2	27.5	1.3	3.5	−64	100	16	1, 16
1040+123 3C 245	1.029	8	34	26.9	26.9	0.0	3.1	−73	−86	13	6, 17
1137+660 3C 263	0.652	49	192	26.7	25.7	−1.0	1.3	110	−68	2	2, 17, 18
1226+023 3C 273	0.158	21	37	26.0	26.9	0.9	~6	−118	−138	20	19, 20, 21
1253−055 3C 279	0.538	15	55	26.3	27.4	1.1	9.2–2.0	−130	−157	27	1, 22, 23, 36
1641+399 3C 345	0.595	21	80	25.9	27.4	1.5	1.4–9.5	−135	−32	~100	1, 3, 24, 25
1642+690	0.751	11	45	26.1	26.8	0.7	7.9	195	168	27	3, 26, 27
1721+343 4C 34.47	0.206	250	539	25.5	25.1	−0.4	3.1	168	163	5	15, 28
1901+319 3C 395	0.635	1	4	26.2	26.7	0.5	15.5	~150	−52	98	10, 29, 35
1928+738	0.302	45	124	25.1	26.4	1.3	7.0	165	164	1	3, 30
2200+420 BL Lac	0.070	15	14	22.9	25.2	2.3	2.4	190	?	?	9, 31, 32
2251+158 3C 454.3	0.859	5.2	22	26.5	27.8	1.3	8.8	−65	−49	16	1, 7, 33

Explanation: (1) Source name. (2) Redshift, z. (3) Angular size (arcsec), θ. (4) Linear size (kpc), l. (5) Logarithm of extended radio luminosity (W Hz^{-1}), $\log L_e$. (6) Logarithm of core radio luminosity (W Hz^{-1}), $\log L_c$. (7) $\log R \equiv \log(L_c/L_e)$. (8) Observed superluminal expansion speed, β_{app}. (9) VLBI position angle (that of the smallest structure) (degrees), PA$_c$. (10) Position angle from the core to the peak of emission in the extended structure (degrees), PA$_e$. (11) Difference in VLBI and large-scale structure position angle (degrees), ΔPA. (12) References for l, β_{app}, PA$_c$, and PA$_e$.

References: 1. Browne et al. 1982. 2. Shone et al., unpublished. 3. Murphy et al., unpublished. 4. Walker 1986. 5. Greybe, A., Shone, D. L., and Porcas, R. W., in preparation. 6. Foley, unpublished. 7. De Bruyn and Schilizzi 1986. 8. Browne and Perley 1986. 9. Antonucci and Ulvestad 1985. 10. Pearson, Perley, and Readhead 1985. 11. Shone, Porcas, and Zensus 1985. 12. Marscher and Broderick 1982. 13. Porcas 1984. 14. Barthel et al. 1986. 15. Barthel, private communication. 16. Shaffer et al., private communication. 17. Hough 1986. 18. Zensus, Hough, and Porcas 1987. 19. Unwin and Biretta 1984. 20. Biretta et al. 1985. 21. Davis, Muxlow, and Conway 1985. 22. Unwin 1986. 23. De Pater and Perley 1983. 24. Biretta, Cohen, and Moore 1986. 25. Biretta, Moore, and Cohen 1986. 26. Pearson et al. 1986. 27. Browne and Orr 1981. 28. Barthel 1984. 29. Waak et al. 1985. 30. Eckart et al. 1985. 31. Mutel and Phillips 1984. 32. Phillips and Mutel 1982. 33. Pauliny-Toth et al. 1984. 34. Porcas 1986. 35. Muxlow, private communication. 36. Cotton et al. 1979.

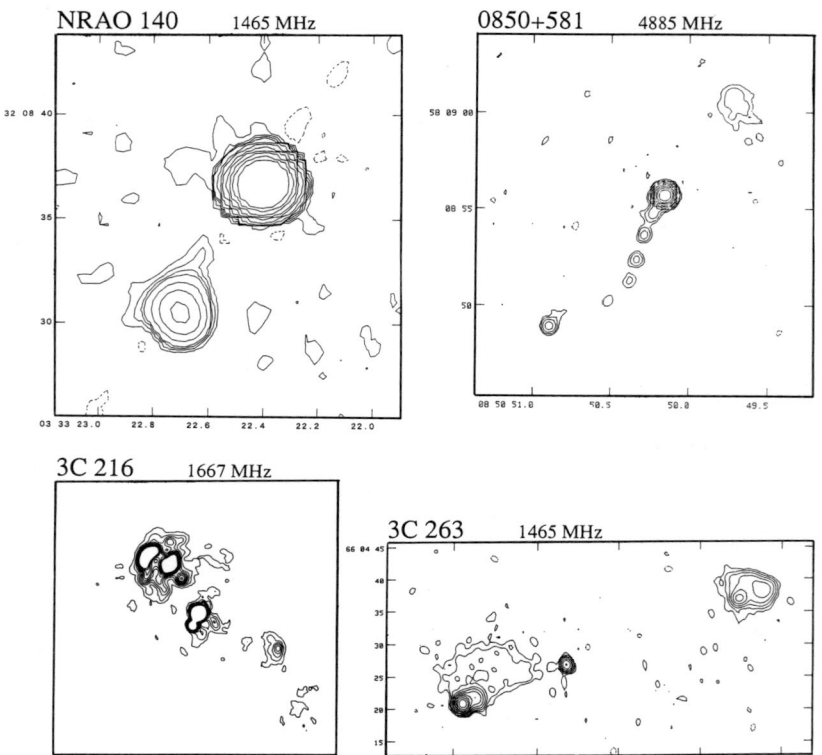

Figure 1a Maps of some of the superluminal sources. NRAO 140: VLA L-band (Murphy, Browne, and Perley, in preparation). 0850+581: VLA L-band (Shone et al., in preparation). 3C 216: MERLIN+EVN L-band (Greybe, Shone and Porcas, in preparation). 3C 263: VLA L-band (Shone et al., in preparation).

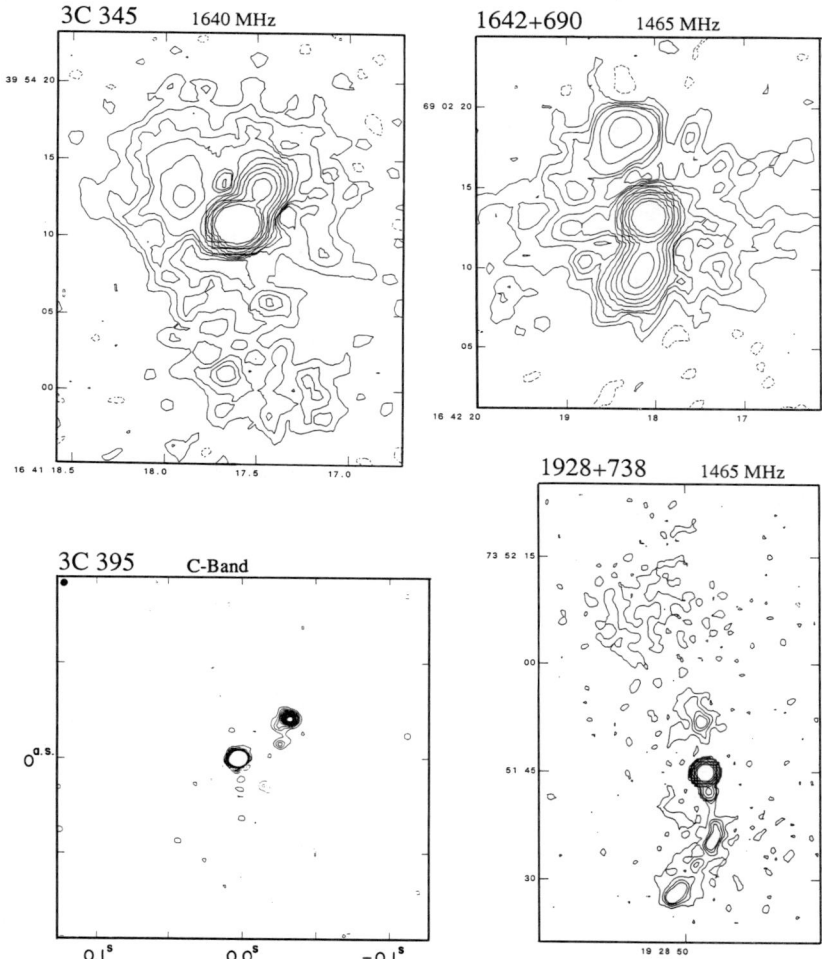

Figure 1b 3C 345: VLA L-band (Murphy, Browne, and Perley, in preparation). 1642+690: VLA L-band (Murphy, Browne, and Perley, in preparation). 3C 395: MERLIN C-band (Muxlow, private communication). 1928+738: VLA L-band (Murphy, Browne, and Perley, in preparation).

the position angle from the core to the southern hotspot are only 12° apart, the kiloparsec jet bends by $\sim 45°$ before entering the hotspot.

3C 216: A compact steep-spectrum source which shows a very complex two-sided structure. The combined MERLIN/EVN map made by Greybe, Shone, and Porcas (in preparation) is shown in Figure 1a. There is a hint of jets on both sides of the core.

4C 39.25: This source has a relatively symmetric arcsecond structure (Browne et al. 1982).

3C 245: An early L-band MERLIN map (Foley 1982; see Hough and Readhead, this Workshop, page 114) is the best available for this source. It shows a jet-like feature pointing to a bright western hotspot. To the east of the core, is a weak diffuse lobe.

3C 263: Another classical double. The VLA C-band map made by Shone et al. (in preparation) is shown in Figure 1a.

3C 273: The "most one-sided" source known (Davis, Muxlow, and Conway 1985; see Zensus and Cohen, this Workshop, page 26).

3C 279: The best maps are those by de Pater and Perley (1983), made with the VLA (see Unwin, this Workshop, page 34).

3C 345: In Figure 1b the L-band VLA map by Murphy, Browne, and Perley is shown. This is in good agreement with the Westerbork map by Schilizzi and de Bruyn (1983). A higher-resolution map showing detail in the arcsecond jet is presented by Browne et al. (1982).

1642+690: In Figure 1b the L-band VLA map of Murphy, Browne, and Perley is shown. Like 3C 345, the core is embedded in a complex extended emission region. The strong elongated component to the south of the core is a curved jet (Browne and Orr 1981).

1721+343: This is the largest known quasar. The best maps are those made with the VLA by Peter Barthel (this Workshop, page 148).

3C 395: This source is very compact. The map shown in Figure 1b was made from MERLIN 5 GHz data (Muxlow, private communication). There is evidence that the core is extended in p.a. $\sim 110°$ which is closer to the 160° p.a. of the VLBI structure than to the p.a. of the arcsecond component ($-52°$). If we assume continuity from the VLBI scale to the arcsecond scale, the jet bends by 98°. No structure on scales in excess of $1''$ is known.

1928+738: Figure 1b shows the VLA L-band map by Murphy, Browne, and Perley. A better L-band map is presented by Simon et al. (this Workshop, page 155).

BL Lac: A VLA L-band map is presented by Ulvestad and Johnston (1984). This shows a diffuse halo of $\sim 20''$ in extent. Schilizzi and de Bruyn (1983) also report the existence of extended emission detected at Westerbork.

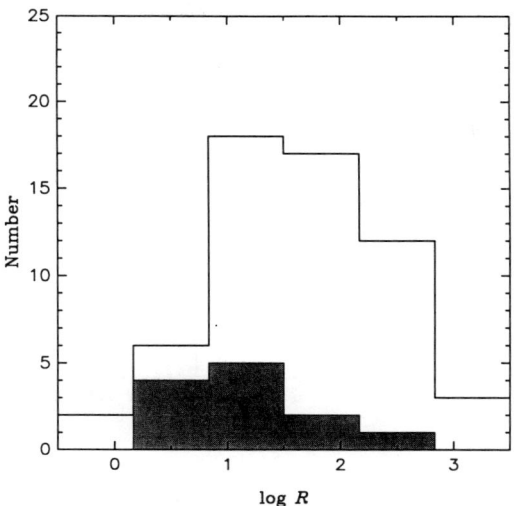

Figure 2 Distribution of R values for superluminal quasars (*shaded*) compared to the distribution for a sample of core-selected quasars (Murphy, Browne, and Perley, in preparation).

3C 454.3: Murphy, Browne, and Perley fail to detect any extended emission other than that from the $5''$ jet (Browne *et al.* 1982; see Pauliny-Toth, this Workshop, page 55). De Bruyn and Schilizzi (1986) present a Westerbork map showing extended structure of scale $\sim 50''$.

3. Statistical Results

Some of the results from the data in Table 1 are summarized below.

(*a*) All superluminal sources have structures on scales $\gtrsim 1''$. In a few cases, there is a well-defined jet (e.g., 0850+581, 3C 454.3), but more commonly the structure consists of one or more extended components (lobes?).

(*b*) Most superluminal sources have extended structure on both sides of the core and often it is clearly double in nature. The one-sided sources are NRAO 140, 3C 273, 3C 395, and 3C 454.3.

(*c*) There is no obvious difference between the radio structures of core-dominated superluminal sources and core-dominated sources in general. With the limited numbers available, it also appears to be true that there is nothing special about the structures of the lobe-dominated superluminal sources.

(*d*) The distribution of values of

$$R = \frac{\text{core luminosity}}{\text{extended luminosity}}$$

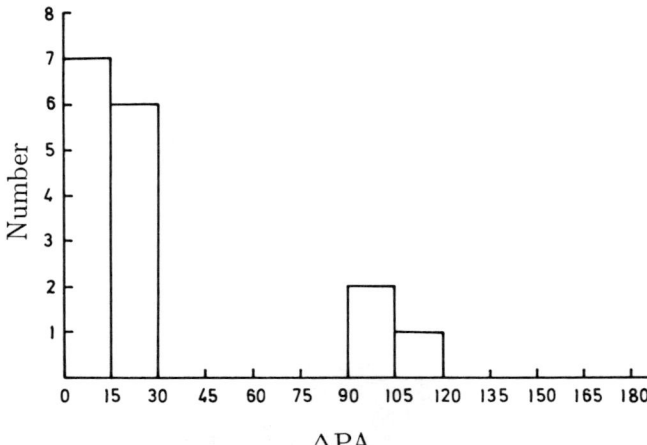

Figure 3 Distribution of difference in position angle between the smallest scale VLBI structure and the line joining the core to the brightest point in the arcsecond structure.

is similar to that found amongst other radio sources. If anything, there is a dearth of superluminal sources with very high values of R (Figure 2).

(e) The superluminal components move towards that part of the extended structure which has the highest surface brightness. Where there is a kiloparsec jet, the VLBI jet either joins up, or looks as if it will. In Figure 3 is plotted the histogram of the difference between the position angle of the smallest VLBI structure, and the position angle of the line joining the brightest point in the extended structure to the core.

(f) Linear sizes of the structures range from a few kiloparsecs to hundreds of kiloparsecs. These sizes and the surface brightnesses are not very different from those found in the general population of radio sources.

Two conclusions can be drawn from these observations. The first is that there is nothing special about the extended radio structure of superluminal sources; both the core-dominated and lobe-dominated superluminal sources are typical representatives of their class. This suggests that nearly all strong radio sources would show superluminal effects if monitored carefully. (The exception to this may be the class of compact VLBI doubles—see the paper in this Workshop by Hodges and Mutel, page 168). The second conclusion is that the relation between the small and large scale asymmetries argues strongly that whatever makes cores one-sided also makes the extended structure asymmetric. Obviously, Doppler boosting is an attractive possibility, but jets could be relativistic and intrinsically one-sided at any particular time.

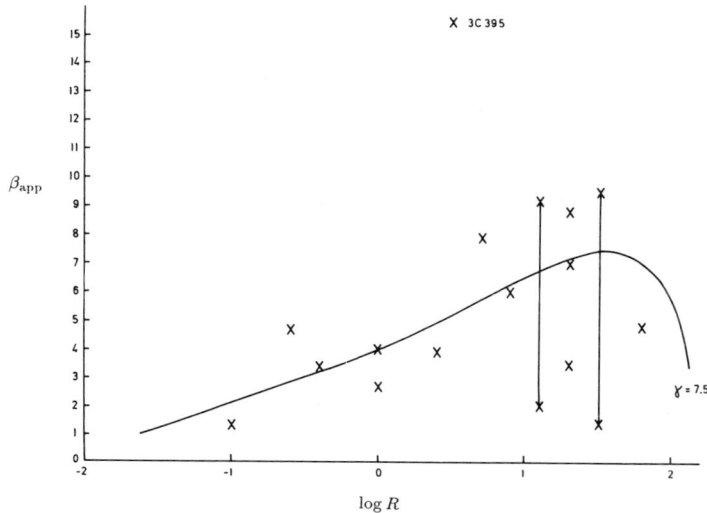

Figure 4 Plot of β_{app} versus $\log R$. The curve is the expected variation of β_{app} with $\log R$ for a source with a Lorentz factor $\gamma = 7.5$ in the unified scheme of Orr and Browne (1982). 3C 345 and 3C 279 are plotted twice and joined by a line to represent the range of reported expansion speeds ($H_0 = 100$ km s^{-1} Mpc^{-1}, $q_0 = 0.5$).

4. Correlations

β_{app} versus R

In Figure 4, I have plotted β_{app} versus R for the quasars from Table 1. (3C 120 and BL Lac have been excluded because radio galaxies and BL Lac objects may be intrinsically different from quasars.) We see a clear trend in the sense that the high-R (core-dominated) sources have the largest values of β_{app}, provided we ignore 3C 395. A linear regression analysis without 3C 395 shows that the correlation between β_{app} and R is significant at better than the 5% level. (With 3C 395, the correlation is not statistically significant.)

For illustration purposes, Figure 4 also shows the variation of β_{app} with $\log R$ expected in the simple beaming scheme of Orr and Browne (1982). The curve is for a Lorentz factor $\gamma = 7.5$ and a minimum value of R, $R_{90} = 0.024$. The general trend is consistent with this kind of model, but a range in Lorentz factor or R_{90} is required to account for all the points. The model plot clearly indicates one possible reason why all superluminal sources found so far (with the exception of 3C 263) have values of $\beta_{\text{app}} \gtrsim 3$. The model predicts that smaller values of β_{app} will only be found in objects with $\log R < -0.5$, and the cores of such objects are rarely strong enough to have been observed with present VLBI sensitivities.

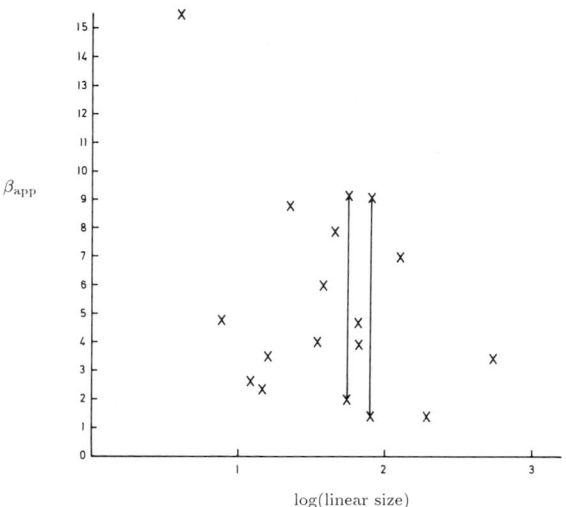

Figure 5 Plot of β_{app} versus the logarithm of the linear size in kiloparsecs.

β_{app} versus Linear Size

In Figure 5, I have plotted β_{app} versus the logarithm of the linear size for the same objects as in Figure 4. No significant trend between β_{app} and size is present. This is not very surprising since the range of projection factors in superluminal quasars probably alters the observed sizes by less than a factor of ten, whereas the range of intrinsic sizes in quasars is known to be at least a hundred. In other words, it appears that the intrinsic spread in linear sizes may mask the trend expected in beaming models. In the next section, I will discuss in detail linear size constraints on beaming models.

5. The Linear Size Problem

It has been pointed out several times (Schilizzi and de Bruyn 1983; de Bruyn and Schilizzi 1986; Barthel *et al.* 1986) that the extended radio structure round superluminal radio sources may be too big for them to have been drawn from the population of "normal" extended radio sources. If deprojected by the "appropriate amount", it is claimed that the superluminal objects lie close to or above the upper envelope of the plot of linear size versus redshift for steep-spectrum quasars. How discrepant the linear sizes are is difficult to quantify since the appropriate deprojection angle is never known. To start with, the measured expansion speed only gives an upper limit to the angle between the direction of motion and the line of sight. A second problem is that this "superluminal angle" is not always the angle one should use to deproject the extended structure. There are often significant misalignments between the direction of motion and the position angle of the outer radio structure.

Despite the above difficulties, I will try to define the magnitude of this linear size problem. The first important step is to select a suitable comparison sample of steep-spectrum quasars. Such a comparison sample should have distributions of redshift and of extended radio luminosity similar to the sample of superluminal sources. This is because, as is well known, there are strong apparent correlations between linear size and redshift and between linear size and luminosity in samples of quasars.

I have produced a comparison sample by selecting for each superluminal two or three quasars that match it as closely as possible in redshift and in luminosity of the extended emission (Table 2). These comparison quasars were selected from one of four well-defined samples of steep-spectrum quasars: the 3C sample of Laing, Riley, and Longair (1983), the Jodrell 966 MHz survey (Shone et al., in preparation), the Bologna 408 MHz sample (Rogora, Padrielli, and de Ruiter 1986; Rogora, Padrielli, and de Ruiter 1987), and a sample of weak steep-spectrum quasars (Browne and Boroson, unpublished). Linear size was ignored in the selection of the comparison objects. However, all the quasars had to have known radio structures which effectively means that all the angular sizes are $> 3''$. The surface brightness sensitivities of the maps of the comparison quasars are typically ≤ 1 mJy arcsec^{-2}. These are therefore of the same quality as the maps of most of the superluminals and thus the sizes of the extended structures of both samples should be subject to a similar sensitivity bias if one exists. In practice, it proved possible to find about two quasars which matched each superluminal to within about 0.1 in redshift and about a factor of two in luminosity.

Histograms of the linear size distributions of the two samples are shown in Figure 6. The superluminal sources are smaller on average than those in the comparison sample. The Kolmogorov-Smirnov test rejects the hypothesis that the two samples are drawn from the same parent distribution at the 10% significance level. If we multiply the size of each superluminal by F, the resulting size distribution is larger than that of the comparison sample at the 5% significance level when F exceeds 3.5.

In practice, we would not expect the same deprojection factor to be appropriate for each superluminal because of the range of observed values for β_{app} and the different ways in which superluminal sources were selected. Furthermore, if the misalignment angle between the parsec and kiloparsec structure is likely to be larger than the angle to the

Table 2 Superluminal Quasars and Comparison Objects

Superluminal quasar	Comparison object	z	$\log L_e$	Size (kpc)
NRAO 140		1.258	25.9	65
	0901+28B	1.121	26.1	107
	1351+31	1.326	26.2	52
3C 179		0.846	26.9	65
	3C 175	0.768	26.7	220
	3C 336	0.927	26.9	93
0850+581		1.322	26.7	65
	1115+536	1.235	26.7	47
	3C 204	1.112	26.7	167
4C 39.25		0.699	26.2	16
	1623+26	0.779	26.1	38
3C 216		0.670	26.6	12
	3C 207	0.684	26.7	51
	3C 254	0.734	26.7	52
3C 245		1.029	26.9	34
	3C 208	1.110	26.9	60
	3C 212	1.049	27.0	47
	1213+53	1.065	27.1	146
3C 263		0.652	26.7	192
	3C 334	0.553	26.3	207
	3C 275.1	0.557	26.5	70
3C 273		0.158	26.0	75
	3C 323.1	0.264	25.8	173
3C 279		0.538	26.3	55
	1742+617	0.523	26.2	152
	1007+417	0.611	26.4	130
3C 345		0.595	25.9	80
	1156+63	0.591	26.0	227
	1434+07	0.681	26.1	174
1642+690		0.750	26.1	45
	1253+10	0.824	26.0	103
	1415+17	0.821	26.1	35
	0839+61	0.862	26.3	121
1721+343		0.206	25.5	539
	3C 249.1	0.311	25.9	115
3C 395		0.635	26.2	4
	3C 455	0.523	26.5	11
	1015+27	0.469	26.0	73
1928+738		0.302	25.1	124
	0911+05	0.303	24.9	263
	1351+26	0.310	24.8	463
	1356+58	0.321	25.2	217
3C 454.3		0.859	26.5	43
	0957+00	0.907	26.4	145
	1011+28	0.899	26.3	59

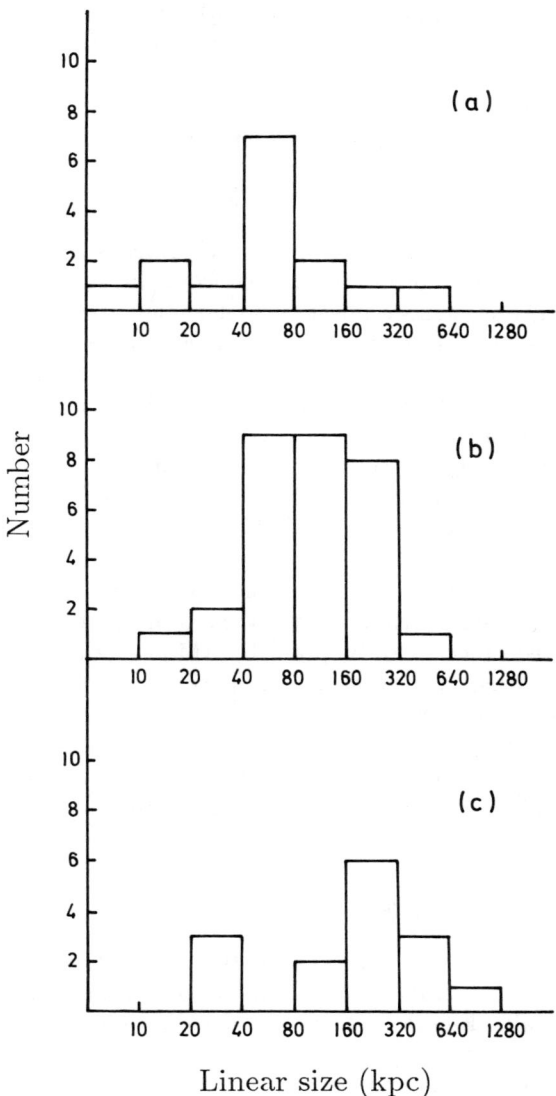

Figure 6 Histograms of linear size distributions. (a) Observed sizes of superluminals (one-sided sizes have been doubled). (b) Sizes of comparison sources (Table 2). (c) Deprojected sizes of superluminals.

line of sight implied by the observed motion, it is this misalignment angle which is appropriate for deprojection. There is, in fact, plenty of evidence that misalignment angles of $\sim 10°$ are quite common. Various indicators of misalignment angles are discussed below.

6. Evidence for Misalignment

Superluminal Quasars

Perhaps the strongest evidence that misalignments are not overwhelmingly important for deprojection comes from the superluminal sources themselves. If the intrinsic misalignment angle between core and outer structure were always greater than the angle made by the core to the line of sight, then we would expect the observed distribution of misalignment angles to be approximately uniform from 0° to 180°. The actual distribution (Figure 3) does contain a few large angles but there is a strong clustering about small angles. Thus, the intrinsic misalignment angles are rarely larger than the angle made by the core motion to the line of sight and so it is this latter angle which should be used for deprojection.

For a single intrinsic misalignment angle and a given core Lorentz factor, the expected distribution of observed misalignments in a sample of core-selected objects can be predicted (Moore et al. 1981; Readhead et al. 1983). Rusk and Rusk (1986) have used this type of analysis on a sample of 43 core-dominated sources. Various combinations of misalignment angle ξ and Lorentz factor γ give reasonable fits (e.g., $\gamma = 5$ and $\xi = 11°$, or $\gamma = 10$ and $\xi = 5°$). Again, this suggests that intrinsic misalignment angles are not often $\gtrsim 10°$.

Lobe-Dominated Quasars

There is generally good alignment between VLBI jets and the outer hotspot in classical double radio sources, though the number of objects studied is small. Figure 7 shows the distribution of angles for a few objects found by cursory inspection of the literature. For these objects, bend angles $> 10°$ seem to be rare. It must be noted that there are probably some selection effects in operation. In particular, the objects studied may have been selected because they are large "well-behaved" doubles and thus one might expect to see good alignment.

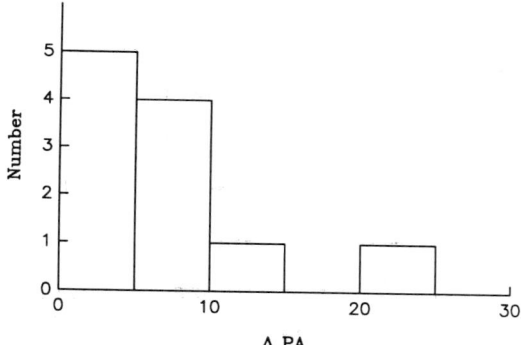

Figure 7 Histogram of the position angle differences ΔPA of parsec and kiloparsec structures. Some objects from Figure 3 are included.

Nonlinearity of Double Sources

A large number of observations show that hotspots and central components of radio sources do not lie on a straight line. Such nonlinearity is clearly a symptom of the misalignment of core and outer structures. Hintzen, Ulvestad, and Owen (1983) find that the median value of the supplement of the angle between the lines joining the lobes to the central component is 9° in a sample of 127 quasars selected at low radio frequency.

For objects with extended radio luminosities matched to the superluminals (e.g., the comparison objects in Table 2 and in the sample of low-luminosity quasars studied by Browne and Boroson), the bend angles are slightly larger. About 13° is typical.

7. The Deprojected Distribution

The various arguments above suggest that misalignment angles of ~ 10° are common. Hence, deprojection factors > 5 should not be applied to superluminals (the finite lateral width of lobes also supports this contention). Therefore, we have adopted the following recipe for deprojection:

(a) When $1.5 \operatorname{arccot}(\beta_{app}) < 10°$, use 10° as the deprojection angle.
(b) For $1.5 \operatorname{arccot}(\beta_{app}) \geq 10°$, use this angle.
(c) Use $H_0 = 100$ km s^{-1} Mpc^{-1} to minimize the discrepancy between observation and prediction.

The reason for using $1.5 \operatorname{arccot}(\beta_{app})$ for the angle to the line of sight is that it is a compromise between $\operatorname{arccot}(\beta_{app})$ which is the angle which minimizes the Lorentz factor of the flow for the observed β_{app}, while $2\operatorname{arccot}(\beta_{\mathrm{app}})$ is the maximum angle to the line of sight even for infinite Lorentz factor.

I have compared the distribution of deprojected linear size with that of the comparison sample (Figure 6). (I have deprojected the comparison sample by a factor $4/\pi$ for this test.) Using the Kolmogorov-Smirnov test, the hypothesis that the two samples are drawn from the same parent population cannot be rejected at the 10% level. However, it only requires H_0 to be reduced to 90 km s^{-1} Mpc^{-1} for the difference to become significant at this level.

8. Discussion

First, I should reemphasize the messiness of the preceding comparison. The statistics are poor, they depend on unknown H_0 and q_0, and we do not know accurately by how much sources bend. In addition, the sources known to be superluminals have been selected for initial observation in no particular systematic way. (For example, 1721+343 was selected because it was the largest known quasar. It may be a surprise that it was found to have an expansion rate of $3.4c$, but it should not come as a surprise that it is larger than any quasar in the comparison sample.) However, in spite of these reservations, I believe that we must work on the hypothesis that the sizes of superluminals are too big for the predictions of the simple unified scheme. This conviction is reinforced by the results obtained by Murphy (Ph. D. thesis, the Victoria University of Manchester, in preparation) who compares the linear sizes of core-selected quasars with those of lobe-selected quasars. Though the linear sizes of the extended emission regions round core-dominated quasars are on average smaller than the sizes of the steep-spectrum lobe-dominated quasars, they are not as small as required by the simple model with average $\gamma \sim 7$. Since values of γ as large as seven, and sometimes larger, are required to explain the superluminal motion, this is telling us the same thing as the analysis of the superluminals themselves. Murphy quantifies the discrepancy by saying that on average a core-to-lobe misalignment of $> 30°$ is required to make the deprojected linear size distribution of the core-dominated sources meet expectations. Though a few such misalignments do exist, this is much larger than the average observed value.

How might beaming models be made more consistent with the linear size data?

(*a*) Make $H_0 = 100$ km s^{-1} Mpc^{-1}.

(*b*) Postulate that relativistic motions occur over a wide range of angles so that the observation of superluminal expansion does not require such good alignment of the source axis to the line of sight.

(c) Look for a missing parent population of sources with large linear sizes. The only possible population is that of radio galaxies.

(d) Postulate that there exists a correlation between linear size and core Lorentz factor amongst quasars. Selecting core-dominated objects would then introduce a bias in favor of large intrinsic linear size which would show up when the observed sizes were deprojected.

In my opinion, the least implausible of these possibilities is (d). There is little evidence that $H_0 > 100$ km s^{-1} Mpc^{-1}. Though an outwardly attractive idea, I think the balance of observational evidence rules out the possibility that a large fraction of high luminosity radio galaxies are quasars with obscured nuclei. (This evidence, concerning narrow line luminosities, luminosities of underlying galaxies, rate of cosmological evolution, etc., is discussed in some detail at the end of Peter Scheuer's talk at this Workshop, page 111.)

Possibility (b) is more difficult to dismiss. One fairly strong argument against it is the narrowness of the observed misalignment angle distribution for superluminals (Figure 3). For model (b) to work, Doppler beaming picks out those bits of a broad cone which are moving at relatively small angles to the line of sight, and it is these bits which display motion. The "bits" may be shocks in a broad beam as postulated by Lind and Blandford (1985), part of a very bent jet as suggested by Peter Scheuer, or individual components shot out of the core with a range of position angles. In all cases, however, in order to make the linear size problem go away, the half-angle of the cone must be 20° to 30°. Of necessity, this half-angle must be larger than the angle to the line of sight deduced from the observed $\beta_{\rm app}$. (If it is not, it will not help the deprojection problem.) This means that the position angle defined by the motion can be up to 20° to 30° away from the true axis of the cone. Since the motion will, in general, be at a small angle to the line of sight, $\leq 30°$, the misalignment between the motion and source axis (kiloparsec axis) when projected on the sky will on average appear very large. This is inconsistent with the relatively narrow distribution for the observed parsec–kiloparsec misalignment angles seen in superluminal sources. (This is exactly the same argument used to dismiss the idea that intrinsic parsec–kiloparsec misalignments can be large enough to solve the linear size problem.)

What about a correlation between linear size and core Lorentz factor? Is there any evidence to support such an idea? Physically, it is quite plausible that faster jets produce bigger sources. Unfortunately, it is a very difficult hypothesis to prove or disprove observationally.

9. Conclusions

(a) All known superluminal sources have extended radio emission.

(b) The core asymmetries are related to asymmetries in the extended structure.

(c) There are no obvious features of superluminal sources which distinguish them from the general population of radio sources.

(d) There is a correlation of β_{app} with the degree of core-dominance, R, as expected for relativistic beaming models.

(e) The linear size problem is not a fundamental difficulty for beaming models. Although the simplest beaming scheme would predict deprojection factors for superluminal quasars which would imply they were too large on average to be drawn from the general population of quasars, this discrepancy can be accounted for by introducing a hypothetical correlation between linear size and core Lorentz factor.

I thank the organizers of the Workshop for financial support, patience, and good company. In discussions, Dave Murphy, Peter Wilkinson, Richard Porcas, Chris Impey, Paddy Leahy, and Robert Laing have made valuable contributions. I am grateful to Dave Murphy, Dave Shone, Tom Muxlow, and Andy Greybe for permission to reproduce their maps.

References

Antonucci, R. R. J., and Ulvestad, J. S. 1985, *Astrophys. J.*, **294**, 158.
Barthel, P. D. 1984, Ph. D. thesis, Rijksuniversiteit Leiden.
Barthel, P. D., Pearson, T. J., Readhead, A. C. S., and Canzian, B. J. 1986, *Astrophys. J. (Letters)*, **310**, L7.
Biretta, J. A., Cohen, M. H., Hardebeck, H. E., Kaufmann, P., Abraham, Z., Perfetto, A. A., Scalise Jr., E., Schaal, R. E., and Silva, P. M. 1985, *Astrophys. J. (Letters)*, **292**, L5.
Biretta, J., Cohen, M., and Moore, R. 1986, in *IAU Symposium 119, Quasars*, ed. G. Swarup and V. K. Kapahi (Dordrecht: Reidel), p. 157.
Biretta, J. A., Moore, R. L., and Cohen, M. H. 1986, *Astrophys. J.*, **308**, 93.
Browne, I. W. A., Clark, R. R., Moore, P. K., Muxlow, T. W. B., Wilkinson, P. N., Cohen, M. H., and Porcas, R. W. 1982, *Nature*, **299**, 788.
Browne, I. W. A., and Orr, M. J. L. 1981, in *Optical Jets in Galaxies*, Proceedings of the second ESO/ESA workshop on the use of the Space Telescope and coordinated ground based research, ESA SP-162, (Noordwijk: European Space Agency), p. 87.
Browne, I. W. A., and Perley, R. A. 1986, *Monthly Notices Roy. Astron. Soc.*, **222**, 149.
Cotton, W. D., Counselman III, C. C., Geller, R. B., Shapiro, I. I., Wittels, J. J., Hinteregger, H. F., Knight, C. A., Rogers, A. E. E., Whitney, A. R., and Clark, T. A. 1979, *Astrophys. J. (Letters)*, **229**, L115.
Davis, R. J., Muxlow, T. W. B., and Conway, R. G. 1985, *Nature*, **318**, 343.
de Bruyn, A. G., and Schilizzi, R. T. 1986, in *IAU Symposium 119, Quasars*, ed. G. Swarup and V. K. Kapahi (Dordrecht: Reidel), p. 203.
de Pater, I., and Perley, R. A. 1983, *Astrophys. J.*, **273**, 64.
Eckart, A., Witzel, A., Biermann, P., Pearson, T. J., Readhead, A. C. S., and Johnston, K. J. 1985, *Astrophys. J. (Letters)*, **296**, L23.
Foley, A. R. 1982, Ph. D. thesis, The Victoria University of Manchester.
Hintzen, P., Ulvestad, J., and Owen, F. 1983, *Astron. J.*, **88**, 709.

Hough, D. H. 1986, Ph. D. thesis, California Institute of Technology.
Laing, R. A., Riley, J. M., and Longair, M. S. 1983, *Monthly Notices Roy. Astron. Soc.*, **204**, 151.
Lind, K. R., and Blandford, R. D. 1985, *Astrophys. J.*, **295**, 358.
Marscher, A. P., and Broderick, J. J. 1982, *Astrophys. J. (Letters)*, **255**, L11.
Moore, P. K., Browne, I. W. A., Daintree, E. J., Noble, R. G., and Walsh, D. 1981, *Monthly Notices Roy. Astron. Soc.*, **197**, 325.
Mutel, R. L., and Phillips, R. B. 1984, in *IAU Symposium 110, VLBI and Compact Radio Sources*, ed. R. Fanti, K. Kellermann, and G. Setti (Dordrecht: Reidel), p. 117.
Orr, M. J. L., and Browne, I. W. A. 1982, *Monthly Notices Roy. Astron. Soc.*, **200**, 1067.
Pauliny-Toth, I. I. K., Porcas, R. W., Zensus, A., and Kellermann, K. I. 1984, in *IAU Symposium 110, VLBI and Compact Radio Sources*, ed. R. Fanti, K. Kellermann, and G. Setti (Dordrecht: Reidel), p. 149.
Pearson, T. J., Barthel, P. D., Lawrence, C. R., and Readhead, A. C. S. 1986, *Astrophys. J. (Letters)*, **300**, L25.
Pearson, T. J., Perley, R. A., and Readhead, A. C. S. 1985, *Astron. J.*, **90**, 738.
Phillips, R. B., and Mutel, R. L. 1982, *Astrophys. J. (Letters)*, **257**, L19.
Porcas, R. W. 1984, in *IAU Symposium 110, VLBI and Compact Radio Sources*, ed. R. Fanti, K. Kellermann, and G. Setti (Dordrecht: Reidel), p. 157.
Porcas, R. W. 1986, in *IAU Symposium 119, Quasars*, ed. G. Swarup and V. K. Kapahi (Dordrecht: Reidel), p. 131.
Readhead, A. C. S., Hough, D. H., Ewing, M. S., Walker, R. C., and Romney, J. D. 1983, *Astrophys. J.*, **265**, 107.
Rogora, A., Padrielli, L., and de Ruiter, H. R. 1986, *Astron. Astrophys. Suppl.*, **64**, 557.
Rogora, A., Padrielli, L., and de Ruiter, H. R. 1987, *Astron. Astrophys. Suppl.*, **67**, 267.
Rusk, R., and Rusk, A. C. M. 1986, *Can. J. Phys.*, **64**, 440.
Schilizzi, R. T., and de Bruyn, A. G. 1983, *Nature*, **303**, 26.
Shone, D. L., Porcas, R. W., and Zensus, J. A. 1985, *Nature*, **314**, 603.
Ulvestad, J. S., and Johnston, K. J. 1984, *Astron. J.*, **89**, 189.
Unwin, S. C. 1986, in *IAU Symposium 119, Quasars*, ed. G. Swarup and V. K. Kapahi (Dordrecht: Reidel), p. 161.
Unwin, S. C., and Biretta, J. A. 1984, in *IAU Symposium 110, VLBI and Compact Radio Sources*, ed. R. Fanti, K. Kellermann, and G. Setti (Dordrecht: Reidel), p. 105.
Waak, J. A., Spencer, J. H., Johnston, K. J., and Simon, R. S. 1985, *Astron. J.*, **90**, 1989.
Walker, R. C. 1986, *Can. J. Phys.*, **64**, 452.
Walker, R. C., Benson, J. M., and Unwin, S. C. 1987, *Astrophys. J.*, **316**, 546.
Zensus, J. A., Hough, D. H., and Porcas, R. W. 1987, *Nature*, **325**, 36.

Feeling Uncomfortable

PETER D. BARTHEL
California Institute of Technology

1. Introduction

Having been in the superluminal business for several years now, and having found a couple of these "action packed" sources myself, I have witnessed an exponential growth in the subject. Between 1980 and 1984 the number of superluminal sources grew slowly from four to seven, but now two years later this last figure has at least doubled. With the growing data base I feel more and more uncomfortable with the simple relativistic beaming model used to explain the various observed phenomena, and with its implications. To communicate and share some of these feelings, thereby finding some relief (I would like to celebrate my own sixtieth birthday someday!), is the main purpose of yhis contribution.

The organization of the paper is as follows. I will start by examining recent observational material using the working hypothesis of the simple relativistic beaming model. This model explains superluminal motion and the apparent small-scale one-sidedness as being due to bulk relativistic motion of the emitting plasma along a direction near the line of sight. This model also attributes the absence of large-scale counterjets in luminous radio sources to Doppler favoritism of the approaching jet over the receding jet. After dealing with some of the recently discovered superluminals and the one-sidedness problem, I present another new superluminal quasar, and point out the problems that these observations pose for the simple relativistic beaming hypothesis. The final paragraph will discuss some ways out of these problems—ways which inevitably complicate matters. I apologize for that.

2. New Superluminal Quasars

Several recently discovered superluminal quasars have fairly unusual properties, including their overall morphology and the absence of variability. Since 1984 the poor superluminal farm of six core-jet sources and one triple source (the black sheep 3C 179) has expanded, and now includes several triple sources, with more (1642+690: Pearson et al. 1986; 1928+738: Eckart et al. 1986; and 0850+581: Barthel et al. 1986) or less (3C 245: Hough and Readhead, this Workshop, page 114; 3C 263:

Zensus, Hough, and Porcas 1987; 4C 34.47: see below) dominant core emission, as well as a compact steep-spectrum source (3C 216: Pearson, Readhead, and Barthel, this Workshop, page 94). Two of these new superluminals, 0850+581 and 3C 216, although definitely core dominated, are not strongly variable (Seielstad et al. 1983). This is in marked contrast with the "classical" superluminal sources, for which strong variability is a characteristic property. It should be noted that this variability in compact radio sources actually led to the development of the relativistic beam model (Rees 1966). In general, the picture posed by the compact steep-spectrum source 3C 216 is not clear (Pearson, Readhead, and Barthel, this Workshop, page 94). Also, the fairly large angular sizes of some of the new superluminal quasars are remarkable. Based on these new data, Barthel et al. (1986) drew attention to the deprojection problem for superluminal quasars with extended morphologies. In that paper we used the angle to the line of sight which would minimize the actual Lorentz factor of the bulk flow in the simple beaming model, $\theta_{\min} = \text{arccot}(\beta_{\text{app}})$, to calculate the deprojected linear sizes for five extended superluminal quasars. The resulting linear sizes put these quasars at the top end of the linear size distribution (Figure 2), and, since the five sources represent a considerable fraction of the superluminal population, this behavior is a problem. We therefore postulated either misalignments between the small and large-scale morphologies, or decoupling of the pattern speed from the bulk plasma flow.

3. One-sided Large-scale Jets

Since the relativistic beaming model requires at least moderately relativistic core velocities ($v_{\text{app}} > 0.9c$ or $\gamma > 2$) to explain superluminal motion, and because of the observed continuity in the one-sidedness from parsec to kiloparsec scales, it has been claimed that the large-scale jets should also be at least moderately relativistic. This would naturally explain the finding that no luminous triple radio source shows any sign of a large-scale counterjet as being due to Doppler favoritism of the approaching jet over the receding one (e.g., Bridle and Perley 1984). The question arises, however, whether relativistic beaming can hide the counterjet in sources which lie in or very near to the plane of the sky. It is an easy (and instructive!) exercise to calculate that for sources within 15° of the sky plane, even for $\gamma \to \infty$, the jet to counterjet ratio cannot exceed five (using a spectral index of the jet $\alpha_{\text{jet}} = -1$, where $S_\nu \propto \nu^\alpha$). Allowing the exponent in the brightness ratio formula to be $(\alpha - 3)$ this figure is eight. The probability for a radio source in a randomly oriented sample to be within 15° of the sky plane is 26%. This explains why Wardle and Potash (1984) were surprised to find eight out of eight

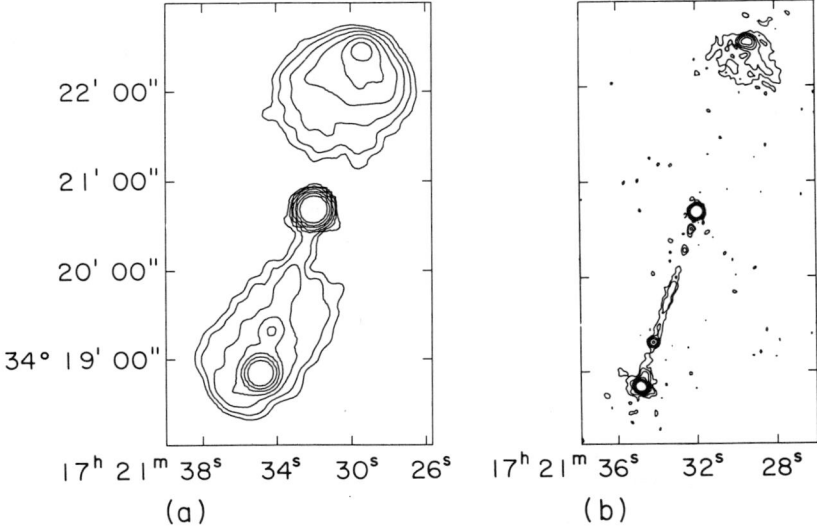

Figure 1 4C 34.47 (1721+343) as mapped with the VLA (Barthel and van Breugel, in preparation): (a) 1.4 GHz, 11″ beam; (b) 5 GHz, 3″.5 beam.

large double-lobed 4C quasars from a complete sample to have one-sided large-scale jets, when from their sizes these sources were expected to lie near the plane of the sky. Equally interesting in this respect is the one-sided jet in 4C 34.47, the largest known quasar (Figure 1). This source has an overall linear size of $560h^{-1}$ kpc ($H_0 = 100h$ km s^{-1} Mpc^{-1}, $q_0 = 0.5$).

Although we again expect it to lie near the plane of the sky, 4C 34.47 displays a one-sided jet, with a brightness ratio exceeding 10 : 1. The extent of these findings is illustrated in Figure 2, which shows linear size distributions for three different samples and for 4C 34.47. The samples include the complete sample of double-lobed 3C quasars (Hough 1986), the Wardle and Potash (1984) sample of eight large 4C quasars, and the five extended superluminal quasars, referred to in the previous paragraph (deprojected sizes for the last sample). The median redshifts for these three samples are 1.04, 0.64, and 0.80; however, recent angular size–redshift data (e.g., Barthel 1984) suggest that redshift effects do not play an important role.

Inspection of Figure 2 shows that we are indeed considering the largest known quasars. Assuming that relativistic beaming hides the large-scale counterjets, we have to conclude that these very large quasars are still at considerable angles from the sky plane. The next paragraph will emphasize this conclusion.

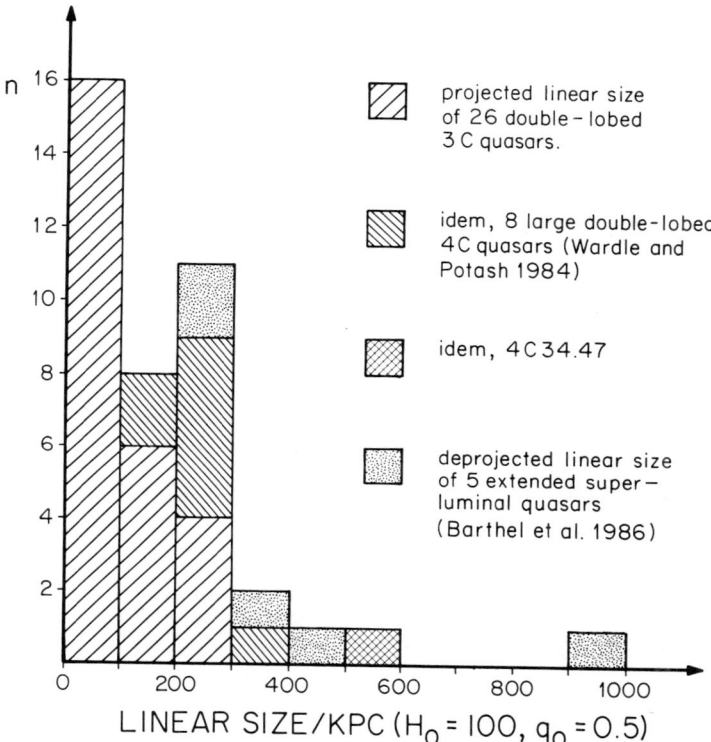

Figure 2 Linear size distributions for quasars (projected and deprojected).

4. Another Superluminal Quasar

The above mentioned largest known quasar, 4C 34.47, has a 5 GHz core flux density of 330 mJy, about 55% of the total 5 GHz flux density. For this reason it belongs in a sample of extended quasars, in which we study the core morphology and evolution using transatlantic VLBI at 5 GHz (Barthel et al. 1984). We have now found that 4C 34.47 almost certainly displays superluminal motion in the core of its triple radio structure. Our two latest 5 GHz maps (epochs 1983.27 and 1986.44) are shown in Figure 3.

The proper motion of the core components is not unambiguous. Although Barthel et al. (1985b) initially assumed that the core coincides with the peak in the 1983.27 map, we are now inclined to identify the northernmost component with the core, for two reasons. First, both the parsec and the kiloparsec scale structure are then one-sided on the same side. Second, the two components south of the assumed core then have the same proper motion. Final identification awaits a

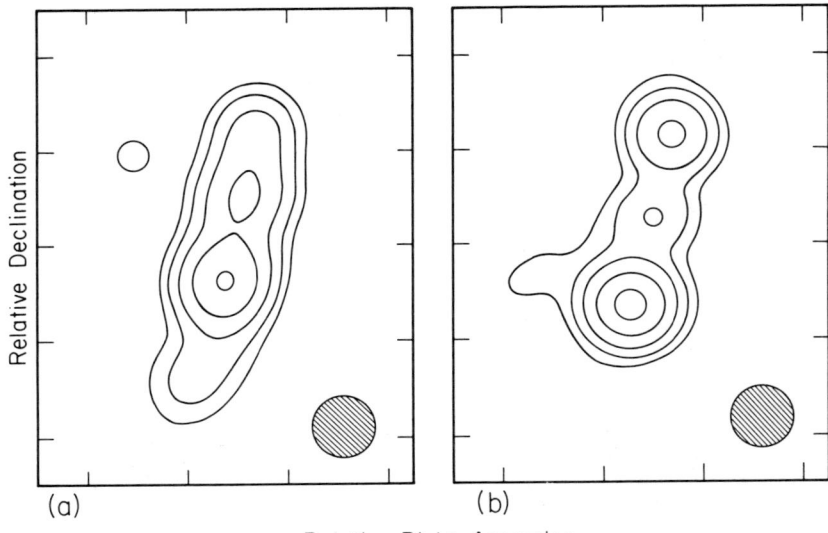

Figure 3 4C 34.47 core at 5 GHz, with 2 mas restoring beam: (a) 1983.27; (b) 1986.44. The scale is 3.2 mas per tick, and identical contour levels have been used: −9, 9, 18, 36, 74, and 135 mJy per beam.

2.8 cm map of the core, which is nearing completion. Assuming that the northernmost component is the core, we measure a proper motion $\mu = 0.36$ mas yr^{-1} or $\beta_{\rm app} = 3.1h^{-1}$ for the two southern components. However, the 1986.44 map is not yet final: a statement on the reality of the southernmost component in the 1983.27 map awaits completion of the 1986 map. A superluminal velocity of $3.1h^{-1}c$ implies in the simple beaming model that this quasar lies within 36° of the line of sight, and minimizing the bulk flow velocity would imply $\theta_{\rm min} = 18°$ (at $\gamma_{\rm min} = 3.3$). As in the other extended superluminal quasars, its deprojected length in the simple beaming model would be very large (Figure 2). Furthermore, the probability for a randomly oriented radio source to lie within a 36° cone is 19%: we would expect at least five quasars to be larger than this largest known quasar.

The inevitable conclusion is that the largest known quasars are at considerable angles from the plane of the sky.

5. What Does This All Mean?

These observational results challenge the simple relativistic beaming picture and raise the following questions.

(a) If the simple relativistic beaming model is correct, we should see quasars with two-sided large-scale jets, even allowing for very high values of γ, unless every quasar is beamed towards us. In the latter case we will have to find their unbeamed counterparts. I already pointed out that several unbeamed sources should be in the megaparsec class, which would rank them among the (rare) giant radio galaxies. Clearly high-dynamic-range mapping of large triple quasars to search for two-sided large-scale jets should continue.

(b) Let us suppose that quasars do have truly one-sided large-scale jets. The tendency of the bright hotspot to be on the jet side (Laing 1984) is relevant in this respect. In that case we are left with the observed superluminal motion in 4C 34.47. Unless the source is a freak (that is to say, we should *not* find similar sources), this again would lead to the question, where are the five or more unbeamed counterparts of 4C 34.47? An obvious test to find out if 4C 34.47 is a freak is to measure proper motion in the core of the radio galaxy 3C 236, the largest known object in the universe (Barthel et al. 1985a).

(c) Modifying the beaming model in order to explain the superluminal motion in extended triple sources (e.g., by allowing for misalignments on small and large scales or by postulating a wide cone angle on the milliarsecond scale and jet collimation further out) may solve part of the deprojection problem, but the question posed by the large-scale one-sidedness—where are the unbeamed quasars?—remains.

(d) A combination of (b) and (c), allowing for truly one-sided jets and a modified beaming model on the small scale, seems able to explain the observations.

A completely different solution of the problems is to postulate that the beaming model in general is not correct. In that case the observations require the following:

(a) the one-sidedness is source-intrinsic, both on small and large scales; this could be caused by flip-flops or asymmetric inefficiency in the energy transport;

(b) the superluminal motion is caused by a different mechanism, such as screen effects (e.g., Miley 1983; de Waard 1986), or oblique shocks (e.g., Marscher, this Workshop, page 280).

Such a model would account for both the apparent one-sidedness and the superluminal motion in sources lying in the plane of the sky. The flow speeds should be at least mildly relativistic ($\gamma \gtrsim 2$) in order to explain the superluminal motion.

There are compelling arguments in favor of the relativistic beaming model, in particular the synchrotron self-Compton argument. It appeared during the Workshop that core-dominated superluminals have on average higher values of β_{app} than do lobe-dominated superluminals, although the data base is still small. I recognize these arguments, but, given the problems mentioned above, I feel that orientation-independent models deserve more attention.

6. Conclusions

With the growing data base it seems harder to reconcile both superluminal motion and jet one-sidedness with a preferred orientation of the radio source with respect to the line of sight. The simple beaming hypothesis will not survive if more very large triple sources with considerable superluminal motion and well aligned small and large scale jets are found. It will also be difficult to maintain the hypothesis if no two-sided jets in large triple sources are detected. The possibilities of orientation-independent models should be investigated in more detail.

In the next couple of years it will become clear whether Wilhelm Busch was right when he wrote *"Aber hier, wie überhaupt, kommt es anders als man glaubt" (Plisch und Plum, 1882)*.

I am grateful to the Dutch fringe fitters (Richard Schilizzi, George Miley, and Jeanette Hooimeijer) and to Wil van Breugel, Tim Pearson, Eugen Preuss, and Tony Readhead for their collaboration. I furthermore thank Charles Lawrence, Tony Readhead, and Richard Schilizzi for a critical reading of the above, and the Workshop organizers for having done a great job.

References

Barthel, P. D. 1984, Ph. D. thesis, Rijksuniversiteit Leiden.
Barthel, P. D., Miley, G. K., Schilizzi, R. T., and Preuss, E. 1984, *Astron. Astrophys.*, **140**, 399.
Barthel, P. D., Miley, G. K., Schilizzi, R. T., and Preuss, E. 1985b, *Astron. Astrophys.*, **151**, 131.
Barthel, P. D., Pearson, T. J., Readhead, A. C. S., and Canzian, B. J. 1986, *Astrophys. J. (Letters)*, **310**, L7.
Barthel, P. D., Schilizzi, R. T., Miley, G. K., Jägers, W. J., and Strom, R. G. 1985a, *Astron. Astrophys.*, **148**, 243.
Bridle, A. H., and Perley, R. A. 1984, *Ann. Rev. Astron. Astrophys.*, **22**, 319.
de Waard, G. J. 1986, Ph. D. thesis, Rijksuniversiteit Leiden.
Eckart, A., Witzel, A., Biermann, P., Pearson, T. J., Readhead, A. C. S., and Johnston, K. J. 1985, *Astrophys. J. (Letters)*, **296**, L23.
Hough, D. H. 1986, Ph. D. thesis, California Institute of Technology.
Laing, R. A. 1984, in *Physics of Energy Transport in Extragalactic Radio Sources*, NRAO Workshop No. 9, ed. A. H. Bridle and J. A. Eilek (Green Bank: National Radio Astronomy Observatory), p. 128.
Miley, G. 1983, in *Astrophysical Jets*, ed. A. Ferrari and A. G. Pacholczyk (Dordrecht: Reidel), p. 99.
Pearson, T. J., Barthel, P. D., Lawrence, C. R., and Readhead, A. C. S. 1986, *Astrophys. J. (Letters)*, **300**, L25.
Rees, M. J. 1966, *Nature*, **211**, 468.
Seielstad, G. A., Pearson, T. J., and Readhead, A. C. S. 1983, *Publ. Astron. Soc. Pacific*, **95**, 842.
Wardle, J. F. C., and Potash, R. I. 1984, in *Physics of Energy Transport in Extragalactic Radio Sources*, NRAO Workshop No. 9, ed. A. H. Bridle and J. A. Eilek (Green Bank: National Radio Astronomy Observatory), p. 30.
Zensus, J. A., Hough, D. H., and Porcas, R. W. 1987, *Nature*, **325**, 36.

The Arcminute Structure of 1928+738

R. S. SIMON, K. J. JOHNSTON
E. O. Hulburt Center for Space Research, Naval Research Laboratory

A. ECKART
Max-Planck Institut für Extraterrestrische Physik

P. BIERMANN, C. SCHALINSKI, A. WITZEL
Max-Planck Institut für Radioastronomie

R. G. STROM
Netherlands Foundation for Radio Astronomy

1. Introduction

There has been a lot of discussion at this Workshop about the relationship between the large-scale radio emission of the superluminal sources and the symmetrical double-lobed radio sources. The question, as yet unresolved, is whether the superluminal sources could be conventional double sources seen end-on, i.e., with their symmetry axes lying close to the observer's line of sight. Here we discuss this question with particular reference to the recently-discovered superluminal source 1928+738 (Eckart et al. 1985; see also the papers presented at this Workshop by Pearson, Readhead, and Barthel, page 94, and by Witzel, page 83). We have found that 1928+738 has two-sided radio structure which extends to 40″ on either side of the compact core, equivalent to a projected linear size of $221h^{-1}$ kpc (assuming $H_0 = 100h$ km s^{-1} Mpc^{-1} and $q_0 = 0.5$). If the superluminal expansion of the core, at $v_{app} = 7.5h^{-1}c$ (Eckart et al. 1985), is interpreted in terms of the standard relativistic jet model (Blandford and Königl 1979), the axis of the small-scale structure must lie within 15.1° [$2\operatorname{arccot}(v_{app}/c)$] from the line of sight. If the large-scale structure is also aligned at this angle, the deprojected size exceeds $700h^{-1}$ kpc.

2. Observations

We have observed the large-scale radio structure of 1928+738 at 20 cm with the VLA and at 49 cm with the WSRT (Johnston et al. 1987). The 20 cm map is shown in Figure 1. This map shows two streams of emission extending out from the central source. The southern stream bends through an angle of about 30° (p.a. −170° to +160°), while the

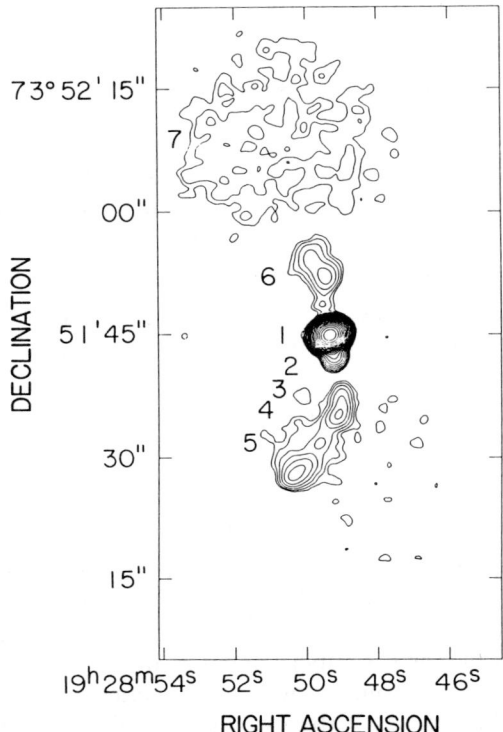

Figure 1 VLA map of 1928+738 at 20 cm wavelength, made using natural weighting to give maximum sensitivity to low-surface-brightness emission. The elliptical Gaussian restoring beam has FWHM $1.8'' \times 1.6''$ in p.a. $-62°$. The compact core has a flux density of 3.35 Jy per beam; the contours are at 2 dB intervals from 2.114 Jy per beam (2 dB below the peak) down to ± 0.53 mJy per beam (38 dB below the peak). The map has a dynamic range of 24000:1.

northern stream ends in a large amorphous region about $15''$ in diameter. Seven "components" (marked 1–7 in Figure 1) can be distinguished. A higher-resolution map, made from the same 20 cm data, is shown in Figure 2. In this map the innermost component (2) is resolved from the compact core (component 1). All the components are unresolved in their shorter dimension with the exception of the northernmost amorphous component 7.

The 49 cm image of 1928+738, with a resolution of $30''$, shows three major components: a dominant central source and two "lobes" spanning about $80''$ in the north-south direction. Figure 3 is an image of the double structure after subtraction of the nuclear component. The southern component appears significantly resolved, and the emission to the west is extended. While the northern lobe corresponds to components 6 and

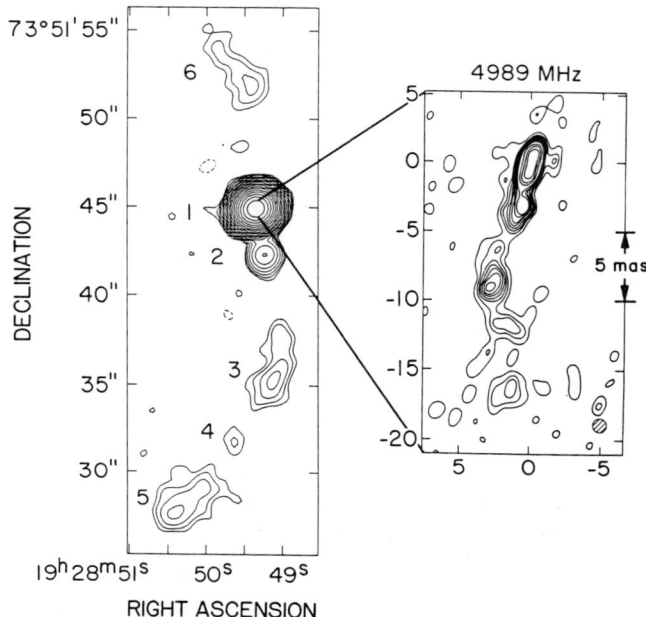

Figure 2 *Left:* full-resolution 20 cm VLA map made with uniformly weighted (u, v) data, excluding the amorphous component 7. The restoring beam has FWHM $1.2'' \times 1.1''$ in p.a. $-54°$. The peak flux density is 3.35 Jy per beam; the contours are at 2 dB intervals from 2.114 Jy per beam down to ± 0.84 mJy per beam. *Right:* the milliarcsecond resolution 6 cm map of 1928+738 from Eckart *et al.* (1985).

7 of Figure 1, the southern lobe probably contains additional flux density that is largely resolved out in the VLA map. Comparison of the 6 cm flux density of components 3, 4, and 5 with the total 20 cm flux density of the southern lobe places a lower limit of -2.4 on the spectral index of this lobe. There may also be extended emission southwest of component 5 that is not seen on the 20 cm map.

3. Discussion

Deprojection

The redshift of this quasar is 0.302 (Lawrence *et al.* 1986) so that $1'' = 2.76h^{-1}$ kpc. Figure 3 shows that the emission is extended over about $80''$, which corresponds to $221h^{-1}$ kpc. From the superluminal motion in the core, we infer that the milliarcsecond-scale jet is aligned within $15.1°$ of the line of sight. If the large scale radio emission is aligned at the same angle to the line of sight, then the deprojected size is $\sim 850h^{-1}$ kpc. In deprojecting the source, though, one must also take into account the width of the lobes, which is independent of the viewing angle. This

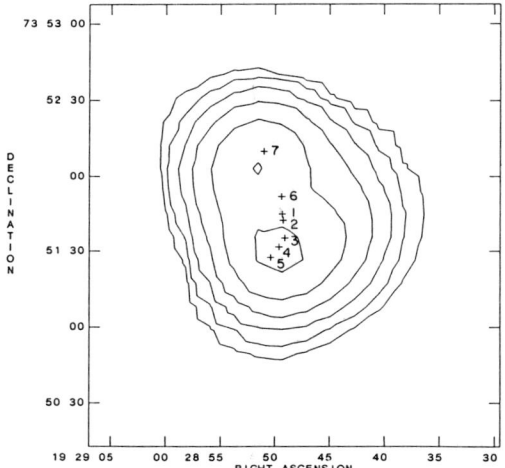

Figure 3 Brightness distribution of 1928+738 at 49 cm after subtraction of the dominant central component. Contour levels are 8.3, 16.5, 33.1, 66.2, 132.3, and 264.6 mJy per beam; the lowest contour is about 0.2% of the peak brightness in the original map. Crosses indicate the positions of components found in the VLA map (Figure 1).

correction reduces the deprojected size by $\sim 20\%$ to $\sim 700h^{-1}$ kpc. Even so, this would make 1928+738 one of the largest known quasars. We should therefore consider the possibility that the axis of the large-scale emission is not aligned with the milliarcsecond jet but lies closer to the plane of the sky.

Typical sizes for large, double-lobed quasars are in the range 150–300h^{-1} kpc. For example, Wardle and Potash (1984) selected a sample of 32 quasars that have clearly-resolved double or triple structure from 4C radio sources in the declination range 20°–40° (Olsen 1970; Schmidt 1974). They mapped the eight quasars in this sample with angular sizes $> 50''$. The apparent major axes of these quasars range from 200 to 400h^{-1} kpc. All eight sources display two-sided emission with one side dominant, and with no counterjet. On the assumption that the quasars in the parent sample are randomly oriented in space, we can estimate that the eight largest quasars lie at angles from the line of sight between 76° and 90° (76° = $\arccos \frac{8}{32}$). Similar apparent sizes (100–350h^{-1} kpc) were found by Burns et al. (1984) for the major axes of high luminosity, edge brightened "classical doubles" (class II; Fanaroff and Riley 1974).

Thus if 1928+738 is not an exceptionally large quasar, its large-scale emission must lie at a larger angle from the line of sight than that inferred from the superluminal motion. For example, taking account of the transverse size of the radio lobes, an angle to the line of sight of 24°

would require a deprojected size $\sim 400h^{-1}$ kpc. It is not impossible, however, that 1928+738 is as large as $700h^{-1}$ kpc, considering that the major axis of the quasar 4C 34.47 is $1h^{-1}$ Mpc (Conway, Burn, and Vallée 1977; Jägers et al. 1982), and that some radio *galaxies* are as large as $4h^{-1}$ Mpc (e.g., 3C 236: Strom and Willis 1980).

The Jet

The milliarcsecond-scale radio emission of 1928+738 is shown in Figure 2 (Eckart et al. 1985). The milliarcsecond jet is directed towards the southern lobe. On this side the arcsecond-scale emission is nearly continuous and presumably represents a continuous jet. The continuity may be used as an indirect argument that the velocity of the jet is continuous out to kiloparsec distances from the core. As in other superluminals, the arcsecond-scale emission close to the compact core is stronger on the side of the milliarcsecond jet. However, 1928+738 is unique among superluminal sources—it also has an apparent counterjet.

Comparison with Other Superluminal Sources

The arcsecond structure of superluminal radio sources has been described by Browne et al. (1982), Browne and Orr (1981), Perley, Fomalont, and Johnston (1982), Schilizzi and de Bruyn (1983), and by several other participants at this Workshop (e.g., Browne, page 129).

Table 1 gives a summary of the large-scale morphology and size of superluminal radio sources. All of the objects are quasars except the radio galaxy 3C 120. The sources 3C 120, 3C 179, 3C 395, 3C 345, NRAO 140, BL Lac, 1642+690, 3C 279, 3C 245, 3C 263, 3C 216, and 4C 34.47 all display emission on both sides of the compact core, like 1928+738. Large extended radio lobes are evident in 3C 179, 3C 120, 3C 279, 4C 34.47, and NRAO 140, while the emission from BL Lac is complex and of low brightness. Thus among the known superluminal sources only 3C 273 does not display emission on both sides.

Table 1 includes the minimum deprojected size of the source assuming that the inclination of the major axis to the line of sight is the same as that inferred from the apparent superluminal motion. The deprojected size of the large-scale emission ranges from 4–$400h^{-1}$ kpc. Three of the sources (3C 120, 4C 34.47, and 1928+738) have apparent major axes in the range of the large quasars observed by Wardle and Potash (1984). The quasar with the largest size is 3C 345; this source also has the smallest angle to the line of sight, 3°. In this case making allowance for the width of the lobes (which we have ignored) would reduce the deprojected size to 400–$800h^{-1}$ kpc. The angles for the other objects

Table 1 Large Scale Structure of Superluminal Sources

Source	Type[a]	z	Angular size (arcsec)	v_{app}/c	θ_{max} (deg)[b]	Apparent size h^{-1} kpc	Deprojected size[c] h^{-1} kpc	Refs.
3C 120	C	0.033	720	2.1	51	370	424	1
NRAO 140	D	1.258	11	5.4	21	61	170	2
3C 179	C/D	0.846	16	4.2	27	80	176	3
3C 216	C	0.669	2	2.5	44	9	13	4
3C 245	D	1.029	9	3.1	36	48	82	5
3C 263	D	0.652	44	1.3	75	199	206	5
3C 273	A	0.158	20	5.3	21	36	100	1
3C 279	C/D	0.538	18	3.5	32	74	140	1
3C 345	D	0.595	20	8.2	3	85	1620	2, 6, 7
1642+690	D	0.751	11	9.3	12.3	53	249	8, 9
4C 34.47	D	0.206	180	3.2	35	406	713	10
1928+738	D	0.302	80	9.1	12.5	235	1090	11, 12
3C 395	D	0.635	1	10.5	10.9	4	21	13, 14, 15
BL Lac	C	0.070	25	2.0	53	22	28	2, 6

[a] A = asymmetric, C = complex, D = double. [b] $\theta_{max} = 2\,\text{arccot}(v_{app}/c)$ for all sources except 3C 345 for which additional constraints are available (Cohen and Unwin 1984). If v/c were the only constraint, $\theta_{max} = 14$ for 3C 345 and the deprojected size is only $> 351 h^{-1}$ kpc. [c] Assumes that large scale emission is misaligned to the line of sight by the angle predicted by the apparent superluminal motion.

References: 1. Walker, Benson, and Unwin 1987. 2. Schilizzi and de Bruyn 1983. 3. Porcas 1984. 4. Pearson, Perley, and Readhead 1985, and Pearson, Readhead, and Barthel, this Workshop, page 94. 5. Hough and Readhead, this Workshop, page 114. 6. Cohen and Unwin 1984. 7. Biretta, Moore, and Cohen 1986. 8. Pearson *et al.* 1986. 9. Browne and Orr 1981. 10. Barthel, this Workshop, page 148. 11. Eckart *et al.* 1985. 12. Johnston *et al.* 1987. 13. Waak *et al.* 1985. 14. Perley, Fomalont, and Johnston 1982. 15. Simon *et al.* 1987.

are greater than 12°, but here again the radio lobes will cause large over-estimates in the deprojected sizes of the sources.

Although 1928+738 is very large compared with the other sources, it is not the only source that has an unreasonably large deprojected size: four of the 14 sources are larger than 400 kpc. This may be taken as evidence that the large scale emission is usually not aligned with the compact emission, and that bending between the core and the outer radio lobes is a common phenomenon.

4. Conclusions

The radio emission at 20 cm and 49 cm from the quasar 1928+738 extends over 80″ ($221 h^{-1}$ kpc), and is centered on the compact core. On one side, the emission is continuous from parsec to kiloparsec scales. If the source is oriented at an angle to the line of sight of $< 15.1°$, as inferred from the apparent superluminal motion, its deprojected size is $\sim 700 h^{-1}$ kpc, which is unusually large in comparison with other quasars. It is thus likely that the source is not aligned with the milliarcsecond emission. Large scale radio structure is probably as common among superluminal sources as it is in other radio quasars, and of comparable physical size.

References

Biretta, J. A., Moore, R. L., and Cohen, M. H. 1986, *Astrophys. J.*, **308**, 93.
Blandford, R. D., and Königl, A. 1979, *Astrophys. J.*, **232**, 34.
Browne, I. W. A., Clark, R. R., Moore, P. K., Muxlow, T. W. B., Wilkinson, P. N., Cohen, M. H., and Porcas, R. W. 1982, *Nature*, **299**, 788.
Browne, I. W. A., and Orr, M. J. L. 1981, in *Optical Jets in Galaxies*, Proceedings of the second ESO/ESA workshop on the use of the Space Telescope and coordinated ground based research, ESA SP-162, (Noordwijk: European Space Agency), p. 87.
Burns, J. O., Basart, J. P., De Young, D. S., and Ghiglia, D. C. 1984, *Astrophys. J.*, **283**, 515.
Cohen, M. H., and Unwin, S. C. 1984, in *IAU Symposium 110, VLBI and Compact Radio Sources*, ed. R. Fanti, K. Kellermann, and G. Setti (Dordrecht: Reidel), p. 95.
Conway, R. G., Burn, B. J., and Vallée, J. P. 1977, *Astron. Astrophys. Suppl.*, **27**, 155.
Eckart, A., Witzel, A., Biermann, P., Pearson, T. J., Readhead, A. C. S., and Johnston, K. J. 1985, *Astrophys. J. (Letters)*, **296**, L23.
Fanaroff, B. L., and Riley, J. M. 1974, *Monthly Notices Roy. Astron. Soc.*, **167**, 31P.
Jägers, W. J., van Breugel, W. J. M., Miley, G. K., Schilizzi, R. T., and Conway, R. G. 1982, *Astron. Astrophys.*, **105**, 278.
Johnston, K. J., Simon, R. S., Eckart, A., Biermann, P., Schalinski, C., Witzel, A., and Strom, R. G. 1987, *Astrophys. J. (Letters)*, **313**, L85.
Lawrence, C. R., Pearson, T. J., Readhead, A. C. S., and Unwin, S. C. 1986, *Astron. J.*, **91**, 494.
Olsen, E. T. 1970, *Astron. J.*, **75**, 764.
Pearson, T. J., Barthel, P. D., Lawrence, C. R., and Readhead, A. C. S. 1986, *Astrophys. J. (Letters)*, **300**, L25.
Pearson, T. J., Perley, R. A., and Readhead, A. C. S. 1985, *Astron. J.*, **90**, 738.
Perley, R. A., Fomalont, E. B., and Johnston, K. J. 1982, *Astrophys. J. (Letters)*, **255**, L93.
Porcas, R. W. 1984, in *IAU Symposium 110, VLBI and Compact Radio Sources*, ed. R. Fanti, K. Kellermann, and G. Setti (Dordrecht: Reidel), p. 157.
Schilizzi, R. T., and de Bruyn, A. G. 1983, *Nature*, **303**, 26.
Schmidt, M. 1974, *Astrophys. J.*, **193**, 505.
Simon, R. S., Hall, J., Johnston, K. J., Spencer, J. H., Waak, J. A., and Mutel, R. L. 1987, *Astrophys. J. (Letters)*, submitted.
Strom, R. G., and Willis, A. G. 1980, *Astron. Astrophys.*, **85**, 36.
Waak, J. A., Spencer, J. H., Johnston, K. J., and Simon, R. S. 1985, *Astron. J.*, **90**, 1989.
Walker, R. C., Benson, J. M., and Unwin, S. C. 1987, *Astrophys. J.*, **316**, 546.
Wardle, J. F. C., and Potash, R. I. 1984, in *Physics of Energy Transport in Extragalactic Radio Sources*, NRAO Workshop No. 9, ed. A. H. Bridle and J. A. Eilek (Green Bank: National Radio Astronomy Observatory), p. 30.

Intrinsic Asymmetry in NGC 6251

DAYTON L. JONES
Jet Propulsion Laboratory

1. Introduction

The compact (parsec-scale) radio sources in active galactic nuclei often appear to be one-sided, even in galaxies with symmetric two-sided radio structure on larger scales (e.g., Readhead and Pearson 1982). This is often explained as a result of bulk relativistic motion, since there is strong evidence that such motion occurs in at least some compact sources (e.g., Cohen and Unwin 1984; Porcas, this Workshop, page 12). This naturally leads to the question of how generally applicable relativistic beaming models are. Specifically, are they a likely explanation of the apparent one-sidedness in compact radio sources which show no evidence of superluminal motion?

The fact that in all sources with one-sided radio jets on both parsec and kiloparsec scales the jets are always found on the same side of the nucleus suggests that the same mechanism must be responsible for the observed asymmetry over the entire length of the jet. This adds importance to the question of how common bulk relativistic motion on parsec scales is, since if it causes the parsec-scale asymmetry it should also be responsible for the kiloparsec-scale asymmetry. At present very little is known about the velocities in large-scale radio jets.

The giant radio galaxy NGC 6251 is a nearly ideal test case, as it has large-scale symmetry (including a faint kiloparsec-scale counterjet), one-sided "core-jet" structure on parsec scales, and sufficient flux density for mapping with the Mark II VLBI system, and it is at a very favorable declination for observations with an array of northern hemisphere telescopes.

2. Observations

The nucleus of NGC 6251 was observed with a VLBI array at 18 cm in March 1983 and April 1985. The first epoch observations were designed to search for a parsec-scale counterjet, and resulted in a very high dynamic range VLBI map (Jones et al. 1986). The detection of such a counterjet would have indicated that the basic jet-producing mechanism is intrinsically symmetric, like the larger scale radio structure. However,

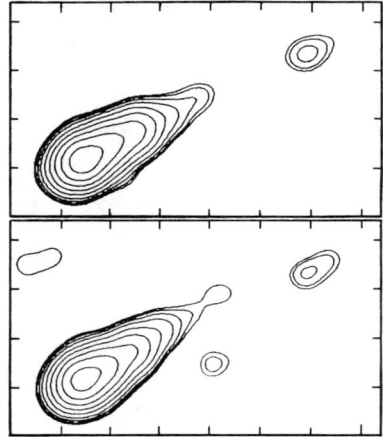

Figure 1 18-cm VLBI maps of NGC 6251 from March 1983 (*top*) and April 1985 (*bottom*) (Jones 1986). In both maps the contours are −1.5, 1.5, 2, 3, 5, 10, 20, 40, 80, and 160 mJy per beam, and the clean beam was circular with FWHM 3.0 mas. The tick marks along the axes are 5 mas apart.

no counterjet was detected by these sensitive observations. It was possible to set a lower limit of 80:1 for the jet-to-counterjet brightness ratio at ±6 mas from the core.

The 1983 map shows a weak but clearly detected feature in the jet about 25 mas from the core. If relativistic beaming were responsible for "hiding" a counterjet which was intrinsically identical to the observed jet, then the transverse motion of this feature (or "knot") should be at least 0.4 mas yr^{-1}. The second-epoch observations were designed to detect this motion if it existed.

In both the 1983 and 1985 experiments, we used large arrays (9–11 telescopes) and observed full (u, v) tracks. Data were recorded with the Mark II recording system, providing a 1.8 MHz bandwidth. In 1983, global fringe-fitting (Schwab and Cotton 1983) was used. This improved the quality of the final map considerably. In 1985 this was found to be unnecessary owing to our use of the sensitive 64-m NASA telescope at Goldstone, which increased the fraction of baselines with high signal-to-noise ratios.

3. Results

Figure 1 shows the maps from both epochs (Jones 1986). The lower noise level on the first-epoch map results from the use of global fringe-fitting.

The measured separations between the main peak and the weaker knot are 24.8 mas for the first epoch and 24.9 mas for the second epoch,

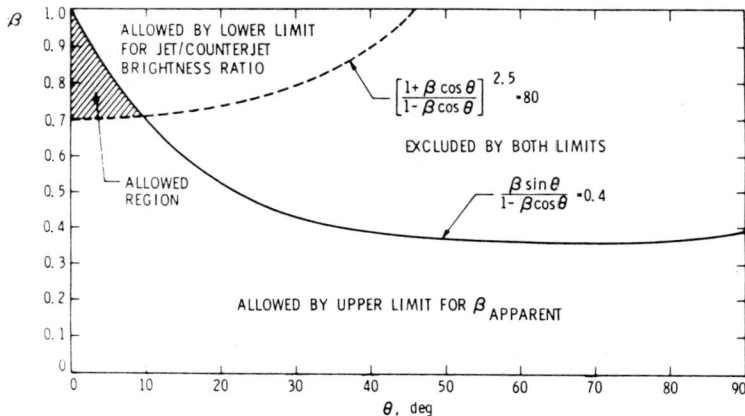

Figure 2 Allowed jet velocity and orientation for a symmetric twin-jet model in which the apparent asymmetry of the parsec-scale radio structure in NGC 6251 is caused by relativistic beaming (Jones 1986).

both with formal errors of ~ 0.1 mas. A more realistic error estimate would be ~ 0.2 times the beam width, or 0.6 mas. This upper limit for the transverse motion of the knot corresponds to an apparent transverse velocity of $\leq 0.3h^{-1}c$ ($H_0 = 100h$ km s^{-1} Mpc^{-1}).

4. Discussion

If we assume that NGC 6251 contains two intrinsically similar jets moving in opposite directions, then our observations constrain the orientation and velocities of the two jets. Figure 2 shows the allowed ranges of β, the bulk velocity of the radio emitting plasma, and θ, the angle between the approaching jet and our line of sight (Jones 1986).

Note that β must be $\geq 0.7c$ and θ must be $\leq 10°$. These particular values assumed $h = 0.75$, but they will not change much for any reasonable value of h. If the large-scale radio axis is aligned within 10° of our line of sight, its total extent would exceed $7.5h^{-1}$ Mpc, making it the largest radio source known. This is certainly possible, but unlikely. It is also possible that the source bends so that the large-scale structure is oriented closer to the plane of the sky, but this bending must occur in a plane which is almost exactly perpendicular to the plane of the sky. Otherwise we would see a bend in the remarkably linear kiloparsec-scale jet. This also seems unlikely.

The large-scale radio structure associated with NGC 6251 differs from that seen around some superluminal sources. It does not appear to be highly distorted by projection effects, and includes two widely separated diffuse lobes and a kiloparsec-scale jet and counterjet which are both very straight near the nucleus. This suggests that the extended

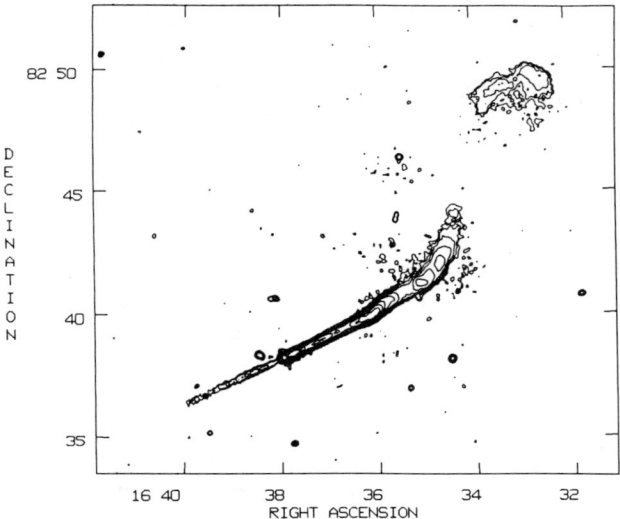

Figure 3 18-cm VLA map of NGC 6251 obtained during the March 1983 VLBI experiment. The contours are 0.03, 0.05, 0.1, 0.2, 0.4, 0.8, 1.6, 3.2, 6.4, 12.8, 25.0, and 50.0% of the peak, which is 506 mJy per beam. The restoring beam was a circular Gaussian with FWHM 10″.

radio source is *not* being seen nearly "end on". Thus, the basic relativistic beaming model does not explain the observed asymmetry in this source.

The other likely explanation for the parsec-scale asymmetry is that it is intrinsic. Since the large-scale structure is basically symmetric (Willis *et al.* 1982), this implies either some sort of "flip-flop" model for the parsec-scale jet (e.g., Rudnick 1982; Rudnick and Edgar 1984) or a difference in the efficiency with which the two jets produce radio emission.

The length of the bright jet seen in VLA maps requires a switching time larger than a million years for "flip-flop" models. This leads to the question of why we see a faint counterjet close to the nucleus (Figure 3). The hypothesis of different efficiencies naturally explains the existence of the counterjet, since in this case energy may be transported to both lobes continuously.

If the large jet-to-counterjet brightness ratio is caused by a difference in the efficiencies of the jets on the two sides, the efficiencies must be determined in the first few parsecs and then remain different throughout the entire length of the jet. The properties of the external medium through which the two jets propagate change greatly over the length of the jets. In particular, the orbital period for gas a few parsecs from the galactic center is orders of magnitude different from the orbital period tens of kiloparsecs from the center. This makes it implausible that the interstellar medium can remain systematically different on different sides of a galaxy over the whole length of a long jet. Thus, I doubt that differences in the medium are responsible for one jet always being brighter than the other in NGC 6251. The brightness difference is more likely to be due to differences in the internal properties of the jets.

It is possible, of course, that the radio knot in this jet does not move with the underlying flow velocity, but this hypothesis differs from the assumption commonly made about the radio knots in other (e.g., superluminal) compact radio sources; and in the one case where we do know the flow velocity—SS 433—the radio knots do move with this velocity. The evidence for bulk relativistic motion in superluminal sources is quite strong, and consequently we are left with two different causes of asymmetry in the brightness of radio jets. This will complicate our efforts to understand the processes occurring in individual sources. Parsec-scale asymmetry may be common in *all* sources with large-scale jets that differ in brightness, including those whose large scale jets are clearly nonrelativistic (e.g., narrow angle tail sources).

5. Summary

At least some of the one sided "core-jet" radio sources in galactic nuclei appear asymmetric for reasons other than relativistic beaming.

I am grateful for the observing support provided by the observatories of the U.S. and European VLBI networks, and thank Tim Pearson for providing most of the software used to analyze these data. This research was supported by the National Research Council through the Resident Research Associate program, and a portion of this research was carried out by the Jet Propulsion Laboratory, California Institute of Technology, under contract with the National Aeronautics and Space Administration.

References

Cohen, M. H., and Unwin, S. C. 1984, in *IAU Symposium 110, VLBI and Compact Radio Sources*, ed. R. Fanti, K. Kellermann, and G. Setti (Dordrecht: Reidel), p. 95.
Jones, D. L. 1986, *Astrophys. J. (Letters)*, **309**, L5.
Jones, D. L., Unwin, S. C., Readhead, A. C. S., Sargent, W. L. W., Seielstad, G. A., Simon, R. S., Walker, R. C., Benson, J. M., Perley, R. A., Bridle, A. H., Pauliny-Toth, I. I. K., Romney, J., Witzel, A., Wilkinson, P. N., Bååth, L. B., Booth, R. S., Fort, D. N., Galt, J. A., Mutel, R. L., and Linfield, R. P. 1986, *Astrophys. J.*, **305**, 684.
Readhead, A. C. S., and Pearson, T. J. 1982, in *IAU Symposium 97, Extragalactic Radio Sources*, ed. D. S. Heeschen and C. M. Wade (Dordrecht: Reidel), p. 279.
Rudnick, L. 1982, in *IAU Symposium 97, Extragalactic Radio Sources*, ed. D. S. Heeschen and C. M. Wade (Dordrecht: Reidel), p. 47.
Rudnick, L., and Edgar, B. K. 1984, *Astrophys. J.*, **279**, 74.
Schwab, F. R., and Cotton, W. D. 1983, *Astron. J.*, **88**, 688.

Willis, A. G., Strom, R. G., Perley, R. A., and Bridle, A. H. 1982, in *IAU Symposium 97, Extragalactic Radio Sources*, ed. D. S. Heeschen and C. M. Wade (Dordrecht: Reidel), p. 141.

Are Compact Doubles Misaligned Superluminals?

MARK W. HODGES
Owens Valley Radio Observatory

ROBERT L. MUTEL
Department of Physics and Astronomy, University of Iowa

1. Introduction

As more multi-epoch VLBI maps of the radio cores of active galactic nuclei become available, it is becoming increasingly clear that superluminal motion is a common feature of many otherwise diverse objects, e.g., quasars, BL Lac objects, and double-lobed radio galaxies. Evidently, the physical conditions which engender superluminal effects are *robust*, i.e., they are operative in a variety of galactic environments. One is naturally led to ask whether there are any classes of compact radio cores which do *not* exhibit superluminal effects.

One such class (perhaps the only one) appears to be the compact double (CD) radio sources (Phillips and Mutel 1982; Pearson and Readhead 1984). The typical CD radio morphology consists of two high brightness ($T_b \sim 10^{10}$ K) components with separations ranging from about ten parsecs to several kiloparsecs. Both components have self-absorbed spectra that peak near 1 GHz. The component angular sizes are consistent with synchrotron self-absorption if the magnetic fields are about 10^{-4} G.

Unlike most superluminal sources, there is little or no variability in either total flux density or polarization. Most importantly, for several sources in which multi-epoch VLBI maps are available, there is no evidence for relative motion at superluminal speeds (Phillips and Shaffer 1983; Readhead, Pearson, and Unwin 1984; Porcas, this Workshop, page 12). A working list of the defining properties of CD sources is given in Table 1.

Multifrequency VLBI maps are now available for seven CD sources (0026+346, 0218+357, 0710+439, 1518+047, 1607+268, 2021+614, and 2050+364), all of which show that the spectra of the two components in each source are similar (2021+614 has complex subcomponent structure: Bartel et al. 1984). There are also eight additional sources for which single frequency maps showing CD structure are currently available. However, they all satisfy the properties listed in Table 1 and are

Table 1 Properties of Compact Double Radio Sources

Morphology	More than 90% of the emission comes from two nearly equal components with separations that are at least three times a component diameter. There is often substructure within each component.
Spectrum	The spectra of both components are similar—each has a self-absorbed synchrotron spectrum with a steep optically thin part.
Size	Overall size is 10–200 mas (about 10–1000 pc). Component sizes from synchrotron-self-absorption calculations are comparable to the measured component sizes.
Polarization	Low ($\lesssim 2\%$).
Variability	Low ($\lesssim 10\%$).
Extended Structure	No arcsecond-scale structure at $\gtrsim$ 3000:1 dynamic range.
Optical ID	Galaxies or empty fields with redshifts between 0.2 and 1.0. Spectra show narrow lines.

therefore *candidate* CD sources. It appears that CD sources comprise perhaps 5% of all active galactic nuclei with bright, milliarcsecond scale cores (Pearson and Readhead 1984). It is not yet clear to what extent the non-morphological properties described in Table 1 are a sufficient condition for CD structure. For example, Hodges, Mutel, and Phillips (1984) found that in a survey of nine previously unmapped sources with "humped" spectra peaking near 1 GHz, only two showed double structure.

A few candidate CD sources identified with quasars have since been discovered to be asymmetrical, i.e., they consist of two components with very different spectra (e.g., 3C 395, Johnston *et al.* 1983; 0153+744, R. L. Mutel, T. Muxlow, and R. B. Phillips, in preparation). So far, all known CD sources are associated with either galaxies or empty fields (e.g., Biretta, Schneider, and Gunn 1985). This is in striking contrast to superluminal sources, more than 50% of which are quasars.

Figure 1 shows a 5 GHz map and component spectra of a typical compact double, the galaxy CTD 93.

2. Genes versus Environment

At first glance, the superluminal sources appear to be a different species than CD sources. The differences could arise either from a different central engine (genetics) or a different ambient medium (environment). We will argue that both superluminals and CD sources could have identical genetics (central engines and beams), but that differences in beam orientation and environment cause the dissimilar radio properties.

Consider genetic differences first. Let us assume that the superluminal sources can be described by the standard relativistic beaming

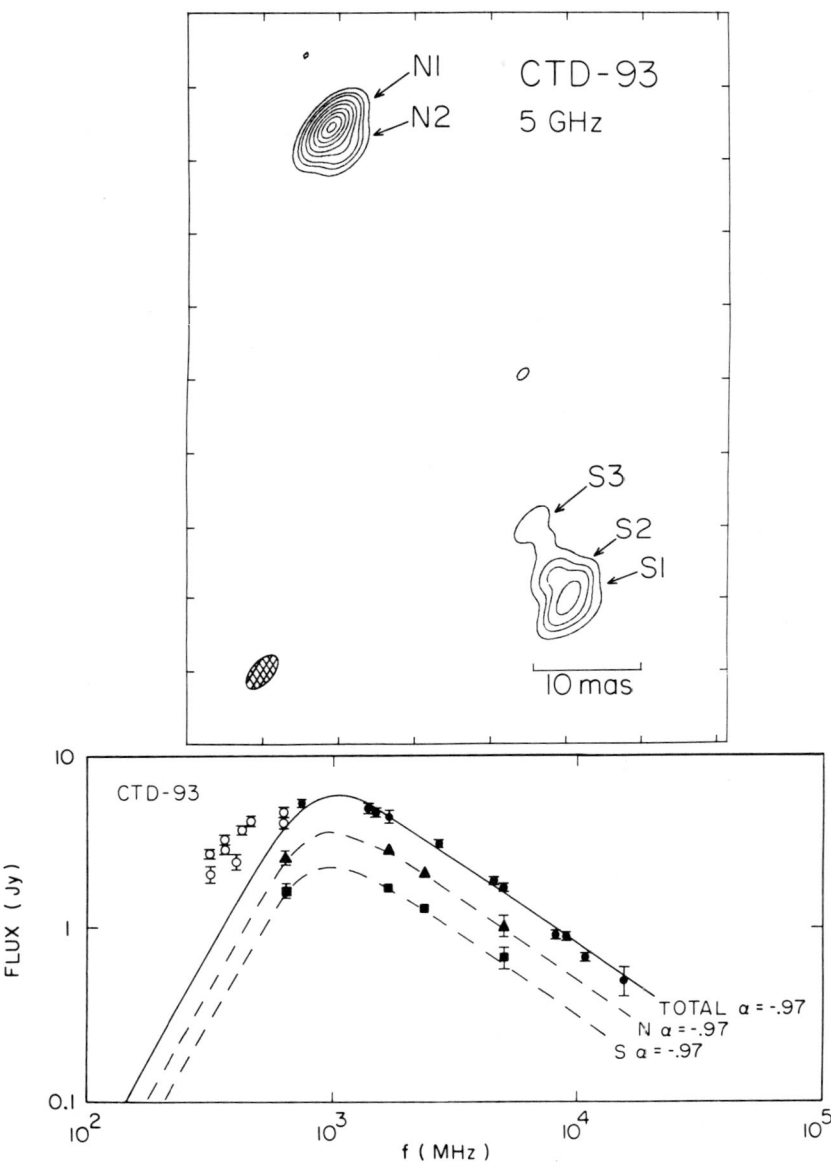

Figure 1 CTD 93: *Top:* hybrid map at 5 GHz. *Bottom:* component spectra and total spectrum (Mutel, Hodges, and Phillips 1985).

model (Blandford and Königl 1979), with two opposed beams emanating from a central source. The rather small subset of such sources that we detect as superluminal sources are simply those whose jet axis happens to be aligned close to the line of sight. The much larger population of misaligned sources will be Doppler dimmed and not easily detected. For the CD sources, we have suggested (Phillips and Mutel 1982; Mutel, Hodges, and Phillips 1985) that the central engine has similar opposed beams, but that the beam axis is near the plane of the sky (like classical radio doubles). The putative core has not yet been detected in any VLBI maps of CD sources, but this is possibly a consequence of the low dynamic range of most VLBI maps. High dynamic range ($\gtrsim$ 100 : 1) VLBI maps are probably needed if the CDs are similar to extended doubles—the ratio of core luminosity to lobe luminosity in extended radio galaxies is $L_{core}/L_{lobe} \lesssim 10^{-2}$ (e.g., Giovannini 1985). Also, since the core (or more appropriately, throat) components of superluminal sources are Doppler boosted by factor $\sim \gamma^3$, they would be severely Doppler dimmed when viewed at a large angle to the line of sight. From these simple considerations, there is no reason to reject the hypothesis of a similar central engine for superluminals and CD sources.

The galactic environment will strongly influence the characteristics of radio jets in all sources, in two distinct ways. First, the density and flow of gas near the central engine determine the fueling rate (Norman and Silk 1983), which in turn influences the jet luminosity and momentum. Second, the ambient medium surrounding the jet influences the degree of bending and bulk velocity of the shocked surfaces at the end of the jet. For radio-loud quasars, Norman and Miley (1984) suggest that the broad-line region and the radio jets strongly interact. The available optical data for the CD sources demonstrate that the gaseous environment of the host galaxy is quite different—there is no evidence of a broad-line region but there are strong forbidden lines characteristic of a kiloparsec-scale narrow-line region in which the radio lobes are presumably embedded. Mutel, Hodges, and Phillips (1985) investigated the confinement of the radio lobes of CD sources in an assumed narrow-line region and concluded that thermal confinement was unlikely. However, ram pressure confinement was plausible if the bulk outflow velocity of the lobes was $v_{CD} \sim 10^{-1.3} c$. This implies a lifetime of $t \sim 10^{4-5}$ yr since emergence from the core. The absence of arcsecond structure in a recent VLA survey of 15 CD sources (R. L. Mutel and Gopal-Krishna, in preparation) argues that the CD phenomenon is not episodic, so that the dynamical time scale represents the true age of the radio source.

Superluminal jets can often be traced continuously from the core to megaparsec scales (e.g., Walker, Benson, and Unwin, this Workshop, page 48). This implies life times of $t_{SL} \gtrsim 10^{7-8}$ yr if the bulk outflow

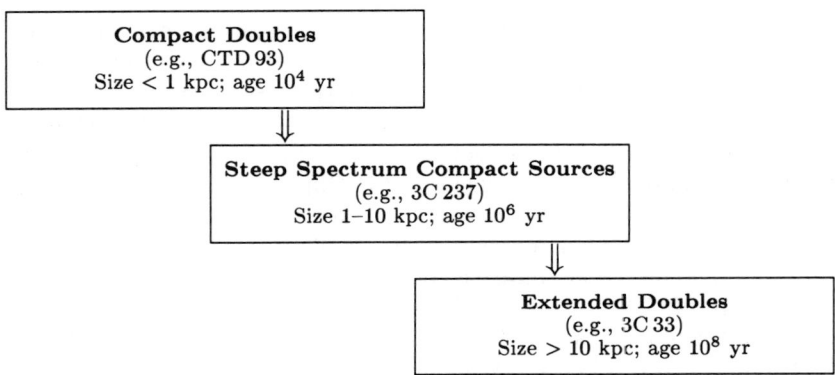

Figure 2 Possible evolutionary sequence from compact doubles to steep-spectrum compact sources to extended double radio sources.

velocity is similar to the CD sources, or $t_{SL} \sim 10^{6-7}$ yr if the bulk flow is relativistic, $v_{bulk} \sim c$ (Bridle and Perley 1984). This difference in source age is unlikely to be a consequence of different environments (or geometry, since the deprojected sizes of superluminals imply that they are older). Rather, it appears that there must be a difference in the rate of fuel supply to the central engine, probably related to the fact that most superluminals have an associated broad-line region.

3. What do CD Sources Evolve Into?

The CD class is unlikely to be *sui generis*—it is almost certainly a phase in the evolution of other classes of sources. The simplest idea is that they represent nascent extended doubles, such as the Fanaroff-Riley class II objects (Phillips and Mutel 1982; Mutel, Hodges, and Phillips 1985; Carvalho 1985). The similarities in radio morphology, parent galaxy, and radio luminosities make this suggestion reasonable. Carvalho (1985) has investigated the possibility of an evolutionary sequence using a simple model for the lobes based on the work of Scheuer (1974), *viz.*, constant energy input from the beams to the lobes, which expand and radiate by synchrotron emission. He finds that the component sizes, fluxes, and separation as a function of time are all consistent with the observations if the expansion velocity is $v_{bulk} \sim 0.2c$, or $t_{CD} \sim 10^4$ yr. The ages of extended doubles are clearly much longer, $t_{FR\,II} \sim 10^{7-8}$ yr. Can we identify the adolescent stage, i.e., sources with a dominant double structure and linear sizes of $10^3 \lesssim l \lesssim 10^{4.5}$ pc? We suggest that at least some of the "steep-spectrum compact" sources (van Breugel 1984; Fanti *et al.* 1985) are good candidates. In particular, sources such as 3C 49, 3C 237, and 3C 241 (Fanti *et al.* 1985) possess all the expected features: dominant double-lobed structures, linear size of a few kiloparsecs, and identification with a moderate-redshift galaxy. In all three sources, high-dynamic-range maps reveal low-brightness features near the dominant lobes. This could indicate that as the lobes expand, less energetic material is swept back and is distorted by interaction with ambient gas flows.

The evolutionary sequence that we have sketched is summarized in Figure 2. An obvious test of this sequence would be to determine the space densities of all three classes of doubles—the ratio of number densities should be approximately 1:10:100 if the expansion velocities are not too dissimilar.

We are grateful to the U.S. VLBI Network for telescope time which has allowed us to study CD sources. This research was supported by the National Science Foundation via grants AST 82-16890 and AST 84-20994 to the University of Iowa.

References

Bartel, N., Shapiro, I. I., Huchra, J. P., and Kühr, H. 1984, *Astrophys. J.*, **279**, 112.
Biretta, J. A., Schneider, D. P., and Gunn, J. E. 1985, *Astron. J.*, **90**, 2508.
Blandford, R. D., and Königl, A. 1979, *Astrophys. J.*, **232**, 34.
Bridle, A. H., and Perley, R. A. 1984, *Ann. Rev. Astron. Astrophys.*, **22**, 319.
Carvalho, J. C. 1985, *Monthly Notices Roy. Astron. Soc.*, **215**, 463.
Fanti, C., Fanti, R., Parma, P., Schilizzi, R. T., and van Breugel, W. J. M. 1985, *Astron. Astrophys.*, **143**, 292.
Giovannini, G. 1985, in *Active Galactic Nuclei*, ed. J. E. Dyson (Manchester: Manchester University Press), p. 93.
Hodges, M. W., Mutel, R. L., and Phillips, R. B. 1984, *Astron. J.*, **89**, 1327.
Johnston, K. J., Spencer, J. H., Witzel, A., and Fomalont, E. B. 1983, *Astrophys. J. (Letters)*, **265**, L43.
Mutel, R. L., Hodges, M. W., and Phillips, R. B. 1985, *Astrophys. J.*, **290**, 86.
Norman, C., and Miley, G. 1984, *Astron. Astrophys.*, **141**, 85.
Norman, C., and Silk, J. 1983, *Astrophys. J.*, **266**, 502.
Pearson, T. J., and Readhead, A. C. S. 1984, in *IAU Symposium 110, VLBI and Compact Radio Sources*, ed. R. Fanti, K. Kellermann, and G. Setti (Dordrecht: Reidel), p. 15.
Phillips, R. B., and Mutel, R. L. 1982, *Astron. Astrophys.*, **106**, 21.
Phillips, R. B., and Shaffer, D. B. 1983, *Astrophys. J.*, **271**, 32.
Readhead, A. C. S., Pearson, T. J., and Unwin, S. C. 1984, in *IAU Symposium 110, VLBI and Compact Radio Sources*, ed. R. Fanti, K. Kellermann, and G. Setti (Dordrecht: Reidel), p. 131.
Scheuer, P. A. G. 1974, *Monthly Notices Roy. Astron. Soc.*, **166**, 513.
van Breugel, W. 1984, in *IAU Symposium 110, VLBI and Compact Radio Sources*, ed. R. Fanti, K. Kellermann, and G. Setti (Dordrecht: Reidel), p. 59.

VLBI Observations of Compact Steep-Spectrum Radio Sources

CARLA FANTI AND ROBERTO FANTI
Dipartimento di Astronomia, Università di Bologna
and Istituto di Radioastronomia CNR, Bologna

1. Introduction

In the study of extragalactic radio sources, the subdivision between sources with flat and steep radio spectra, classically related to the subdivision between compact and extended structures (Miley 1980; Kellermann and Pauliny-Toth 1981), does not account for all the possible morphological classifications. There exists a third category of objects which, though unresolved by conventional interferometry (e.g., Cambridge Five-Kilometre Telescope), do display the overall straight steep spectrum typical of the extended radio sources up to 5–10 GHz. This implies that they do not contain a dominant compact core. Whether or not this is a peculiarity with respect to the other radio sources is not clear yet. Generally these sources are referred to as Compact Steep-Spectrum (CSS) radio sources or Steep-Spectrum Cores (SSC).

The advent of instruments capable of very high resolution, like the VLA, MERLIN, and especially VLBI, has allowed several of these sources to be mapped in great detail, revealing a wide variety of structures (e.g., Wilkinson *et al.* 1984, 1986; van Breugel, Miley, and Heckman 1984; Fanti *et al.* 1985; Pearson, Perley, and Readhead 1985; Spencer *et al.*, in preparation). Sometimes they resemble shapes that we see in the more extended objects, e.g., doubles (Figure 1; Fanti *et al.* 1985) or core-jets (Figure 2). But quite often they are distorted and wind up, suggesting that we might be seeing objects oriented close to the line of sight, where small intrinsic distortions are enhanced by projection effects (Figure 3; but see also Fanti *et al.* 1986; Wilkinson *et al.* 1984, 1986). If relativistic beaming plays a role, we would expect to observe Doppler-boosted radio cores in these objects.

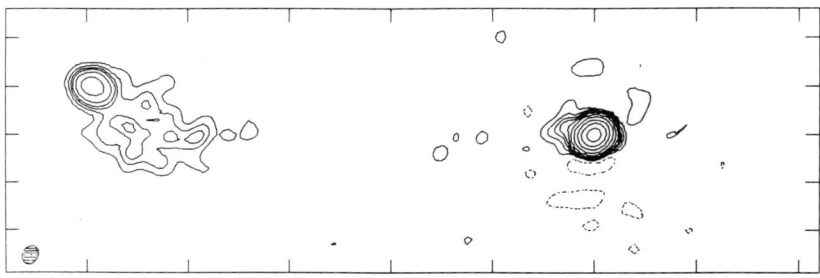

Figure 1 EVN+MERLIN map of the galaxy 3C 49 at 18 cm. The tick separation is 200 mas in right ascension. A 6 cm VLBI map (in preparation) shows that the unresolved component to the west has a complex structure.

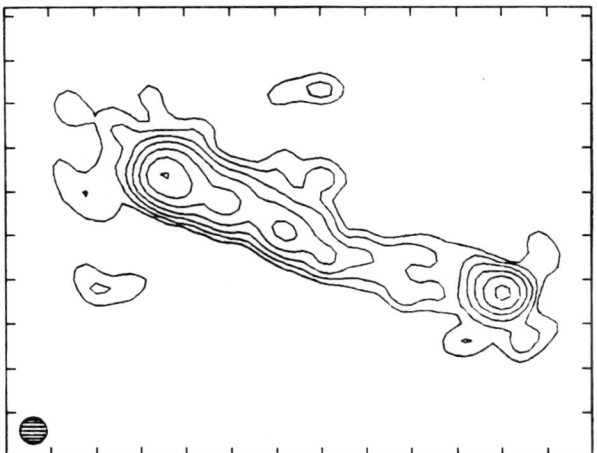

Figure 2 EVN map of the quasar 3C 138 at 18 cm. The tick separation is 50 mas. The unresolved component to the south-west is resolved at 6 cm and is likely to contain the source core.

2. Linear diameter statistics

Both the observed prominent distortions and the small projected linear sizes ($\lesssim 10$ kpc with $H_0 = 100$ km s^{-1} Mpc^{-1}) of CSS radio sources indicate that many of these objects may actually be more extended radio sources oriented towards us. Statistical considerations, however, suggest that this is the case only for a minority of CSS radio sources. We consider the sources in the 3CR catalogue mapped at 5 GHz by Jenkins, Pooley, and Riley (1977), and assume that they are randomly oriented. This assumption is reasonable because the sources mostly have steep radio spectra, and hence are not likely to undergo dominant relativistic boosting in their cores. We can model an intrinsic source diameter distribution which, after projection on the plane of the sky, reproduces

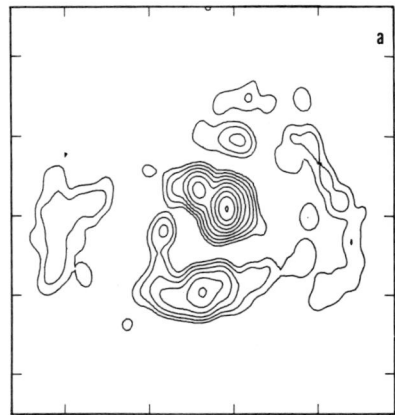 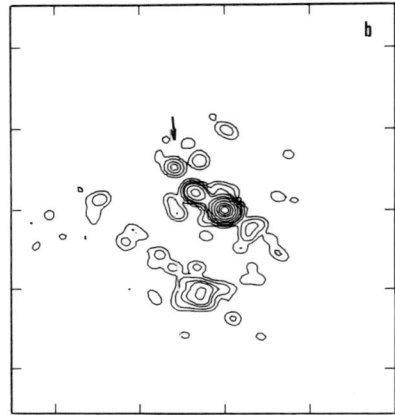

Figure 3 VLBI maps of 3C 119 (quasar?) at 18 cm (a) and 6 cm (b). The component indicated by the arrow is the inverted-spectrum core. The structure recalls a distorted jet seen end-on. The tick separation is 50 mas in both maps.

the observed diameter distribution of this sample (Figure 4a). The comparison between the "true" diameter and the "projected" diameter distributions (Figure 4b) shows that no more than about six among 25 objects ($\sim 25\%$) with projected linear size < 10 kpc can be larger objects seen close to the line of sight. This excess also implies a small bias in the orientation of these sources (more small angles to the line of sight than in a random distribution), that should be included in the statistical model.

The evidence is therefore that the majority of CSS radio sources are intrinsically small. Presumably they are small because they are young, because they are inefficient in transporting the radiating material far from its origin (to form extended lobes), or because they are trapped within the core of the underlying optical object by a dense interstellar medium (van Breugel, Miley, and Heckman 1984; Fanti *et al.* 1985). But whether the CSS radio sources are a separate class of object or whether they represent a stage in the evolution of the extended steep-spectrum or the compact flat-spectrum radio sources is still a matter of debate.

3. Core-to-extended Emission Ratio

If projection is not the major factor responsible for the observed shapes and sizes of CSS sources then the absence of a dominant, Doppler-boosted core is no longer a difficulty in the context of relativistic beaming, and we are allowed to consider the population of small sources (< 10 kpc) as a whole, regardless of their overall spectrum.

Some of the CSS radio sources are known to or suspected to show superluminal motion (e.g., 3C 147, 3C 216), but most of them have

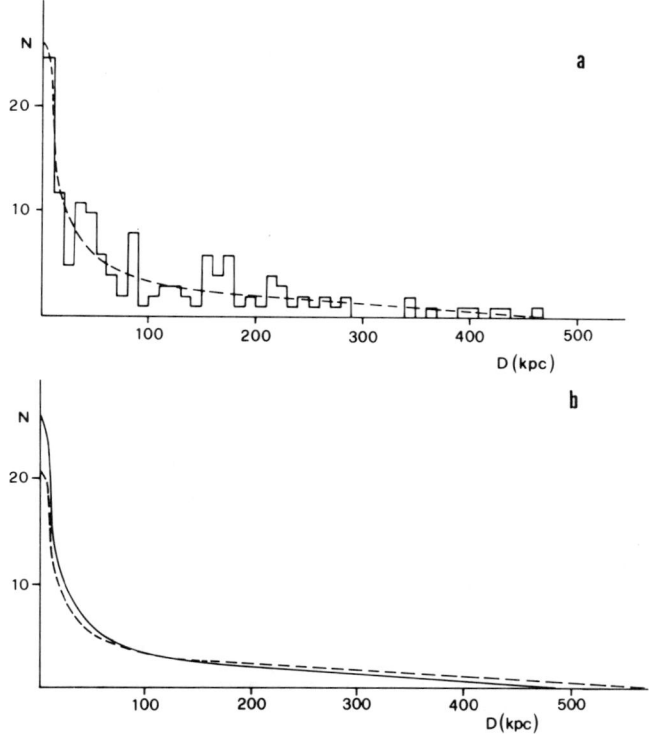

Figure 4 (a) Observed linear diameter distribution for 3CR radio sources. The curve represents the distribution derived from the model by projection (see text). (b) Assumed intrinsic (*broken line*) and projected (*solid line*) diameter distribution.

a core-to-extended emission ratio (R) similar to that of their larger cousins. To analyze this point in detail, we have examined all the sources in the 3CR catalogue with $\log P_{178} \geq 26$, independent of their radio spectrum, and have compared the distribution of R for the small radio sources to that for all the 3CR radio sources. This is now possible since enough structural information is available for a large number of the smallest radio sources in the catalogue. In order to avoid objects which might have been included in the catalogue only because of significant boosting in their core, we considered only those objects whose flux density in extended structures is above the 3CR flux density limit. In practice this excludes only 3C 454.3.

There are known systematic differences in R between quasars and radio galaxies, e.g., (a) quasars have an R almost two order of magnitudes greater than galaxies of similar radio power, and (b) among galaxies, R is anticorrelated with $\log P$ of the extended regions (Fanti

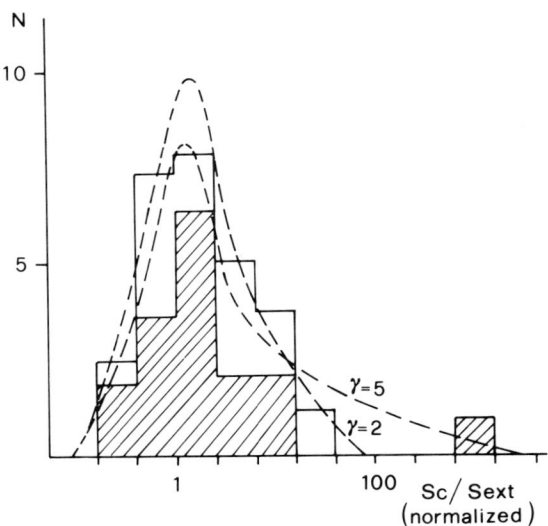

Figure 5 Core-to-extended emission ratio for CSS sources normalized to population median value (see text). Hatched areas represent quasars. The curves represent the expected distributions for two values of γ.

et al. in preparation; Feretti et al. 1984). Therefore, to combine all the data without producing an artificial spread in the distribution, we normalized the value of R for each object to the median R value of the corresponding class (galaxy or quasar in the appropriate range of P). The overall distribution thus has a normalized median value $R = 1$.

The resulting distribution of normalized R for the small-diameter objects is shown in Figure 5. The median value of R is 1.3, different from unity at the 2σ level. This is consistent with the slight bias in the orientation of the small sources and with the hypothesis of a measurable relativistic effect in their cores. No significant difference seems to be present between quasars and radio galaxies. The curves plotted in Figure 5 represent the expected statistics of R for two values of the Lorentz factor γ (corrected for the orientation bias). There is no compelling evidence that either value is a better fit to the data, although a maximum likelihood test favors $\gamma \approx 2$. A distribution of Lorentz factors is probably required. In any case the distribution of R for the small-diameter sources does not differ significantly from that observed for the more extended objects in the 3CR catalogue.

4. Conclusions

Our present conclusions are that small (CSS) radio sources in the 3CR catalogue are a population of truly small radio sources, randomly oriented in the sky. The relative proportion of weak and strong cores among them does not differ significantly from that in the 3CR population, and therefore there is no need to include any special feature in the relativistic beaming theories to justify their existence. Finally, the reason for their small size has to be sought not in projection effects but, for instance, in the properties of their core environment.

References

Fanti, C., Fanti, R., Parma, P., Schilizzi, R. T., and van Breugel, W. J. M. 1985, *Astron. Astrophys.*, **143**, 292.
Fanti, C., Fanti, R., Schilizzi, R. T., Spencer, R. E., and van Breugel, W. J. M. 1986, *Astron. Astrophys.*, **170**, 10.
Feretti, L., Giovannini, G., Gregorini, L., Parma, P., and Zamorani, G. 1984, *Astron. Astrophys.*, **139**, 55.
Jenkins, C. J., Pooley, G. G., and Riley, J. M. 1977, *Mem. Roy. Astron. Soc.*, **84**, 61.
Kellermann, K. I., and Pauliny-Toth, I. I. K. 1981, *Ann. Rev. Astron. Astrophys.*, **19**, 373.
Miley, G. 1980, *Ann. Rev. Astron. Astrophys.*, **18**, 165.
Pearson, T. J., Perley, R. A., and Readhead, A. C. S. 1985, *Astron. J.*, **90**, 738.
van Breugel, W., Miley, G., and Heckman, T. 1984, *Astron. J.*, **89**, 5.
Wilkinson, P. N., Booth, R. S., Cornwell T. J., and Clark, R. R. 1984, *Nature*, **308**, 619.
Wilkinson, P. N., Kus, A. J., Pearson, T. J., Readhead, A. C. S., and Cornwell, T. J. 1986, in *IAU Symposium 119, Quasars*, ed. G. Swarup and V. K. Kapahi (Dordrecht: Reidel), p. 165.

VLBI Observations of the Suspected Superluminal 3C 371

KEVIN R. LIND

California Institute of Technology
and National Radio Astronomy Observatory

1. Introduction

The low dynamic range of typical VLBI maps introduces a model uncertainty which impedes detailed understanding of the underlying physics, especially of such phenomena as superluminal motion. It has not yet been determined whether the underlying flow patterns in parsec-scale jets are continuous or discontinuous; VLBI images can be interpreted equally well either as steady outflow with brightness-enhanced "knots" or as the emission of diffusive isolated "plasmoids". The theoretical distinction is crucial. If the emission originates from isolated packets of plasma, then the fluid velocity (beaming) and the pattern velocity (subluminal or superluminal motion) should be nearly identical; if the emission originates from shocks or other hydrodynamic structures within a continuous jet, these velocities may differ considerably.

An observational solution to this problem would be the detection of an underlying continuous jet. Suitable sources for such a study should be at low redshift in order to obtain high linear resolution, e.g., redshift $z \leq 0.5$, which gives $\geq 7.5h^{-1}$ pc mas^{-1} for $H_0 = 100h$ km s^{-1} Mpc^{-1}. The sources must also be strong enough, at least 1 Jy at 5 GHz, to allow the detection of faint structure. Examination of the flux-density limited sample of Pearson and Readhead (1984) revealed a number of good candidates; the best of these was 3C 371, with $z = 0.051$, $\delta = 70°$, and $S_{5\ \mathrm{GHz}} = 1.86$ Jy.

2. Previous Observations

3C 371 (1807+698) is a nearby N galaxy located in a small cluster of galaxies (Sandage 1966; Miller 1975). The optical emission from this object can be separated into a thermal component associated with a moderately luminous elliptical galaxy ($m_v = 16$), and a nonthermal pointlike continuum source centered on the galaxy, which is optically polarized and variable at optical and shorter wavelengths, similar to the core of a BL Lac object (Oke 1967; Miller 1975; Oke 1978; Worrall et al. 1984, Staubert et al. 1986). Radio observations showed that 3C 371 is one-sided and core-dominated above 1.5 GHz at all resolution scales (Perley, Fomalont, and Johnston 1980; Browne and Orr 1981; Perley 1982; Pearson and Readhead 1984; van der Laan et al. 1984; Wrobel and Lind, in preparation). 3C 371 was described as a "likely candidate for superluminal motion" by Readhead, Pearson, and Unwin (1984).

3. VLBI observations: May 1985

3C 371 was observed for 24^h on 1985 May 28–29 at 5 GHz, using the Mark II recording system. A total of 12 stations participated in this experiment, five from the European VLBI Network (EVN) plus one in Metsähovi, Finland, and six from the U.S. VLBI Network. The (u,v) coverage was exceptionally uniform and complete, despite failures at three stations. The dynamic range in the final maps is of order 400:1, only a factor of three or so above the thermal noise level.

The data were mapped in the AIPS package. In addition to the full resolution map, a tapered map was produced to enhance underlying extended structure (Figure 1). In the following, I will refer to the components labeled in these maps with subscripts t (for tapered) and u (for untapered). These maps show several interesting features.

(a) The structure is one-sided. The apparent counterjet (CJ) in the untapered map is not visible on the tapered map, and is probably an artefact, due to closure phase errors on the longest triangles (Linfield 1986).

(b) The jet-to-counterjet brightness ratio of the tapered map is about 30:1, in the inner part of the jet, and 10:1 at the distance of component C_t.

(c) The jet is unusually straight. The component position angles differ by no more than 8°, and in fact all but two components line up to within 3°. The jet opening angle—defined as the angle subtended by the first contour level down from the peak of component F_u as measured from the core CC—is on the order of 5°, small enough that a bend of a few degrees should be detectable.

(d) Both maps show two bends of order 30°–40° in the jet, associated with components A_t, B_t, and C_t (or C_u, D_u, and E_u), and B_t, C_t, and D_t. This suggests that the measured variations in the component position angles are not due to large-scale misalignment or radial expansion, but are caused by semi-periodic deviations ("wiggles") in the jet. The *existence* of these wiggles is not an artefact of the analysis, although details are questionable. The wiggles appear to increase in wavelength from 5 mas to about 20 mas, and in amplitude from a fraction of a milliarcsecond to possibly 3 mas. This would be the expected appearance

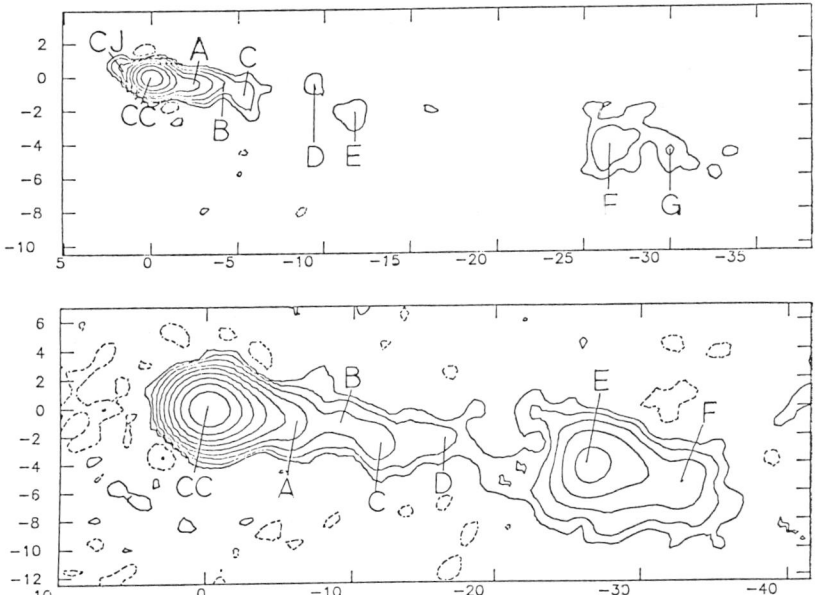

Figure 1 VLBI maps of 3C371 at 5 GHz (May 1985): *Top:* full resolution, circular beam with FWHM 1 mas. Contour levels are −0.4, 0.4, 0.8, 1.6, 3.2, 6.4, 12.8, 25.6, 51.2% of the peak brightness (843 mJy per beam). Total map flux density is about 1.6 Jy. *Bottom:* tapered resolution map; the visibilities were tapered using a circular Gaussian, resulting in an effective circular beam of FWHM 2.5 mas. Contour levels are −0.4, −0.2, 0.2, 0.4, 0.8, 1.6, 3.2, 6.4, 12.8, 25.6, 51.2% of the peak brightness (1094 mJy per beam). Total map flux density is about 1.6 Jy.

of a jet undergoing helical oscillations. In this case, the brightening at the end of the jet (E_t) and the apparent recollimation (F_t) could be explained as a shock developing where the oscillation goes nonlinear. It would be premature to suggest a physical explanation of the wiggles without a second-epoch observation of comparable sensitivity, and preferably identical coverage.

The general appearance of the source is that of a wiggling jet, decreasing monotonically in surface brightness out to D_t. Beyond this point, there is a gap, followed by a bright region (component E_t) and apparent recollimation of the presumed outflow (component F_t). The apparent drop in brightness at the end of the jet could be an artefact due to field-of-view limitations of the maps.

4. Comparison with 1982.9 Map

For comparison with the above results, I have reanalyzed the 5 GHz VLBI data of 3C 371 of December 1982 (Pearson and Readhead 1984)

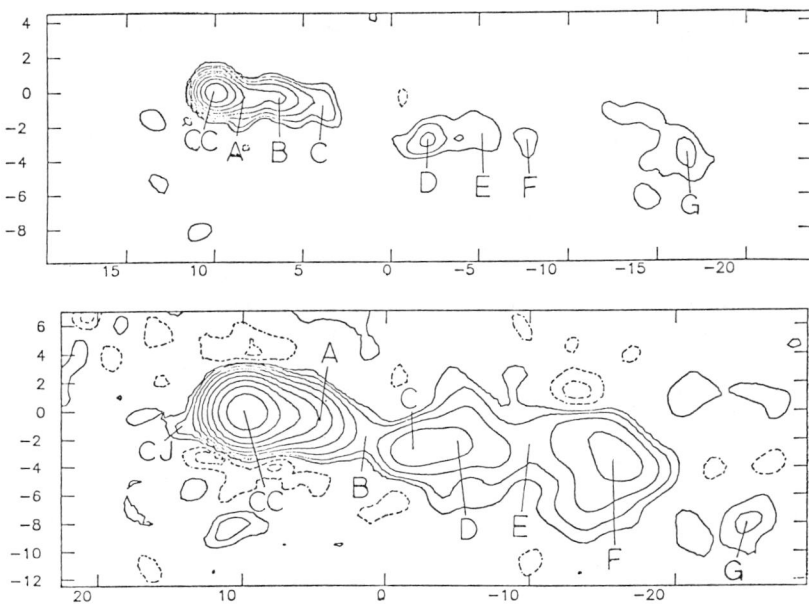

Figure 2 December 1982 VLBI maps of 3C 371 (5 GHz): *Top:* full resolution, circular beam with FWHM 1 mas. Contour levels are −0.4, 0.4, 0.8, 1.6, 3.2, 6.4, 12.8, 25.6, 51.2% of the peak brightness (840 mJy per beam). Total map flux density is about 1.6 Jy. *Bottom:* tapered map; the visibilities were tapered using a circular Gaussian, resulting in an effective circular beam of FWHM 2.5 mas. Contour levels are −0.4, −0.2, 0.2, 0.4, 0.8, 1.6, 3.2, 6.4, 12.8, 25.6, 51.2% of the peak brightness (1079 mJy per beam). Total map flux density is about 1.6 Jy.

using the AIPS package. The results are excellent (Figure 2): the rms noise level achieved is 1.5 mJy, only 50% above the rms noise level in the May 1985 maps. These maps show essentially the same features as the May 1985 maps and confirm the jet-to-counterjet ratio (at least 30:1 in both maps), the straightness of the jet, the opening angle of about 7°, and the existence of wiggles (two bends on the order of 30°) in the inner jet. The jet opening angle may be increased by the wiggles. The tapered map suggests that the misaligned components in the untapered map are actually peaks in a more-or-less continuous wiggling jet of wavelength on the order of 10 mas. The apparent recollimation at the end of the jet is not visible at this epoch.

5. Discussion

The N galaxy 3C 371 has an extended VLBI jet which appears to be the base of the arcsecond scale jet (Wrobel, this Workshop, page 186). The jet orientation can be traced over many resolution scales, and the misalignment between the milliarcsecond jet and the arcsecond jet is explained by the bending seen in the MERLIN map of Browne and Orr (1981). If the jet asymmetry is assumed to be due to beaming, rather than intrinsic, a weak limit may be put on the jet speed and orientation from the limit on the ratio of jet-to-counterjet brightness of 30:1; this corresponds to $\beta \cos \theta \geq 0.56$, or $\gamma \geq 1.24$ and $\theta \leq 56°$.

The VLBI jet is straight to within $\eta \sim 3°$, with most of the deviations in position angle attributable to wiggles with local bending angles of order 30°. This straightness is remarkable even without amplification of the bending angle by projection effects, considering that there are apparent local bends of order ten times the apparent overall bending angle. Such large bends in a jet would normally be associated with instabilities, which should disrupt the flow. In fact, the high surface brightness in the outermost VLBI component, which in conjunction with the surrounding emission region accounts for 12% of the compact flux, may be indicative of a shock or instability in the jet. This contention is supported by the apparent recollimation of the flow behind the emission peak (visible in the tapered map), which may be indicative of a shock followed by recollimation. Even if this is an instability, it still may not disrupt the flow sufficiently to explain the large wiggles in the jet. Furthermore, the apparent smoothness and the large percentage of low-brightness extended flux in the jet (e.g., region D_u–E_u), suggest a relatively strong underlying jet component. This also supports the contention that the underlying flow is stable, since strong shocks in the jet should give the jet a more irregular appearance, especially if they were beamed (Jones et al. 1986).

A rough estimate of relativistic motion may be made from brightness changes in the two VLBI maps, if it is assumed that these occur in the component complexes coherently across at least a fraction of each complex, where that fraction is given by the fractional change in the total flux from the complex. This is fairly conservative, as the appearance in the maps is that the entire complex changes, and not just a fraction of it. In the region of the central component (B_t, C_t, and D_t at both epochs), the flux density dropped from 200 mJy to 65 mJy. This drop in flux density of 65% in 2.5 yr over a region of linear extent 10 mas corresponds to an estimated rate of change of 0.65×10 mas/2.5 yr $= 2.6$ mas yr^{-1}, or about $6h^{-1}c$. The outer component complex (E_t and F_t in both epochs) increased in flux density from 82 mJy to 185 mJy over a distance of about 13 mas, giving an estimated rate of change of 0.55×13 mas/2.5 yr $= 2.9$ mas yr^{-1}, or about $7h^{-1}c$. This is comparable to the estimate of $10h^{-1}c$ made on the basis of X-ray to radio flux (Worrall et al. 1984).

I plan to make further VLBI observations at the same frequency with nearly identical coverage to confirm the above results and to determine whether there are superluminal changes present in this faint and less "traditional" object.

References

Browne, I. W. A., and Orr, M. J. L. 1981, in *Optical Jets in Galaxies*, Proceedings of the second ESO/ESA workshop on the use of the Space Telescope and coordinated ground based research, ESA SP-162, (Noordwijk: European Space Agency), p. 87.
Jones, D. L., Unwin, S. C., Readhead, A. C. S., Sargent, W. L. W., Seielstad, G. A., Simon, R. S., Walker, R. C., Benson, J. M., Perley, R. A., Bridle, A. H., Pauliny-Toth, I. I. K., Romney, J., Witzel, A., Wilkinson, P. N., Bååth, L. B., Booth, R. S., Fort, D. N., Galt, J. A., Mutel, R. L., and Linfield, R. P. 1986, *Astrophys. J.*, **305**, 684.
Linfield, R. P. 1986, *Astron. J.*, **92**, 213.
Miller, J. S. 1975, *Astrophys. J. (Letters)*, **200**, L55.
Oke, J. B. 1967, *Astrophys. J. (Letters)*, **150**, L5.
Oke, J. B. 1978, *Astrophys. J. (Letters)*, **219**, L97.
Pearson, T. J., and Readhead, A. C. S. 1984, in *IAU Symposium 110, VLBI and Compact Radio Sources*, ed. R. Fanti, K. Kellermann, and G. Setti (Dordrecht: Reidel), p. 15.
Perley, R. A. 1982, *Astron. J.*, **87**, 859.
Perley, R. A., Fomalont, E. B., and Johnston, K. J. 1980, *Astron. J.*, **85**, 649.
Readhead, A. C. S., Pearson, T. J., and Unwin, S. C. 1984, in *IAU Symposium 110, VLBI and Compact Radio Sources*, ed. R. Fanti, K. Kellermann, and G. Setti (Dordrecht: Reidel), p. 131.
Sandage, A. 1966, *Astrophys. J.*, **145**, 1.
Staubert, R., Brunner, H., and Worrall, D. M. 1986, *Astrophys. J.*, **310**, 694.
van der Laan, H., Zieba, S., and Noordam, J. E. 1984, in *IAU Symposium 110, VLBI and Compact Radio Sources*, ed. R. Fanti, K. Kellermann, and G. Setti (Dordrecht: Reidel), p. 9.
Worrall, D. M., Puschell, J. J., Bruhweiler, F. C., Miller, H. R., Rudy, R. J., Ku, W. H.-M., Aller, M. F., Aller, H. D., Hodge, P. E., Matthews, K., Neugebauer, G., Soifer, B. T., Webb, J. R., Pica, A. J., Pollock, J. T., Smith, A. G., and Leacock, R. J. 1984, *Astrophys. J.*, **278**, 521.

VLA Polarimetry of the Active Galaxy 3C 371

J. M. Wrobel
New Mexico Institute of Mining and Technology

1. Introduction

Radio imaging with arcsecond resolution of VLBI core-jet sources can be used to examine the case for continuity of jet-like features between parsec and circumgalactic scales (cf. 3C 120: Walker, Benson, and Unwin, this Workshop, page 48). Furthermore, polarimetry of such sources allows investigation of the dominant magnetic field topologies as a function of linear offset from the central "engine". Examination of these continuity and field topology issues is essential for an understanding of how energy is channeled from the nuclear region to the circumgalactic environment.

The active galaxy 3C 371 (1807+698) is particularly well suited for such investigations. It is identified with an N galaxy at $z = 0.05$ (Sandage 1966), implying $1'' = 0.7h^{-1}$ kpc for $H_0 = 100h$ km s^{-1} Mpc^{-1} and $q_0 = 0.5$. Its radio emission is dominated by a compact core (CC) smaller than $1''$ ($0.7h^{-1}$ kpc). VLBI imaging of the CC at 6 cm shows a core-jet structure 5 mas ($4h^{-1}$ pc) in extent in p.a. $-97°$ (Pearson and Readhead 1981; Readhead, Pearson, and Unwin 1984), plus an outer component along a similar p.a. and separated from the VLBI core by ~ 30 mas or $\sim 20h^{-1}$ pc (Lind, this Workshop, page 180). Superluminal motion of these VLBI-scale components has been suspected but not demonstrated (Readhead, Pearson, and Unwin 1984; Worrall et al. 1984).

Two secondary radio structures are adjacent to the compact core: a weak component A is offset by $\sim 3''$ ($2h^{-1}$ kpc) in p.a. $-116°$ and shows a smooth connection back to the CC; and a very weak component B is offset by $\sim 22''$ ($15h^{-1}$ kpc) in p.a. $-92°$ (Perley, Fomalont, and Johnston 1980; Browne et al. 1982; van der Laan, Zieba, and Noordam 1984; Pearson, Perley, and Readhead 1985). These arcsecond-scale components are surrounded by a radio halo of largest extent $\sim 4\rlap{.}'8$ ($\sim 200h^{-1}$ kpc) elongated in p.a. $-85°$ (Ulvestad and Johnston 1984; Simon, Spencer, and Johnston 1987). Figure 1 shows the relative locations of components CC, A, and B, and the enveloping halo.

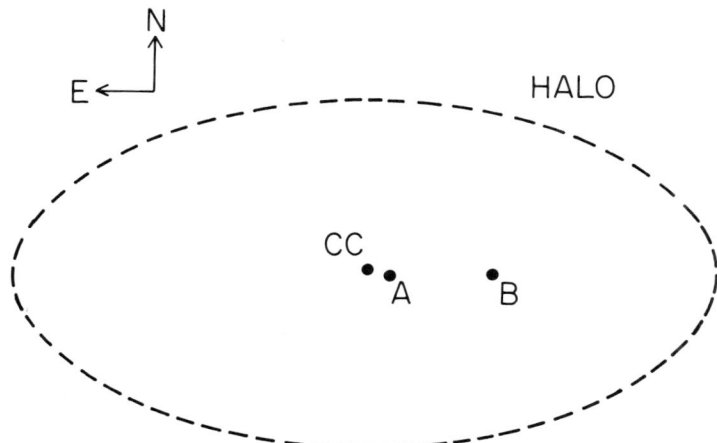

Figure 1 A sketch identifying the kiloparsec-scale components of 3C 371 (not to scale).

Miller (1975) finds a high, variable optical polarization percentage for 3C 371, and suggests that this source is a BL Lac object embedded in an elliptical galaxy. Worrall *et al.* (1984) explain the continuum emission at infrared to X-ray wavelengths as relativistically beamed synchrotron self-Compton emission. Staubert, Brunner, and Worrall (1986) present an EXOSAT spectrum and report substantial X-ray variations on time scales as short as 25 min.

2. VLA Polarimetry

3C 371 was observed during two separate VLA sessions. During the 6-cm session the VLA was in B configuration and operating in a phased-array mode (Clark 1981) as a station for a VLBI experiment (Lind, this Workshop, page 180). A scheme designed to monitor the RCP *vs.* LCP phase stability of the reference antenna, a prerequisite for accurate VLA polarimetry (Bignell 1982), was successfully employed. Observations of 3C 371 were also made during an A-configuration snapshot program designed to determine rotation measures of core-jet VLBI sources (Wrobel *et al.*, in preparation). These data were acquired in normal continuum mode at multiple wavelengths within the VLA's 20-cm window.

3. Results and Discussion

The new VLA imaging provides linear polarization information on three components of 3C 371, namely the CC and components A and B (Figure 1). The high-resolution images at five wavelengths (not shown;

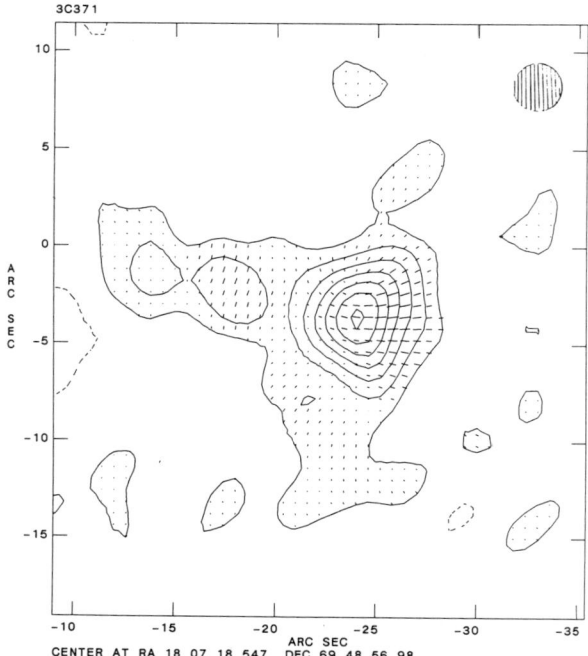

Figure 2 Tapered (FWHM = $2''.5$, *shaded circle*) VLA cleaned map of component B at 6 cm. Origin is at the location of the compact core (CC). Scale is $1'' \approx 0.7h^{-1}$ kpc. Total intensity contour interval CI = 0.05% of CC peak (not shown). Equally-spaced contours between -4 and 10 times the CI are plotted. Lengths of bars are proportional to linearly polarized intensity and orientation of bars gives E-field direction.

FWHM = $1''.6 = 1.1h^{-1}$ kpc) separate the unresolved CC from component A. The low-resolution image of Figure 2 (6 cm, FWHM = $2''.5 = 1.8h^{-1}$ kpc) resolves component B and accurately locates its peak relative to the CC (offset $24'' = 17h^{-1}$ kpc in p.a. $-98°$). Images of component B in the 20-cm band with matching resolution have not yet been made.

Values of the E-field orientation (χ_λ) measured at the total intensity peak of component A are plotted as crosses in Figure 3 as a function of λ^2. A least-squares-fit yields a rotation measure $RM_A = 16 \pm 1$ rad m^{-2} which agrees well with that obtained by Rudnick and Jones (1983) from lower resolution VLA images at 18 and 20 cm in which components A and CC were partially blended. Further, this rotation measure is similar to those of three steep-spectrum extragalactic sources (Simard-Normandin, Kronberg, and Button 1981) whose positions on the sky are within about 10° of 3C 371, which is at Galactic longitude $l^{II} = 100°$ and latitude $b^{II} = +29°$. Thus, it is likely that the rotation measure of

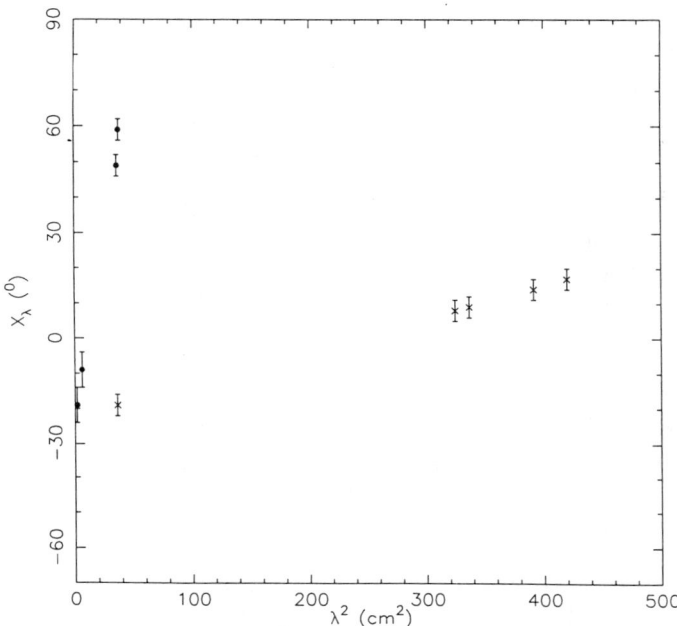

Figure 3 χ_λ as a function of λ^2 for 3C 371. Data are from the present work and from Rudnick (private communication). *Filled circles*: compact core (CC); *crosses*: component A.

component A primarily originates within the disk of the Milky Way.

Figure 2 shows that component B has a sharp leading edge, extends back toward the CC, has a width of a few kiloparsecs, and reaches its highest linearly polarized intensity on its leading edge. Furthermore, a circumferential projected magnetic field is inferred for B if it is optically thin and has a rotation measure similar to that of A. In all these respects, B resembles the hot spots in powerful double radio sources (e.g., Laing 1981). However, if component B is emitting isotropically, then its 6-cm power ($\sim 10^{23} h^{-2}$ W Hz^{-1}) is about two orders of magnitude lower than in those hot spots. In the beam model of powerful double radio sources, a hot spot is assumed to be associated with the "working surface" where the beam impinges on the ambient gas (e.g., Begelman, Blandford, and Rees 1984). It is worth noting that component B resembles the outermost jet components of the compact, core-dominated, superluminal source 3C 273 (Perley 1984; Flatters and Conway 1985).

The CC is linearly polarized by $< 0.4\%$ (2σ) at the four wavelengths observed in the 20-cm region. Plotted as filled circles in Figure 3 are the polarization position angles of the CC, based on the present work at 6 cm and on Rudnick's unpublished VLA polarimetry at 1, 2, and

6 cm (FWHM $\leq 3'' \sim 2h^{-1}$ kpc, epoch 1983 April). (For these latter 6-cm data, the blending of the CC and component A are not severe: the linearly polarized emission is dominated by the CC and the E-field p.a. measured for component A by Rudnick agrees with that plotted in Figure 3.) If the rotation measure for component A also applies to the CC, the difference in the rotation-measure-corrected χ_λ between 2 and 6 cm is $\Delta^{\rm corr}_{6-2} \sim 60°$. Such a value is unusually high for compact radio sources as a class, for which $\Delta^{\rm corr}_{6-2} \leq 20°$ is typically found (Jones et al. 1985). Below, three possible ways of producing the CC's large $\Delta^{\rm corr}_{6-2}$ are presented and criticized.

(a) *The CC rotation measure $RM_{\rm CC} \gg RM_{\rm A}$*, meaning there is a large rotation measure local to the CC: $RM_{\rm local} \sim +200\text{--}300$ rad m^{-2}. Such a high local rotation measure would make 3C 371 unique among compact extragalactic radio sources since for such objects Rudnick and Jones (1983) find $|RM_{\rm local}| \leq 30$ rad m^{-2}.

(b) *χ_λ twists because the VLBI core-jet structure twists, and a wavelength-dependent opacity changes the core-jet appearance.* VLBI polarimetry of a number of compact cores shows that the polarized emission is primarily associated with optically thin (jet) components (Cotton et al. 1984; Wardle et al. 1986). If this is also true of 3C 371, then this second case appears unlikely, since the elongation p.a. of the parsec-scale VLBI jet at 2.8 cm (Unwin, private communication) differs by $< 10°$ from that of the longer VLBI jet at 6 cm (Pearson and Readhead 1981; Readhead, Pearson, and Unwin 1984).

(c) *χ_λ varies with position in the VLBI core-jet source, and a wavelength-dependent opacity changes the core-jet appearance.* At a fixed wavelength in a given source, different VLBI-scale components can exhibit different E-field orientations (Cotton et al. 1984; Wardle and Roberts 1986; Wardle et al. 1986). Again assuming that the VLBI-scale polarized emission of 3C 371 is mainly associated with the jet, a mechanism which causes a large observed difference in the E-field orientation along the jet must be invoked. Possible mechanisms abound. Wardle and Roberts (1986) discuss the case of magnetic field configuration changes along a VLBI jet, induced by adiabatic expansion, transverse shear, or internal shocks. Another intriguing explanation is related to the suspected superluminal nature of 3C 371 (Readhead, Pearson, and Unwin 1984; Worrall et al. 1984), namely relativistic aberration (Blandford and Königl 1979). If the parsec and few-parsec scale jet components move close to the line of sight with relativistic, but *different*, speeds, then the apparent direction from which these components are viewed will differ. This could cause differences in the observed E-field position angles between these two scales, i.e., between 2 and 6 cm. Different relativistic speeds for the VLBI jet components of 3C 371 would not make this radio source unique, as such differences are observed in the superluminal source 3C 345 (Biretta, Moore, and Cohen 1986).

Linear polarization VLBI observations of 3C 371 should be undertaken to establish which, if any, of the mechanisms described above is responsible for the CC's large Δ^{corr}_{6-2}.

4. Conclusions

New VLA polarimetry of the active galaxy 3C 371 reveals that the rotation measure of component A ($2h^{-1}$ kpc offset from the compact core) is low and probably primarily of Galactic origin; and that the morphology, size, and inferred magnetic field configuration of component B ($17h^{-1}$ kpc offset) resemble those of hot spots in extended, double extragalactic radio sources. Rudnick's simultaneous polarimetry of the compact core (private communication) shows that this component exhibits an unusually large difference in the E-field p.a. χ_λ between 6 and 2 cm. This behavior may be caused by very pronounced source inhomogeneities on milliarcsecond or parsec scales.

I am pleased to acknowledge the financial support of the National Science Foundation, through grant AST 82-10259 to the Caltech VLBI group, and thank the VLA for providing computing facilities. Rick Perley and Craig Walker provided constructive comments on a draft manuscript. NRAO is operated by Associated Universities, Inc., under contract with the NSF.

References

Begelman, M. C., Blandford, R. D., and Rees, M. J. 1984, *Rev. Mod. Phys.*, **56**, 255.
Bignell, R. C. 1982, Lecture No. 6 in *Synthesis Mapping*, Proceedings of the NRAO–VLA Workshop held in Socorro, New Mexico, June 21–25, 1982, ed. A. R. Thompson and L. R. D'Addario (Green Bank: National Radio Astronomy Observatory).
Biretta, J. A., Moore, R. L., and Cohen, M. H. 1986, *Astrophys. J.*, **308**, 93.
Blandford, R. D., and Königl, A. 1979, *Astrophys. J.*, **232**, 34.
Browne, I. W. A., Orr, M. J. L., Davis, R. J., Foley, A., Muxlow, T. W. B., and Thomasson, P. 1982, *Monthly Notices Roy. Astron. Soc.*, **198**, 673.
Clark, B. G. 1981, *How to Use the VLA for VLB*, internal document, National Radio Astronomy Observatory.
Cotton, W. D., Geldzahler, B. J., Marcaide, J. M., Shapiro, I. I., Sanromá, M., and Rius, A. 1984, *Astrophys. J.*, **286**, 503.
Flatters, C., and Conway, R. G. 1985, *Nature*, **314**, 425.
Jones, T. W., Rudnick, L., Aller, H. D., Aller, M. F., Hodge, P. E., and Fiedler, R. L. 1985, *Astrophys. J.*, **290**, 627.
Laing, R. A. 1981, *Monthly Notices Roy. Astron. Soc.*, **195**, 261.
Miller, J. S. 1975, *Astrophys. J. (Letters)*, **200**, L55.
Pearson, T. J., Perley, R. A., and Readhead, A. C. S. 1985, *Astron. J.*, **90**, 738.
Pearson, T. J., and Readhead, A. C. S. 1981, *Astrophys. J.*, **248**, 61.
Perley, R. A. 1984, in *IAU Symposium 110, VLBI and Compact Radio Sources*, ed. R. Fanti, K. Kellermann, and G. Setti (Dordrecht: Reidel), p. 153.
Perley, R. A., Fomalont, E. B., and Johnston, K. J. 1980, *Astron. J.*, **85**, 649.

Readhead, A. C. S., Pearson, T. J., and Unwin, S. C. 1984, in *IAU Symposium 110, VLBI and Compact Radio Sources*, ed. R. Fanti, K. Kellermann, and G. Setti (Dordrecht: Reidel), p. 131.
Rudnick, L., and Jones, T. W. 1983, *Astron. J.*, **88**, 518.
Sandage, A. 1966, *Astrophys. J.*, **145**, 1.
Simard-Normandin, M., Kronberg, P. P., and Button, S. 1981, *Astrophys. J. Suppl.*, **45**, 97.
Simon, R. S., Spencer, J. H., and Johnston, K. J. 1987, *Astron. J.*, submitted.
Staubert, R., Brunner, H., and Worrall, D. M. 1986, *Astrophys. J.*, **310**, 694.
Ulvestad, J. S., and Johnston, K. J. 1984, *Astron. J.*, **89**, 189.
van der Laan, H., Zieba, S., and Noordam, J. E. 1984, in *IAU Symposium 110, VLBI and Compact Radio Sources*, ed. R. Fanti, K. Kellermann, and G. Setti (Dordrecht: Reidel), p. 9.
Wardle, J. F. C., and Roberts, D. H. 1986, *Can. J. Phys.*, **64**, 434.
Wardle, J. F. C., Roberts, D. H., Potash, R. I., and Rogers, A. E. E. 1986, *Astrophys. J. (Letters)*, **304**, L1.
Worrall, D. M., Puschell, J. J., Bruhweiler, F. C., Miller, H. R., Rudy, R. J., Ku, W. H.-M., Aller, M. F., Aller, H. D., Hodge, P. E., Matthews, K., Neugebauer, G., Soifer, B. T., Webb, J. R., Pica, A. J., Pollock, J. T., Smith, A. G., and Leacock, R. J. 1984, *Astrophys. J.*, **278**, 521.

Milliarcsecond Polarization of Superluminal Sources

DAVID H. ROBERTS AND JOHN F. C. WARDLE
Department of Physics, Brandeis University

1. Introduction

The polarization of the radiation from quasars and other active galactic nuclei contains a wealth of information. The emission mechanisms, source structure, relativistic particle spectrum, thermal particle environment, relativistic motion of the radiating plasma, and the order and orientation of the magnetic field, all play roles in determining the polarization structure of the source. Here we review linear polarization measurements of three bright sources made at 5 GHz and briefly discuss their implications for the physics of extragalactic jets on the parsec scale.

2. Observations

The observations were made during 1981 December and 1982 December using antennas of the US VLBI Network: Haystack Observatory ("K", 37 m), NRAO Green Bank ("G", 43 m), Owens Valley ("O", 40 m), and the phased-up NRAO Very Large Array ("Y", equivalent to about 100 m). To improve the sensitivity through increased bandwidth, and to make it convenient to record two senses of polarization, the data were acquired using the Mark III recording system. At Green Bank and the VLA, seven 2-MHz channels of right circular polarization (RCP) and seven 2-MHz channels of left circular polarization (LCP) were recorded simultaneously; at Haystack and Owens Valley only LCP was available. The data were cross-correlated using the Mark III correlator at Haystack Observatory. Calibration of fringe amplitudes was done in the standard way (using noise tube measurements and published antenna gain curves) at K, G, and O, and using internal interferometry at Y.

The critical part of VLBI polarization measurements is the determination of the antenna instrumental polarizations. Our method (described by Roberts et al. 1984) is based on observations of an unresolved polarization calibrator over a wide range of parallactic angles. The sources OQ 208 (1981) and 0106+013 (1982) were used as such calibrators. The instrumental terms of each antenna were determined to better than 0.5%, and ranged from 11% (G, LCP, in 1981) to 1.5% (Y, LCP, 1981 and 1982). The absolute position angle of polarization was calibrated by comparing observations of several sources on our shortest baseline (KG) to integrated values communicated by H. Aller and L. Molnar (1981) or measured by us at the VLA (1982).

Maps of the total intensity structure of the sources (Stokes parameter I) were made from the calibrated parallel-hand (LL) data using the hybrid mapping algorithm of Readhead and Wilkinson (1978). An automatic byproduct of this was a set of relative antenna phase gains for each coherent scan (LCP). These were combined with the $R - L$ phase difference of each dual-polarization station to produce a set of amplitude- and phase-calibrated cross-hand fringes. (The $R - L$ phase difference at the VLA was determined from the relative phase of the RR and LL fringes on the baseline GY, assuming that $R - L$ was constant at G. This assumption is supported by the fact that it led to consistent polarization calibrations, and by consistent calibrator polarization position angles throughout each observation.) Maps of the complex polarization (Stokes parameter $P = Q + iU = pe^{2i\chi}$, where $p = mI$ is the polarized flux density, m the fractional linear polarization, and χ the position angle of the electric vector) were made by Fourier transformation of the cross-fringes (RL and LR; Conway and Kronberg 1969). The I and P maps are assured of good registration through the common antenna phase calibration. The resulting "dirty" P maps were then deconvolved using a complex "clean" and restored using elliptical Gaussian beams. The slightly different resolution of the P and I maps is due to the missing RCP receivers at K and O.

3. The Quasar 3C 345

Our maps of the quasar 3C 345 ($z = 0.595$; $3.8h^{-1}$ pc mas^{-1} for $H_0 = 100h$ km s^{-1} Mpc^{-1}, $q_0 = 0.5$) are shown in Figure 1 (Wardle et al. 1986). The displacement of the centroids between the I and P maps indicates that the polarization of the optically thick core (component D) is substantially less than that of the jet. Summation of the clean components shows that $m < 1\%$ for D; the inner and outer knots (C3 and C2) have polarization parameters (m, χ) equal to (11%, 22°) and (6%, 83°), respectively.

If the Faraday rotation of the milliarcsecond jet is small (as suggested by the integrated rotation measure of 23 rad m^{-2}), then the magnetic field direction at C3 is very closely aligned along the jet, while that at C2 is nearly transverse to the jet. This may represent the evolution of the magnetic field under the expansion of the jet, a twisted field (Königl and Choudhuri 1985), changing relativistic aberration (Bland-

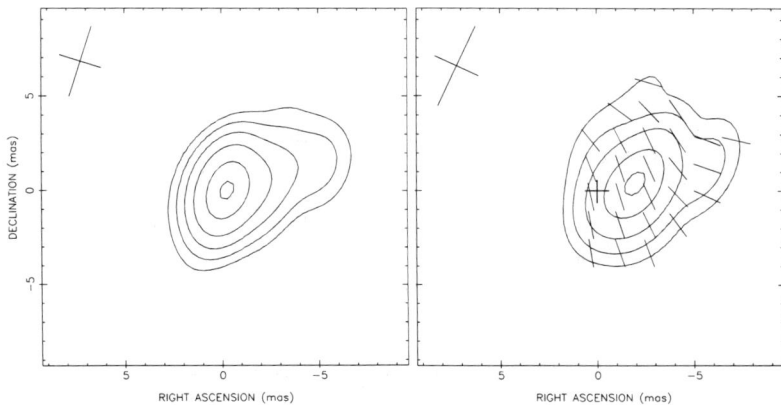

Figure 1 3C 345 at 5 GHz, epoch 1981.9 (from Wardle et al. 1986). *Left:* Total intensity map; contours are −4, 4, 8, 16, 32, 64, and 95% of peak (6407 mJy per beam); beam size is 3.9 × 2.3 mas, p.a. −17.5°. *Right:* Complex polarization map. The plane of the electric vector is shown by the line segments and the location of the compact I core (component D) is shown by the cross. Contours of polarized intensity are 8, 16, 32, 64, and 95% of peak (194 mJy per beam); beam size is 4.7 × 2.5 mas, p.a. −24.9°.

ford and Königl 1979), optical depth effects, or a different rotation measure for the knots. Only a series of multi-frequency observations can decide among these and other possibilities. At arcsecond resolution the jet in 3C 345, like other powerful radio jets, has a magnetic field parallel to the jet.

The relatively high fractional polarization in the milliarcsecond jet is expected for a transparent synchrotron source. However, the polarization of the core is much smaller than expected in an optically-thick, Faraday-thin synchrotron source. For example, the integrated polarization of a Blandford-Königl inhomogeneous jet model for the core of 3C 345 is 30–50%, depending on the exact Doppler and Lorentz factors adopted. Possible depolarization mechanisms include a disordered magnetic field, a non-power-law electron energy distribution, a non-synchrotron emission mechanism, internal Faraday depolarization due to thermal or mildly relativistic particles in the jet, and external Faraday depolarization due to gas associated with the narrow line region.

4. The Galaxy 3C 120

The galaxy 3C 120 ($z = 0.033$, $0.45h^{-1}$ pc mas^{-1}) has a very complex radio structure on scales from milliarcseconds to arc minutes (Walker, Benson, and Unwin, this Workshop, page 48). Our I map (Figure 2) is in good agreement with the maps by these authors—we detect the core and the first three knots (B, C, and D) in the jet (see Walker et

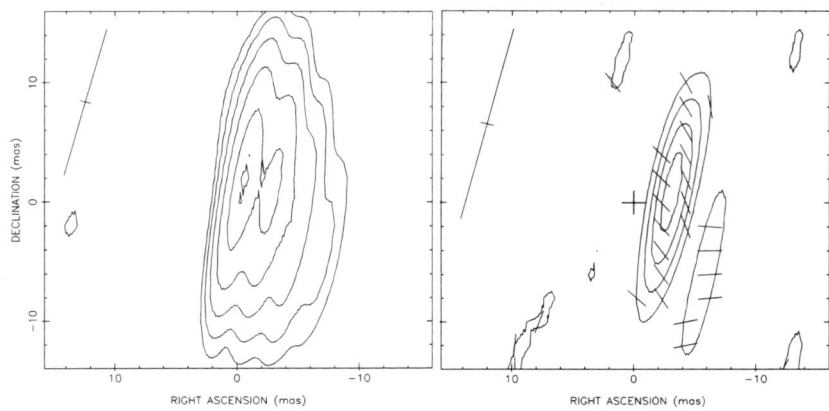

Figure 2 3C 120 at 5 GHz, epoch 1982.9. *Left:* Total intensity map. Contours are −4, 4, 8, 16, 32, 64, and 95% of the peak (981 mJy per beam); beam size is 13.5 × 0.9 mas, p.a. −15.1°. *Right:* Complex polarization map. The plane of the electric vector is shown by the line segments. The location of the compact I core is shown by the cross. Contours of polarized intensity are 30, 50, 70, and 90% of the peak (23 mJy per beam); beam size is 17.5 × 1.1 mas, p.a. −14.3°.

al. 1984). The P map has a peak intensity which is only 2.3% of that of the I map; comparison of the two maps shows the core to be < 1% polarized, and the knots to have fractional polarizations of about 11, 3, and 3%, respectively. However, the correspondence between the I and P structures is not very tight. The integrated polarized flux in the P map is only 24 mJy at $\chi = -22°$, while that obtained simultaneously at the VLA was 177 mJy at 171°; thus most of the polarized flux density seen by the VLA is missing from the VLBI map. The VLA polarization maps by Walker, Benson, and Unwin (1987) show electric vectors at $\chi \simeq 180°$ for both the core and the arcsecond jet. Our P map is not of sufficient quality to be certain that the electric vectors change their position angles down the jet (as is the case in 3C 345).

5. The BL Lac Object OJ 287

OJ 287 ($z = 0.306$; $2.8h^{-1}$ pc mas^{-1}) has strong and highly variable polarization at centimeter wavelengths. Its VLBI structure is very compact, and it is often used as a calibrator. Our maps of OJ 287 made at two epochs a year apart are shown in Figures 3 and 4. The I maps each contain a point source with a slight extension to the southwest. The P maps, however, show a striking and variable structure. The simplest models which fit simultaneously the I and P data consist of three unresolved components: a moderately polarized core (C), a strongly polarized inner component (SW1), and a weakly polarized outer component (SW2). The positions and polarized fluxes of C and SW1 are

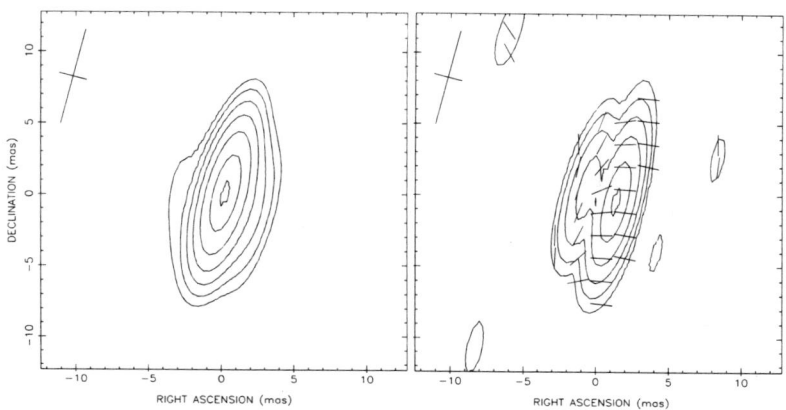

Figure 3 OJ 287 at 5 GHz, epoch 1981.9. Contours are −2, 2, 4, 8, 16, 32, 64, and 95% of the peak (3067 mJy per beam); beam size is 6.9 × 2.0 mas, p.a. −14.6°. *Left:* Total intensity map. *Right:* Complex polarization map, with contours of polarized intensity and the plane of the electric vector shown by the line segments. Contours are 4, 8, 16, 32, 64, and 95% of the peak (117 mJy per beam); beam size is 6.9 × 2.0 mas, p.a. −14.6°.

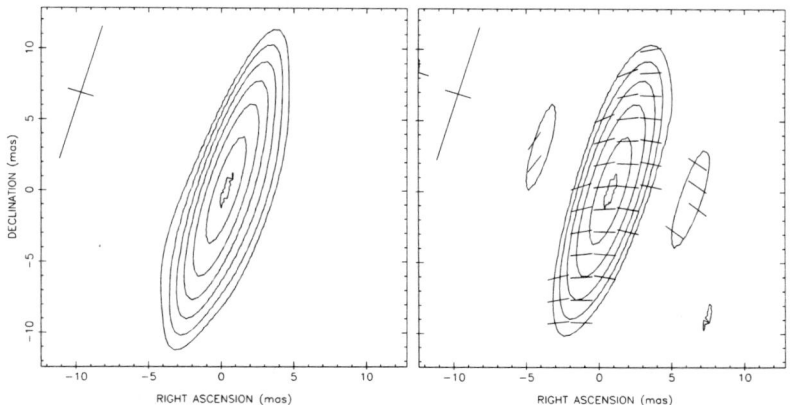

Figure 4 OJ 287 at 5 GHz, epoch 1982.9. *Left:* Total intensity map. Contours are −2, 2, 4, 8, 16, 32, 64, and 95% of the peak (2932 mJy per beam); beam size is 9.9 × 1.9 mas, p.a. −17.5°. *Right:* Complex polarization map, with the plane of the electric vector shown by the line segments. Contours of polarized intensity are 4, 8, 16, 32, 64, and 95% of the peak (148 mJy per beam); beam size is 9.9 × 1.9 mas, p.a. −17.5°.

Table 1 Three-Component Models of OJ 287

Component	$\Delta\alpha$ (mas)	$\Delta\delta$ (mas)	I (mJy)	p (mJy)	χ (deg)	m (%)
		Epoch 1981.9				
C (Core)	...	...	2910	98	−20	3.4
SW1 (Jet?)[a]	−0.88	−0.43	283	180	+83	64
SW2 (Jet?)[b]	−2.08	−1.20	244	6	+65	2.5
		Epoch 1982.9				
C (Core)	...	...	2830	129	−81	4.5
SW1 (Jet?)[c]	−1.16	−0.48	421	95	+78	23
SW2 (Jet?)[d]	−2.55	−0.82	114	4	+71	3.5

[a] Separation $r = 0.98$ mas at $\theta = -116°$ with respect to the core.
[b] Separation $r = 2.40$ mas at $\theta = -120°$ with respect to the core.
[c] Separation $r = 1.26$ mas at $\theta = -113°$ with respect to the core.
[d] Separation $r = 2.68$ mas at $\theta = -108°$ with respect to the core.

determined by the P data, while the presence of SW2 is required to produce the closure phases on the longest I baselines. The parameters of these models are given in Table 1.

The model parameters show several striking features. (a) The polarization of SW1 was as high as 64%, very nearly the maximum possible for a homogeneous synchrotron source. (b) Both SW1 and SW2 moved away from C by 0.28 mas in one year; this corresponds to apparent transverse motion at $3.3h^{-1}c$. (c) The electric vectors in SW1 maintained their orientation during this motion, while the fractional polarization dropped by a factor of two. (d) The core C is moderately polarized, and this polarization is variable in both fraction and electric field vector direction. (Complete details may be found in a paper in preparation by Roberts, Wardle, and Gabuzda.)

6. Conclusions

Although these maps represent only a first glimpse of the milliarcsecond polarization of active galactic nuclei, we have already obtained valuable information about the physical conditions on parsec scales. The cores of 3C 345 and 3C 120 have low polarization; if this is due to depolarization by emission-line gas, it will provide a valuable probe of this region. The core of the BL Lac object OJ 287 is significantly polarized, which may indicate a lack of depolarizing gas (or magnetic field). The high polarization in the "jet" of OJ 287 implies a highly ordered magnetic field. Both the core and jet show polarization variability, and detailed studies of their evolution will provide valuable clues to their structure, evolution, and environment. Superluminal motion may be suggested by the polarization data on OJ 287.

We thank the US VLBI Network for generous allocations of observing time, and the staff of the Mark III Correlator Facility at Haystack Observatory for their help. We are especially grateful for the assistance of Leslie Brown and Denise Gabuzda (Brandeis) and Alan Rogers (Haystack). This work was supported by the NSF under grants AST 83-15945 and AST 85-19529 (DHR) and AST 84-1863 (JFCW).

References

Blandford, R. D., and Königl, A. 1979, *Astrophys. J.*, **232**, 34.
Conway, R. G., and Kronberg, P. P. 1969, *Monthly Notices Roy. Astron. Soc.*, **142**, 11.
Königl, A., and Choudhuri, A. R. 1985, *Astrophys. J.*, **289**, 173.
Readhead, A. C. S., and Wilkinson, P. N. 1978, *Astrophys. J.*, **223**, 25.
Roberts, D. H., Potash, R. I., Wardle, J. F. C., Rogers, A. E. E., and Burke, B. F. 1984, in *IAU Symposium 110, VLBI and Compact Radio Sources*, ed. R. Fanti, K. Kellermann, and G. Setti (Dordrecht: Reidel), p. 35.
Walker, R. C., Benson, J. M., Seielstad, G. A., and Unwin, S. C. 1984, in *IAU Symposium 110, VLBI and Compact Radio Sources*, ed. R. Fanti, K. Kellermann, and G. Setti (Dordrecht: Reidel), p. 121.
Walker, R. C., Benson, J. M., and Unwin, S. C. 1987, *Astrophys. J.*, **316**, 546.
Wardle, J. F. C., Roberts, D. H., Potash, R. I., and Rogers, A. E. E. 1986, *Astrophys. J. (Letters)*, **304**, L1.

The Low Frequency Variability of Extragalactic Radio Sources: a Relativistic Effect or Galactic Scintillation?

R. FANTI AND L. GREGORINI
Dipartimento di Astronomia, Università di Bologna
and Istituto di Radioastronomia CNR, Bologna

L. PADRIELLI
Istituto di Radioastronomia CNR, Bologna

S. SPANGLER
Department of Physics and Astronomy, University of Iowa

1. Introduction

The discovery of low frequency variability (LFV) of extragalactic radio sources (e.g., Hunstead 1972) opened a set of problems of considerable magnitude for our understanding of the underlying physics of these objects. Under the assumption that the variability is an intrinsic phenomenon, the short time scales and the implied brightness temperatures pose severe theoretical problems that have been discussed by several authors (e.g., Jones and Burbidge 1973).

Two main "intrinsic" solutions to the problems have been advanced in the literature: (*a*) relativistic motion in the galactic nuclei may boost the incoherent synchrotron radiation, producing apparent brightness temperatures in excess of the 10^{12} K limit; or (*b*) coherent processes may be responsible for the phenomenon. The former hypothesis has been favored in the absence of evidence for diffractive scintillation in the interstellar medium (Dennison and Condon 1981).

However, "extrinsic" explanations have also been suggested, either in the form of absorption by a medium of variable opacity (by W. B. McAdam: see Marscher 1979; Condon *et al.* 1979), or as an interstellar medium propagation effect. The latter is referred to as "slow scintillation" or "refractive scintillation" (Shapirovskaya 1978; Rickett, Coles, and Bourgois 1984). The refractive scintillation hypothesis has received wide acceptance in view of the new results on the slow variability of pulsars (Sieber 1982).

Low frequency VLBI observations are required to distinguish the two possible causes of the LFV phenomenon.

The intrinsic explanation based on relativistic beaming (which requires very large Lorentz factors) predicts two features: (a) continuity between the low frequency and the high frequency variability; and (b) structural changes at superluminal speeds associated with large low frequency variations.

The extrinsic explanation (Rickett, Coles, and Bourgois 1984) leads us to expect correlations between the variability parameters and the galactic coordinates, interstellar medium parameters, and source size.

It is quite likely that both intrinsic and extrinsic processes are at work in these sources. We may then ask what fraction of the variability is explained by the relativistic model, what Lorentz factors are required, how many sources are described instead by the scintillation model, what parameters are required of the interstellar medium, and so on.

2. Multifrequency Observations and Structure Studies

It is now known from multifrequency variability observations (e.g., Altschuler et al. 1984; Padrielli et al. 1987) that $\sim$ 15–25% of the low frequency variables show continuity in the variability from high to low frequencies, in a manner qualitatively similar to that expected from the standard expanding synchrotron model (van der Laan 1966), albeit modified by relativistic effects. We regard these "class C" sources as "bona fide" intrinsic variables. There are six such sources in the Bologna sample (3C 120, 0605−08, 3C 279, 1510−08, 3C 345, and BL Lac), plus a few other candidates (Padrielli et al. 1987). The remaining sources of the sample are either of "class L", varying only at low frequency ($\nu \lesssim 1.4$ GHz), or of "class U", varying at both high and low frequencies but with no correlation between the two (see Spangler and Cotton 1981). The prototype of class L is DA 406 (1611+34) and that of class U is 3C 454.3.

Strong support for sources of class C being intrinsic variables with associated relativistic effects comes from structure studies. Four of the class C sources listed above are well-known superluminal sources. Other important information comes from a two-epoch VLBI experiment (Romney et al. 1984; Padrielli et al. 1986) carried out for 21 sources of the Bologna sample. This subsample contained four sources of class C, two of class L, and five of class U. Among sources of class C, three varied significantly at 408 MHz between the two epochs of the VLBI observations, and two showed expansion at rates in agreement with the relativistic model. Among sources of classes L and U, three varied by large amounts between the two epochs without showing significant expansion, although expansion might have been expected if the variability was due to relativistic effects.

These findings, therefore, reinforce the interpretation that the C sources are intrinsically variable and superluminal, while the variability of the L and U sources is likely to be extrinsic and probably due to slow (refractive) scintillation.

3. The Refractive Scintillation Model

We are presently analyzing the properties of the LFV sources of the Bologna sample in the light of the predictions of a slow refractive scintillation model. This contribution is a progress report; a more detailed discussion is in preparation.

The basic relations of this model have been derived by Rickett (1986):

$$m = 0.5\theta_s/(\theta_s^2 + \theta_I^2)^{1/2},$$
$$\tau = L \csc |b| (\theta_s^2 + \theta_I^2)^{1/2}/1.7v,$$
$$\theta_s = \theta_0 (\csc |b|)^{1/2} \lambda^2,$$

where m is the fractional rms variability, τ is the variability time scale, b is the galactic latitude, L is the thickness of the scattering medium (assumed to be 500 pc), v is the speed at which the irregularities move (assumed to be 50 km s^{-1}), θ_I is the source angular size, θ_s is the scattering angular size, and θ_0 is a parameter depending on the medium characteristics (assumed to be 8 mas).

A stringent test of the scintillation hypothesis would require knowledge of the intrinsic source size θ_I, which is rarely available. However, a combination of the three equations gives the following relation:

$$m\tau(\sin|b|)^{3/2} \sim 6 \left(\frac{L}{500 \text{ pc}}\right) \left(\frac{50 \text{ km s}^{-1}}{v}\right) \left(\frac{\theta_0}{8 \text{ mas}}\right) \% \text{ yr},$$

which can be used to test the scintillation model. The two quantities m and τ are deduced from observations, while the right-hand side depends only on the properties of the interstellar medium.

The time scale τ is evaluated statistically by means of both autocorrelation functions and structure functions. The two methods give results in excellent agreement. It is found that at least 30%, and perhaps as many as 50%, of the sources show two distinct time scales, which we will call the "short" and the "long" time scales. The rms flux variability associated with each of the two has been evaluated, and two variability indices m_s and m_l have been computed (rms divided by the average flux). We stress the following difficulties: (a) the average flux may also come from a non-varying component, so that m is a lower limit to the true fractional variability; and (b) if the two time scales are due to two

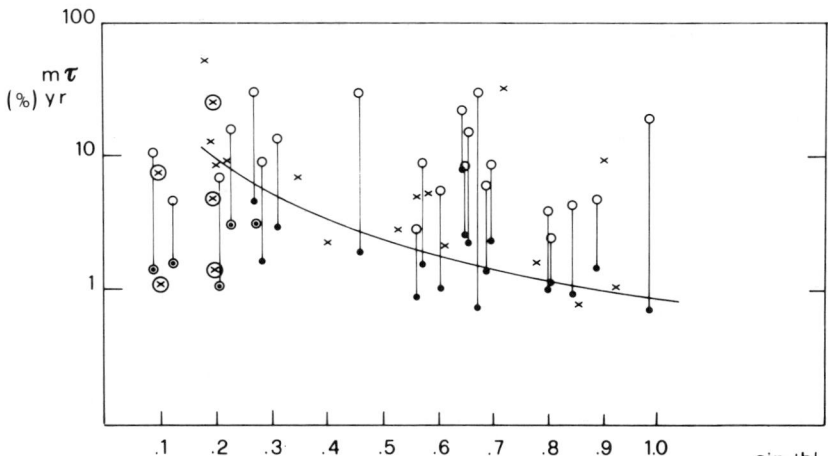

Figure 1 $m\tau$ plotted versus $\sin|b|$ (see text for definitions). The solid line represents $m\tau = 1.2(\sin|b|)^{-3/2}$. *Symbols:* ×: sources with a single time scale; • and ○: short and long time scales of the same source. The points for sources with $|b| < 5°$ in the galactic anticenter direction are enclosed in circles.

different components, we will be unable to normalize each of the two rms fluxes by the corresponding component flux, because of our ignorance of the low frequency structure.

Figure 1 shows a plot of the product $m\tau$ versus $\sin|b|$. Different symbols are used for the single-time-scale sources and for the short and long τ's of the two-time-scale sources. The lower part of the distribution (the short τ's and the τ's of the single-time-scale sources) follows the $(\sin|b|)^{-3/2}$ law predicted by the refractive scintillation theory, with a dispersion of $\pm \log(m\tau) \sim 0.4$. We note that the short variability $m\tau$'s of every class lie on this band, but there are some noticeable exceptions: the sources DA 406 and 1117+14 (see below), and sources with $|b| < 5°$ in the anticenter direction. The latter anomaly is not unexpected, since the path through the interstellar medium may well be much less than $L/\sin|b|$ in this region of the sky, owing to the finite size of the Galaxy.

We believe that this relation represents the scintillation effect predicted by the refractive model of Rickett (1986). Note, however, that the average of $m\tau(\sin|b|)^{3/2}$ is about 1, instead of about 6 as expected for the parameters quoted by Rickett. This requires some modification of either L, v, or θ_0. We have used the very few source sizes measured at low frequency (Wilkinson *et al.* 1979: for 3C 345, 3C 454.3, CTA 102, and 3C 286, at 610 MHz) to try to calibrate the above parameters, using the short time scale m and τ. We find $L \sim 700$ pc and $\theta_0 \sim 2$ mas. These values are not in disagreement with those currently reported in the literature ($L = 500$ pc, $\theta_0 = 4$–8 mas), although our value of θ_0

is slightly smaller. Maps at low frequency with better resolution are needed to improve the estimates of these parameters.

What about the slower variability? Figure 1 does not show (although it does not exclude) a clear relation between m, τ, and $\sin|b|$. A good fraction of the sources with longer time scales are the "bona fide intrinsic" variables described above. However, one-third of them are either L or U. Is this again refractive scintillation from a somewhat larger component? Or, are these a new set of intrinsic variables? It is perhaps too early to make any statement, although the similarity of their $m\tau$ products to those of the "bona fide intrinsic" sources is tantalizing.

A few more points should be made regarding refractive scintillation. All "bona fide intrinsic" sources are expected to be small, so they should exhibit refractive scintillation as well as intrinsic variability. We have already identified the former with the long time scale and the latter with the short one. However, one source, 1510−08, does not have a short time scale short enough to be identified with refractive scintillation. This cannot easily be attributed to an anomaly of the interstellar medium in that direction, as two other nearby sources (1504−16 and 1524−13) show the fast (scintillation) variability. Furthermore, BL Lac, firmly believed to be an intrinsic variable, shows one time scale only. However, with our parameters the scintillation time scale can be similar to the intrinsic one if the source has a size of 5 mas, which in turn implies a scintillation index smaller than the true variability by a factor of order two.

On the other hand, two of the single-time-scale sources (DA 406 and 1117+14) are definitely varying too slowly or varying too much ($m \sim 33$ and $m \sim 10$, respectively), as compared to the probable scintillators. Again, their excessively slow variability cannot easily be attributed to an anomaly of the medium in that area of the sky, since they are close enough to other sources which show the proper "scintillating" behavior. Why do they scintillate so slowly? Or, if they are intrinsic variables, why do we not also detect scintillations?

4. Conclusions

Our analysis suggests the following results:

(*a*) There is a component of variability in the LFV sources which is an effect of propagation through the Galaxy, possibly refractive scintillation as discussed by Rickett (1986), and which accounts for the shorter time scale variation. However, our results require a revision of some of the parameters of the interstellar medium, probably the scattering angle.

(*b*) In multifrequency and structure studies, at least 15–25% of the LFV sources show an indication of intrinsic variability, which produces the long time scales. In these cases, the brightness temperatures computed from the observed time scales go up to a few times 10^{14} K, but do not reach the ridiculous values of 10^{16} K or so that were derived from the short time scales which we now attribute to refractive scintillation.

(c) In addition to scintillation, several sources in the L and U classes show a longer time scale variability whose nature is at present not understood.

(d) Some sources show an anomalous scintillation behavior that is also not understood.

References

Altschuler, D. R., Broderick, J. J., Condon, J. J., Dennison, B., Mitchell, K. J., O'Dell, S. L., and Payne, H. E. 1984, *Astron. J.*, **89**, 1784.
Condon, J. J., Ledden, J. E., O'Dell, S. L., and Dennison, B. 1979, *Astron. J.*, **84**, 1.
Dennison, B., and Condon, J. J. 1981, *Astrophys. J.*, **246**, 91.
Hunstead, R. W. 1972, *Astrophys. Letters*, **12**, 193.
Jones, T. W., and Burbidge, G. R. 1973, *Astrophys. J.*, **186**, 791.
Marscher, A. P. 1979, *Astrophys. J.*, **228**, 27.
Padrielli, L., Aller, M. F., Aller, H. D., Fanti, C., Fanti, R., Ficarra, A., Gregorini, L., Mantovani, F., and Nicolson, G. 1987, *Astron. Astrophys. Suppl.*, **67**, 63.
Padrielli, L., Romney, J. D., Bartel, N., Fanti, R., Ficarra, A., Mantovani, F., Matveyenko, L., Nicolson, G. D., and Weiler, K. W. 1986, *Astron. Astrophys.*, **165**, 53.
Rickett, B. J. 1986, *Astrophys. J.*, **307**, 564.
Rickett, B. J., Coles, W. A., and Bourgois, G. 1984, *Astron. Astrophys.*, **134**, 390.
Romney, J., Padrielli, L., Bartel, N., Weiler, K. W., Ficarra, A., Mantovani, F., Bååth, L. B., Kogan, L., Matveenko, L., Moiseev, I. G., and Nicolson, G. 1984, *Astron. Astrophys.*, **135**, 289.
Shapirovskaya, N. Ya. 1978, *Astron. Zh.*, **55**, 953. English translation: *Soviet Astron.*, **22**, 544.
Sieber, W. 1982, *Astron. Astrophys.*, **113**, 311.
Spangler, S. R., and Cotton, W. D. 1981, *Astron. J.*, **86**, 730.
van der Laan, H. 1966, *Nature*, **211**, 1131.
Wilkinson, P. N., Readhead, A. C. S., Anderson, B., and Purcell, G. H. 1979, *Astrophys. J.*, **232**, 365.

Superluminal Motion in CTA 102

LARS B. BÅÅTH
Onsala Space Observatory, Sweden

1. Introduction

Flux variations in compact continuum radio sources were first reported in 1965 (Sholomitskii 1965). From the time scale of the variability Sholomitskii concluded that the linear size of CTA 102 (2230+114) was 0.1 pc, while an angular size of 0″.01 was derived from the spectral turnover (Slish 1963). This led Sholomitskii to suggest that CTA 102 might be a Galactic object. CTA 102 was later identified as a quasar with a redshift of 1.037 which, assuming the redshift to be cosmological, placed the source at a considerably larger distance. The "causality argument" used by Sholomitskii to get the linear size is still valid though, and the varying component must therefore have a much smaller angular extent, making up only a small fraction of the source while emitting the major part of its radio flux. The brightness temperature of a component of such a small size exceeds the limit where the inverse Compton process should cause the X-ray flux to be very much higher than is actually observed. For this reason, low frequency variability has often been referred to as "Superluminal Flux Variation" (Romney et al. 1984).

The high brightness temperature and the flux variations at high frequencies (> 1 GHz) have both been successfully explained with the beaming model using moderately high Lorentz factors ($\gamma \sim 5$–10). The disagreement between the minimum size implied by the Compton limit and the maximum size implied by the time scale of the variations is larger at lower frequencies, and a number of theories have been suggested to explain the flux variations at low frequencies. Basically, the variations can be either intrinsic to the source, extrinsic, or a combination of both. The intrinsic models include an extrapolation of the beaming model, where the variations have the same cause and explanation as for the superluminal radio sources. The extrinsic models include noncosmological redshifts, again making CTA 102 a Galactic object, and interstellar scintillations or "interstellar weather".

Slow, refractive scintillation caused by the interstellar medium is very likely to play a role in the flux variations particularly at low frequencies (Rickett, Coles, and Bourgois 1984; Fanti et al., this Workshop, page 200). However, such a source also has to contain compact components at low frequencies in order to make scintillation effective. This suggests that low frequency variables as a group represent good candidates for superluminal motion.

2. Observations

The low frequency variables show variations, correlated or uncorrelated, either at low and at high frequencies, or at low frequencies only. The sources which show uncorrelated variations at high and low frequencies, or vary only at low frequencies, are usually quiescent in a band 1–2 GHz, often referred to as the intermediate-frequency gap (Dennison et al. 1984). In order to look for structural changes in such objects we started a VLBI monitoring project at 932 MHz, just below the intermediate frequency gap. This gave us the maximum possible resolution at a frequency where there is still hope to observe the components responsible for the flux variations. Observations were made at six epochs during 1983–1984 with the following antennas: Kiruna, Sweden; Arecibo, Puerto Rico; Green Bank and Owens Valley, USA; and at the last three epochs also Jodrell Bank, England. A total of 15 sources were observed, in the same range of sidereal time at all six epochs. The (u,v)-plane coverage yields a beam that is only slightly larger in the north-south direction than in the east-west direction. We have, therefore, in order to make interpretation easier, used a circular 5 mas beam to restore the cleaned maps. This corresponds to 10% super-resolution in the north-south direction, which in the case of our uncomplicated maps does not introduce any ambiguity in the structure.

The mapping of the first epoch is nearly completed and we present maps for the objects 2147+145, DA 406 (1611+343), 1422+202, and CTA 102. CTA 102 was the first source to be mapped at more than one epoch. This seemed appropriate since it was the source in which low frequency variability was first detected. We show maps only for the three best epochs. The other three epochs were hampered by various problems, including a lovely northern light at Kiruna and recorder problems at Green Bank.

3. Results

The "optically quiet quasar" 2147+145 is shown in Figure 1a. The map at 932 MHz is similar to the map made by Cotton et al. (1984) at 1660 MHz.

Our sources do not all have the same morphology. A few are compact with an unresolved component and a more extended halo, e.g.,

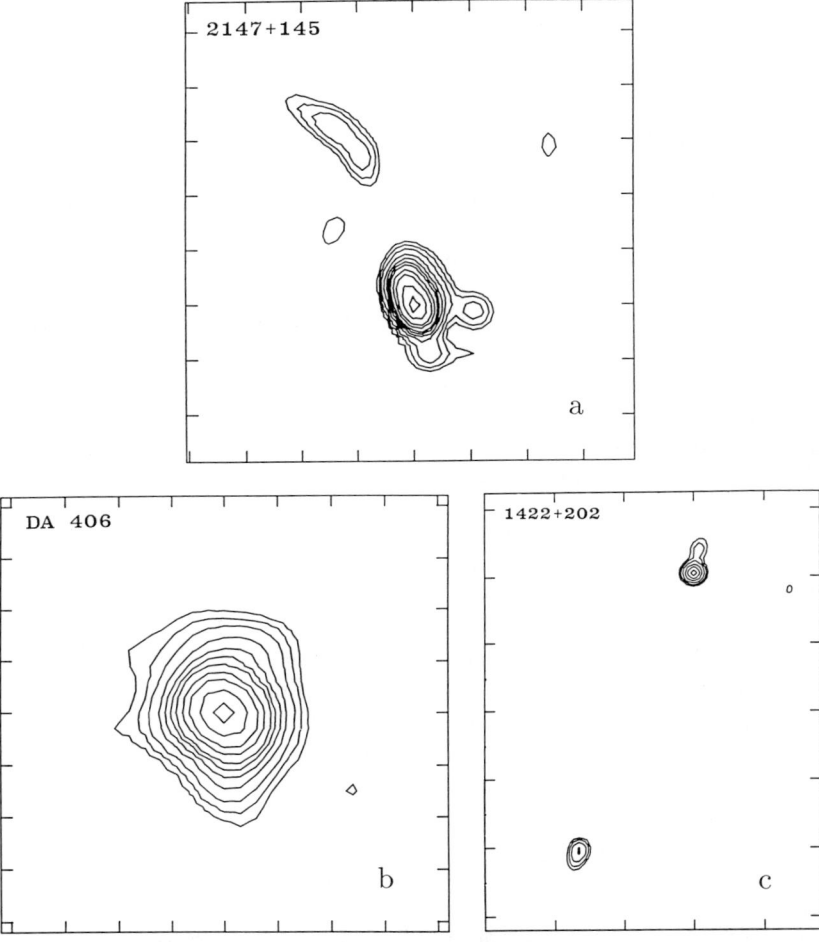

Figure 1 VLBI maps of three sources at 932 MHz, convolved with a Gaussian restoring beam with FWHM 5 mas. The contour levels in each map are at −5, 5, 7, 10, 15, 20, 25, 30, 40, 50, 70, and 90% of the peak flux density. (a) 2147+145: peak flux density 0.61 Jy per beam; scale is 10 mas per tick. (b) DA 406 (1611+343): peak flux density 2.26 Jy per beam; scale is 5 mas per tick. (c) 1422+202: peak flux density 0.28 Jy per beam; scale is 20 mas per tick.

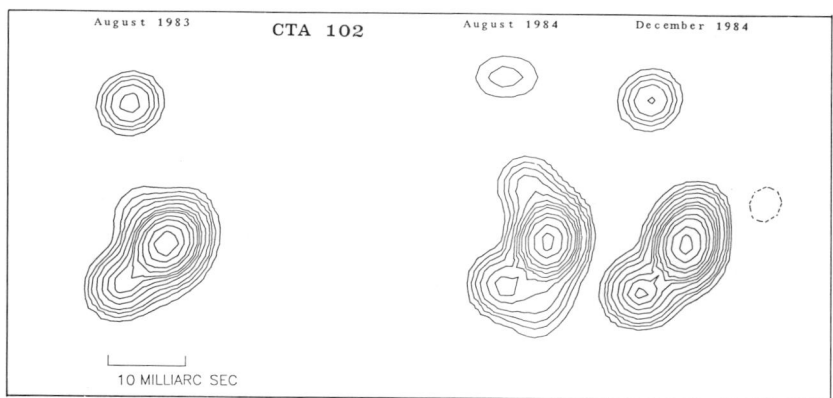

Figure 2 VLBI maps of CTA 102 (2230+114) at 932 MHz obtained at three epochs. The maps are convolved with a Gaussian restoring beam with FWHM 5 mas. The contour levels are at −5, −3, 3, 5, 7, 10, 15, 20, 25, 30, 40, 50, 70, and 90% of 2 Jy per beam.

DA 406 (Figure 1b). DA 406 has never shown any VLBI structure and may be a good candidate for flux variations due to interstellar scintillation. Most of the sources are doubles, e.g., 1422+202 (Figure 1c), with separations varying from 20 to 100 mas. 1422+202 has also been observed with MERLIN revealing a 10″ double structure (F. Mantovani et al., in preparation). However, the use of the global fringe-fitting and self-calibration techniques makes it impossible to decide in which of the two arcsecond lobes the VLBI double is situated. EVN observations at 1660 MHz indicate that both lobes contain compact structure.

Figure 2 shows maps of CTA 102 obtained from the three epochs that we have analyzed so far—1983 August, 1984 August, and 1984 December. The second-epoch map is of poorer quality than the other two, and the north-eastern component in particular is rather poorly determined in that map. The distance between the central component and the south-eastern component was 7.91 ± 0.10 mas, 8.56 ± 0.2 mas, and 8.87 ± 0.1 mas at the three epochs. The separation thus increased at 0.65 ± 0.15 mas yr^{-1}. The corresponding apparent separation speed is $(18 \pm 4)h^{-1}c$. The relative position angle was constant within the error bars. The apparent superluminal motion along a constant position angle suggests that the flux variations in CTA 102 are intrinsic rather than due to interstellar scintillation. Interstellar scintillation would be expected to play only a minor role at the galactic latitude of CTA 102, $-38°.6$. A backward extrapolation of the apparent motion at 0.65 mas yr^{-1} indicates that the two components (we cannot tell which of them, if either, is the core) overlapped around 1972. CTA 102 underwent several large outbursts of total flux density around 1970 (Padrielli 1982), but there is an unfortunate lack of data between 1970 and 1972.

4. Conclusion

CTA 102 is a superluminal source with an apparent expansion speed of $(18 \pm 4)h^{-1}c$. CTA 102 might not be typical of the low frequency variables, but the fact that the very first source we have mapped at several epochs has turned out to be superluminal suggests that beaming effects play an important role at frequencies below the intermediate frequency gap as well as at high frequencies.

References

Cotton, W. D., Owen, F. N., Geldzahler, B. J., Johnston, K., Bååth, L., and Romney, J. 1984, *Astrophys. J. (Letters)*, **277**, L41.
Dennison, B., Broderick, J. J., O'Dell, S. L., Mitchell, K. J., Altschuler, D. R., Payne, H. E., and Condon, J. J. 1984, *Astrophys. J. (Letters)*, **281**, L55.
Padrielli, L. 1982, in *Low Frequency Variability of Extragalactic Radio Sources*, ed. W. D. Cotton and S. R. Spangler (Green Bank: National Radio Astronomy Observatory), p. 1.
Rickett, B. J., Coles, W. A., and Bourgois, G. 1984, *Astron. Astrophys.*, **134**, 390.
Romney, J., Padrielli, L., Bartel, N., Weiler, K. W., Ficarra, A., Mantovani, F., Bååth, L. B., Kogan, L., Matveenko, L., Moiseev, I. G., and Nicolson, G. 1984, *Astron. Astrophys.*, **135**, 289.
Sholomitskii, G. B. 1965, *Astron. Zh.*, **42**, 673. English translation: *Soviet Astron.*, **9**, 516.
Slish, V. I. 1963, *Nature*, **199**, 682.

Imaging Superluminal Sources: Prospects for the Next Decade

PETER N. WILKINSON

University of Manchester, Jodrell Bank

1. Introduction

The development ten years ago of VLBI imaging techniques (Readhead and Wilkinson 1978; Cotton 1979) using what has now come to be known as self-calibration is one of the main reasons why we are able to have this Workshop at all. It wasn't until reliable maps of 3C 273 and 3C 345 were produced (e.g., Pearson et al. 1981) which clearly showed "blobs" of emission moving away from a bright "core" that the skeptics were finally convinced that superluminal motion was a reality and was not merely a failure of the observers' ability to interpret their data correctly.

We have seen during the course of this Workshop that VLBI maps have revealed superluminal motion in the cores of a wide range of objects; the phenomenon is clearly rather common, but equally clearly we are still "light years" away from understanding it! A significant part of the problem lies in the maps themselves—with a few honorable exceptions they are still primitive, and so watching the apparent motion of a few disconnected "blobs" remains the order of the day. Given the wide range of astrophysical questions raised by the observations of milliarcsecond jets in general and of superluminal motion in particular the importance of obtaining better and better images should be obvious—it is not just an end in itself.

What are the major problems which might be solved with better data? In brief they seem to me to be as follows:

(a) How and where are the beams formed? Interesting scales to study are those of accretion disks around massive black holes, i.e., 10^{16-17} cm.

(b) Are there really two beams? Obviously one needs to look very hard for counterjets close to the core.

(c) What is the physics of the subsequent flow? Is the beam confined or ballistic, i.e., how does one account for the bends and wiggles? And related to this: Is superluminal motion a pattern motion of relativistic shocks or does it represent the real bulk motion of "plasmons"? Could the beam be much wider (because of Doppler beaming) than the apparent cone angle and hence could collimation be occuring on the scale of hundreds of parsecs (typically 0.1 arcsec)? Does the motion remain superluminal out to the kiloparsec scale or does the beam slow down fairly early on?

Question (a) can only be tackled in the powerful, distant objects with the highest possible resolution, using millimeter-wavelength VLBI and QUASAT; however one should note that in some nearby objects, such as M87, observations are already probing scales of 10^{-2} pc (Spencer and Junor 1986). To pursue questions (b) and (c), however, we most certainly require high fidelity, high dynamic range images of the jets—preferably in all Stokes parameters. Not only that, we also need them at many frequencies, at closely spaced epochs, and with positional information at the 100-microarcsec level from relative astrometry. This is a big wish list but there is reason to be optimistic—progress is being made in all these areas now. With the advent of the full VLBA in the early 1990s, even more rapid progress can be expected.

2. Guesstimates

In some of these areas, one can make quantitative estimates of the capability we shall have when we hold the Workshop to celebrate Marshall's seventieth birthday. These estimates are based on the last 20 years' experience of aperture synthesis and, at least to a factor of two or so, are rather firmly based. Let us assume that we shall have at our disposal an array of 20 or more telescopes based on the VLBA and the EVN.

Sensitivity

Assuming 64 MHz bandwidths and 30 K system temperatures, thermal noise levels will be around 10 μJy after a 12-hour integration. This figure is applicable to observations at $\sim$5 GHz and below. At 22 GHz, noise levels several times higher, say 30 μJy, can be expected. At 90 GHz, collecting areas are much less and system temperatures much worse than at 5 GHz, and my guess is that noise levels of 200 μJy are the best that one can hope for. At the lower frequencies then, VLA-like sensitivities will be available, but in order to exploit them in imaging complex sources, we shall need to achieve very high dynamic ranges. I hope that, by now, the goal of "getting down to the thermal noise level" has permeated all VLBI heads.

Dynamic Range

This general topic has been addressed recently by Wilkinson (1983) and Cornwell and Wilkinson (1984), and, strictly in the context of the VLA, by Perley (1986). All that we need to note here is that sidelobe-like errors at the level of 10^{-5} of the peak brightness are certainly sky-plane representations of errors in the visibility data—nowadays errors in the imaging algorithms themselves are rather small. As is well known, these visibility errors come in two forms, telescope-related and baseline-related, and the trick is to recognize and correct them in a systematic way. Self-calibration works beautifully for the telescope-related errors (Pearson and Readhead 1984), and recently aficionados of the various arrays have begun to learn how to cope with the more awkward baseline-related terms. As a result of their efforts, the current status of "best-efforts" dynamic range can be roughly summarized as:

Connected-element arrays (WSRT, VLA): $\sim 10^5 : 1$
Radio-linked arrays (MERLIN): $\sim 10^4 : 1$
Tape-recording arrays (Global VLBI): $\sim 10^3 : 1$

It is as well to note what this means in terms of the residual errors in amplitude (ΔA) and phase ($\Delta \phi$):

Dynamic range	ΔA (%)	$\Delta \phi$ (degrees)
$10^5 : 1$	< 0.02	< 0.01
$10^4 : 1$	< 0.2	< 0.1
$10^3 : 1$	< 2	< 1

In terms of dynamic range, then, VLBI is currently about ten years behind connected-element arrays and about five years behind MERLIN—but it is keeping up in relative terms. The dynamic range in our first VLBI maps in 1976 was about 20 : 1, so we have improved by about a factor 50 in the last decade. Now that most of the problems in VLBI imaging are appreciated, and with MERLIN to use as a stalking horse, we should beat that performance in the next decade and get to a dynamic range of $10^5 : 1$. However, to achieve this goal we shall need to pay very careful attention to all aspects of the data path, starting with the polarization purity of the receivers and the bandpasses of the filters and continuing with high fidelity recording and playback through a well-understood correlator. Polarization impurity, for example, has received little attention for intensity imaging, but if the source is polarized, it can give rise to baseline-related offsets that vary slowly (with parallactic angle). I believe that the new level of professionalism which the VLBA will engender will be vital if the $10^5 : 1$ goal is to be reached.

How Far Can Jets Be Imaged?

Given a dynamic range of $10^5 : 1$, how far from the core can a jet in a "standard" superluminal source be imaged? Walker's analysis of the many images of the 3C 120 jet is most useful here (Walker 1986; and this Workshop, page 48). He shows that the surface brightness (Σ) falls off with distance along the jet (r) roughly as $r^{-2.4}$. From his data, which have all been scaled to 5 GHz, one can estimate the surface brightness in a 1 mas beam (corresponding to the longest earth baselines) as follows:

Distance along jet, r (mas)	Brightness, Σ (μJy per beam)
2	6×10^5
20	2400
200	10

Thus we could, in principle, follow the jet in 3C 120 out to about 200 mas (i.e., 200 beams or 100 pc) until it disappears into the noise.

At other frequencies, the jet flux scales as ν^α (with $\alpha \sim -0.65$) and the beam area scales as ν^{-2}; given the expected sensitivities, one can make the following estimates of the length of the jet one will see in 3C 120:

Frequency (GHz)	Jet length (mas)	Number of beams
1.6	700	230
5.0	200	200
22.0	30	120
90.0	1.5	25

A few comments on these numbers are worthwhile. First, polarization will only be measurable less than one-third of the way out at full resolution. Secondly, bright knots may be traced further out than these numbers imply. Thirdly, at 1.6 and 5 GHz arrays of more than 20 telescopes will be needed to make reliable images containing so many filled pixels. At 22 GHz, fewer telescopes will suffice, whilst at 90 GHz as few as six telescopes can do the job. At the highest frequencies, the onus falls on making sensitivity improvements—there are enough telescopes to make the images. Fourthly, for stronger sources like 3C 273, one could in principle image the jet even further out than for 3C 120 but the required dynamic range ($10^6 : 1$) and the number of synthesized beams along the jet (> 1500) are, even for an optimist like me, out of reach in the next decade.

Fidelity

One must again emphasize the difference betweeen dynamic range—which refers basically to the level of the residual sidelobes off-source and which for a given level of data error only reduces as the square root of the number of baselines (i.e., roughly as the number of telescopes N)—and the fidelity of the reconstruction on-source. This depends on the (u,v) coverage, i.e., on $N(N-1)/2$, although no one has established the dependence properly, as well as the particular deconvolution algorithm in use (e.g., Dulk et al. 1984). Reproducibility of subtle features will be vital for testing specific theories related to jets. For example, the shock model for particle acceleration in the extended jet in 3C 273 (Meisenheimer and Heavens 1986) can be tested by comparing accurate maps at two frequencies, but only if they are reliable at the 0.1 arcsec level, i.e., $1/200^{th}$ of the jet length.

To achieve this fidelity, arrays of 20 or more telescopes will probably be needed whatever the "flavor of the month" in deconvolution algorithms turns out to be. However, another possibility for obtaining the necessary (u,v) coverage is currently emerging. Multi-frequency synthesis, which Tim Cornwell, John Conway, and I are currently working on, uses observations over a $\pm$ 5–10% range of frequencies to fill in gaps in the (u,v) plane. This bandspread works well for relatively modest arrays of around ten telescopes, as long as one can solve simultaneously for spectral index as well as the intensity distribution at an intermediate frequency. Progress in this area is rather rapid and promising results on simulated data have been achieved. The next step is to test the algorithms on real data from MERLIN and the VLA—the data are already in hand.

3. Summary

I have ignored polarization and astrometry in this discussion, but as regards dynamic range and map fidelity, I am reasonably confident that in ten years time VLBI arrays will be achieving what the VLA is achieving now for extended jets, viz., a dynamic range of 10^5 : 1 and excellent fidelity over 100–200 beams. Such maps will be similar in quality to the VLA maps of the NGC 6251 jet (Perley, Bridle, and Willis 1984). This will mean that we shall be working with images 10–20 times longer (in parsecs) than we are now. This must surely lead to a more profound understanding of the superluminal phenomenon.

References

Cornwell, T. J., and Wilkinson, P. N. 1984, in *Indirect Imaging*, ed. J. A. Roberts (Cambridge: Cambridge University Press), p. 207.
Cotton, W. D. 1979, *Astron. J.*, **84**, 1122.
Dulk, G. A., McLean, D. J., Manchester, R. N., Ostry, D. I., and Rogers, P. G. 1984, in *Indirect Imaging*, ed. J. A. Roberts (Cambridge: Cambridge University Press), p. 355.
Meisenheimer, K., and Heavens, A. F. 1986, *Nature*, **323**, 419.
Pearson, T. J., and Readhead, A. C. S. 1984, *Ann. Rev. Astron. Astrophys.*, **22**, 97.
Pearson, T. J., Unwin, S. C., Cohen, M. H., Linfield, R. P., Readhead, A. C. S., Seielstad, G. A., Simon, R. S., and Walker, R. C. 1981, *Nature*, **290**, 365.
Perley, R. A. 1986, in *Synthesis Imaging*, NRAO Workshop No. 13, ed. R. A. Perley, F. R. Schwab, and A. H. Bridle (Green Bank: National Radio Astronomy Observatory), p. 161.
Perley, R. A., Bridle, A. H., and Willis, A. G. 1984, *Astrophys. J. Suppl.*, **54**, 291.
Readhead, A. C. S., and Wilkinson, P. N. 1978, *Astrophys. J.*, **223**, 25.
Spencer, R. E., and Junor, W. 1986, *Nature*, **321**, 753.
Walker, R. C. 1986, *Can. J. Phys.*, **64**, 452.
Wilkinson, P. N. 1983, in *Techniques d'Interférométrie à Très Grande Base*, Proceedings of an International Conference on Very Long Baseline Interferometry Techniques organized by the Centre National d'Études Spatiales (Toulouse: Cepadues-Éditions), p. 375.

A Different Perspective on Superluminal Sources

LAWRENCE RUDNICK
Department of Astronomy, University of Minnesota

> Take the Levites [the priests, or *Cohens*] instead of all the first-born... and as for the redemption of the *two hundred and threescore and thirteen* of the first-born...
> *Numbers*, C3:45. Proof of a Conspiracy.

1. Introduction

Something is moving relativistically in superluminal sources, but there are good reasons to be skeptical of any more detailed claims about their nature. I will begin with a pessimistic appraisal of our ability to study such sources and then try to place them in the more general context of compact radio sources (Section 2). In Section 3, I will highlight some of the interesting information we have gained from studies of broadband spectra, variability, and polarization. Finally, in Section 4, I will vaguely suggest a different way of looking at active nuclei, which might provide some useful insights into their nature.

The reader is cautioned that this is more a stream of consciousness than a coherent review—see Wiita (1985) and the conference proceedings edited by Jones (1986), Giuricin *et al.* (1986), Sitko (1986), Dyson (1985), and Miller (1985) for a more general perspective on active nuclei. I have also ignored a number of important areas, and am often remiss in citing either evidence or the genesis of ideas.

A Pessimistic Outlook

One major difficulty encountered in the study of active galactic nuclei is that *anything can happen*. Historically, by focusing on one or two simple properties, we have been able to define different classes of active nuclei, including Seyfert, Markarian, starburst, LINER, N, broad-line radio and narrow-line radio galaxies, quasars (quiet and noisy), and BL Lac objects, and the overlapping classes of high polarization QSOs, optically violent variables, and blazars. Exploring the connection between these classes has been a frustrating task. Advocates of unified models feel obligated to address this question; others often hope we can solve our problems without addressing it. Nonetheless, it is useful to remind

ourselves about the ways in which these different types of objects might be related. For example, they could be:
1. *identical*, except that they are *anisotropic* and we view them from different angles. Such anisotropy could arise, for example, from an opaque accretion disk or relativistic beaming; or
2. *identical*, except that they have either stochastic or regular variations in their behavior, which change their observed properties into those of another class; or
3. *based on the same physical process*, but differing in some critical parameter, such as the central mass or accretion rate; or
4. *evolutionary stages* of the same physical objects; or
5. *different*.

Each of these possibilities deserves consideration. Ruling out all but one of them will be a long and dirty job, but somebody's got to do it!

A second major problem is that the radio-emitting electrons probably do not represent the energetically (dynamically?) dominant component of active nuclei. In terms of luminosity (νF_ν), the submillimeter or far infrared radiation is more important. Optical depth in the radio regime adds its complications, perhaps hiding the most important structures. Finally, there may be an underlying, non-radiating fluid, as is often invoked for large-scale jets.

The third class of problems results from the probable beaming, namely, that what we *do* see is likely to be misleading. This argument has been made most compactly by Phinney (1985), who points out that different observers will see different components and regions of a relativistic jet, perhaps reaching quite different conclusions regarding its nature. This brings us full circle, requiring a coherent overall model for these sources, and not just explanations of the details seen in components and outbursts.

2. Where do the Superluminals Fit?

Over the past 5–10 years, a considerable amount of progress has been made in distinguishing (phenomenologically) the various classes of active nuclei. Surprisingly little effort has gone into seeing how this relates to superluminal behavior. Perhaps now that a much wider range of structural variations are being measured with VLBI, observers will take a keener interest in correlating these with other information.

As an *Ansatz* for this paper, I will continue to duck the issue, and will speak about the class of blazars as if they were coincident with the class of superluminally expanding sources. A more responsible approach is taken by Chris Impey, in this Workshop (page 233).

The blazars can be readily distinguished from other active nuclei (including other compact "flat-spectrum" radio sources) by a number of properties in addition to their VLBI behavior:

1. They are strong radio variables, although the episodic nature of these variations requires long-term observation for proper classification.

2. They are strong millimeter emitters, which has led to their classification as "flat spectrum" sources. This "flat spectrum" moniker actually serves us poorly (see below), and should be reserved for the few sources with truly flat radio spectra (e.g., 0735+178).

3. They have radio spectra that are not power laws, and are not dominated by narrow peaks, but are gently curved or "complex."†

4. They have high radio luminosities ($> 10^{24}$ W Hz^{-1}), as distinguished from Seyferts, radio quiet QSOs and most nuclei of (low-frequency selected) radio galaxies.

5. They have bright optical counterparts (QSOs or BL Lac objects, dominating any host galaxy), with comparable radio and optical luminosities (νF_ν).

6. Their X-ray luminosity correlates with the radio/optical emission [for radio-selected blazars (Ledden and O'Dell 1985), but see section 3, below].

7. They have strong (> 1–2%), variable, radio polarization.

8. They have strong ($> 3\%$), variable, optical polarization.

Just as importantly, there are a number of properties which should, but do not, help us distinguish blazars from other sources:

1. The presence of strong, extended radio emission.‡

2. Their infrared luminosities are not distinct, as a population. Nor are the infrared colors, even in the 10–100 μm region, an indicator of blazar activity (Neugebauer et al. 1986), although some low detection percentages are still a problem.

3. Their optical luminosities are not distinct, as a population, although the optical continua tend to be smoother and steeper (Oke, Neugebauer, and Becklin 1970; Rieke and Lebofsky 1979; Moore and Stockman 1984).

4. Their emission-line equivalent widths are similar to other sources

---

† I agree with R. Blandford's admonition, and suggest that we drop the term "humped" radio spectrum, and use instead "narrow peak" spectrum, to label this distinct class of active nuclei (Rudnick and Jones 1982; Phillips and Mutel 1982).

‡ This is a whole separate, interesting area, which is unfortunately not being treated in this Workshop. The reader is referred to Ulvestad and Antonucci (1986) and references therein. At the Workshop, I also showed, but am not ready to print, a provoking diagram by Jeff Pedelty et al. which shows that any combination of core and extended luminosities may be found, but that different radio/optical selection criteria will isolate regions of this full parameter space.

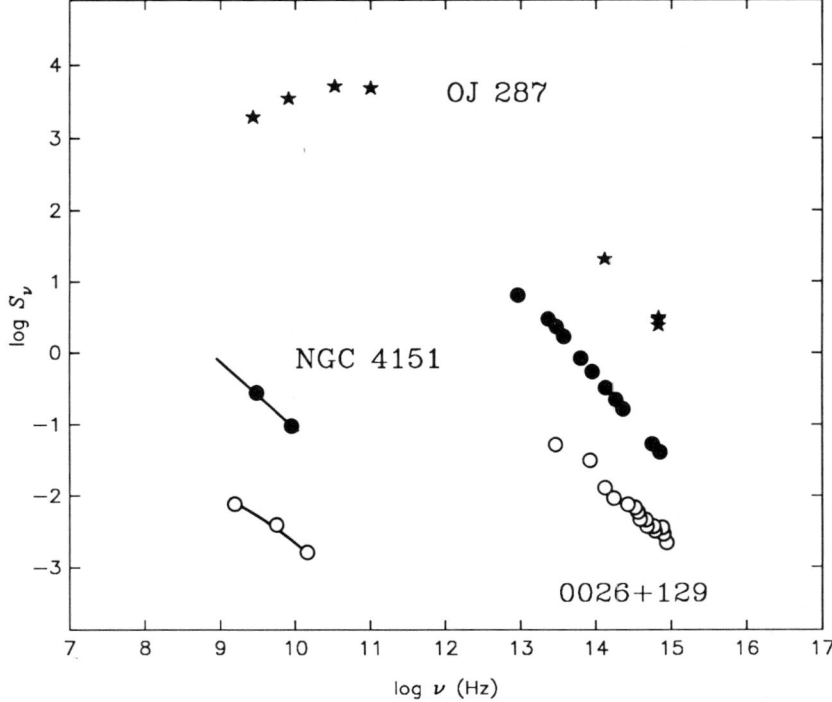

Figure 1 Comparison of broadband spectra for different types of compact nucleus. Blazar OJ 287 from Rudnick et al. (1978). Seyfert galaxy NGC 4151 and radio quiet QSO 0026+129 from Rudnick, Sitko, and Stein (1984). The spectra are plotted with displaced zero levels.

(Moore and Stockman 1984) although emission lines are sometimes absent in BL Lac objects.

5. The class of X-ray selected blazars (Stocke et al. 1985; Ledden and O'Dell 1985) shows that strong X-ray emission is *not* a reliable indicator of strong nuclear radio emission.

These non-correlating properties need to be documented carefully, as critical data for the various "unification" possibilities discussed above.

Compared to the efforts of multi-frequency, multi-epoch VLBI mapping, it is a fairly simple undertaking to obtain the complementary broadband information which has proven so useful in distinguishing various classes of active nuclei. This is often not done systematically, despite the fact that spectra, optical and radio polarizations, variability, and luminosity have all proven to be useful discriminators (Rudnick and Jones 1982; Moore and Stockman 1984; Saikia, Swarup, and Kodali 1985; Landau et al. 1986; Rudnick, Jones, and Fiedler 1986). It is now incumbent upon us to tap this kind of information by assuring that our

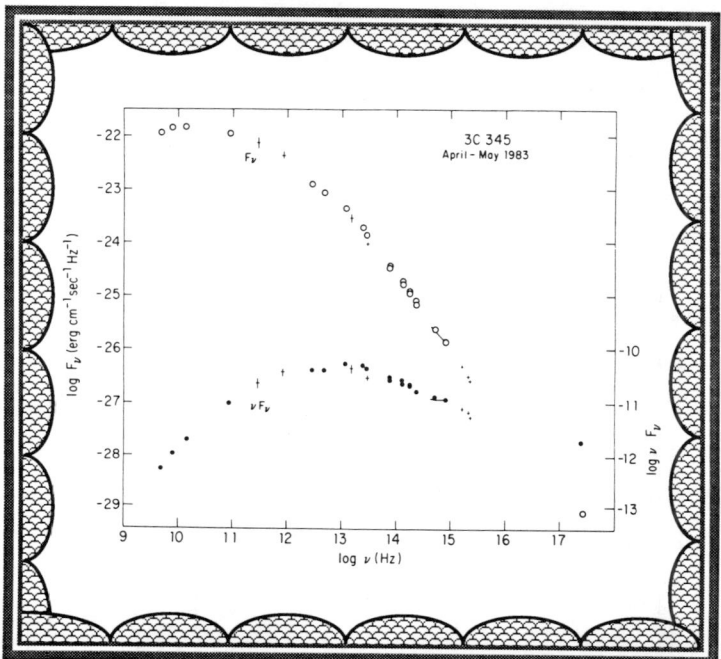

Figure 2   3C 345's well-sampled broadband spectrum, adapted from Bregman *et al.* (1986).

broadband and VLBI programs are better coordinated, as in the project described at this Workshop by Arno Witzel (page 83).

3. Measuring "Point" Sources

Broadband Spectra

Early discussions of the broadband (radio–infrared–optical) spectra of blazars noted a characteristic radio–optical spectral index of -0.7, although there was speculation about how much this value was due to the selection of the samples. However, there are other classes of objects with much different radio/optical indices, such as Seyferts and radio quiet QSOs (Figure 1).

Evidence for the *continuity* of the broadband spectra came from identification of a characteristic "break frequency" (Jones *et al.* 1981; Landau *et al.* 1983) and the equivalent, but more graphic, match between radio and infrared extrapolated fluxes (Ennis, Neugebauer, and Werner 1982). Recently, data from the Kuiper Airborne Observatory and IRAS have allowed us to fill in the annoying gap in the broadband spectra (Harvey, Wilking, and Joy 1982), with a delightful result. For

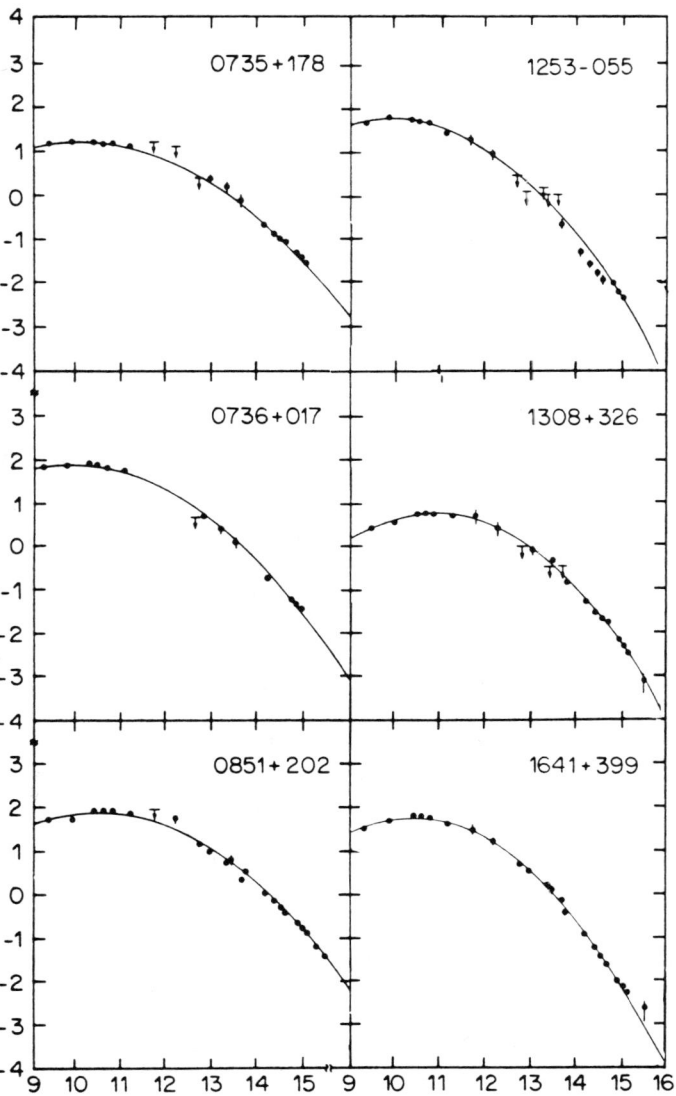

Figure 3 Broadband spectra of a sample of blazars with curves showing best fit parabolae ($\log S_\nu$ versus $\log \nu$ in Hz). Adapted from Landau et al. 1986.

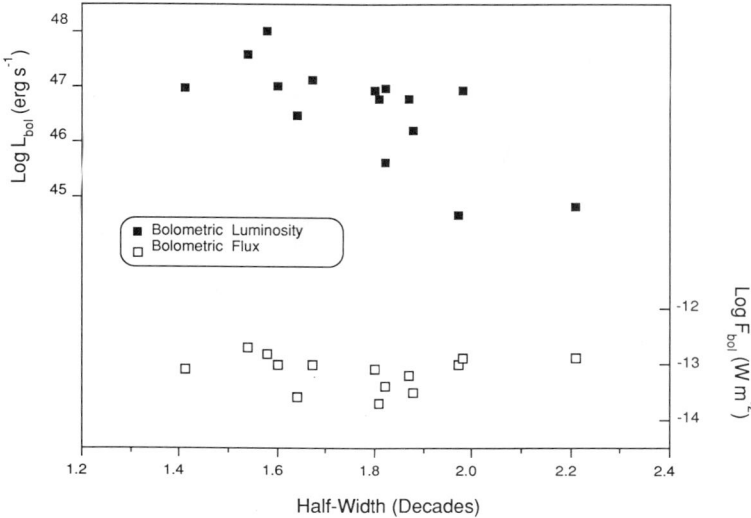

Figure 4 Correlations between spectral parameters for a sample of blazars. Note that a correlation is only apparent when the (physically relevant) luminosity is used, not the observed flux. Adapted from Jones, Rudnick, and Landau (1986).

blazars, the spectra can now be seen to be smoothly varying from the radio through the ultraviolet. Figure 2 shows this beautiful result for 3C 345 (Bregman et al. 1986).†

Broadband observations of a modest sample of blazars led Landau et al. (1986) to conclude that not only were the fluxes smoothly varying, but so were the local spectral slopes; *parabolae*, the simplest second-order curves, provided a surprisingly good fit to the global spectra (Figure 3). Deviations from the parabolae were $\sim 10\%$, representing real departures from the parabolic shape as well as measurement errors and variability. Broadband parabolic spectra can arise from parabolic electron energy distributions, or from standard energy gain/loss equilibria, combined with opacity effects starting in the submillimeter region. Landau et al. found no evidence for "breaks" in the spectra, nor are any expected theoretically, even for sharply cut off electron populations. In the rest-frame, the peak luminosity of blazars (νF_ν) always occurs in the infrared ($13 < \log \nu_{\rm peak} < 14.5$).

As striking as the simple parabolic nature of the spectra was, even more surprising was the appearance of a correlation between the fit parameters for the active sources (Figure 4; see also Figure 3 of Landau et al. 1986). This strongly suggests that there is a common regulat-

---

† The X-ray fluxes and slopes typically don't fit a simple extrapolation—so I'll ignore them! But see, e.g., Worrall (1986) and Urry (1986).

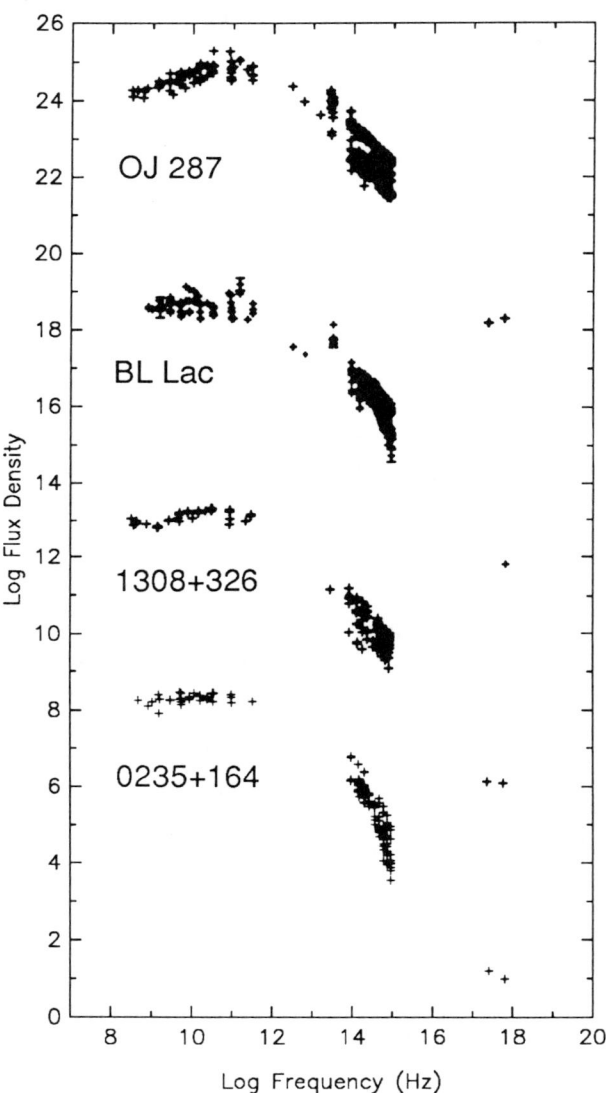

Figure 5 Broadband flux measurements of variable sources. Adapted from Kidger and Beckman (1985).

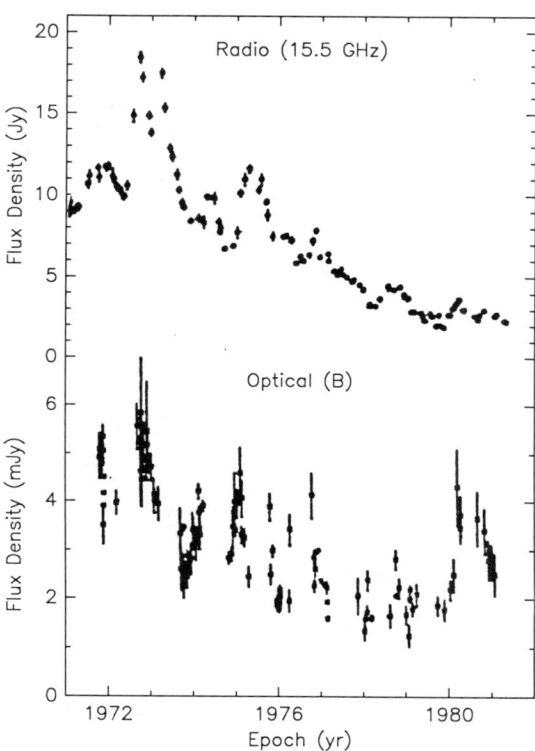

Figure 6 Optical and radio "light curves" for 3C 120, showing correlated outbursts. Adapted from Balonek (1986).

ing mechanism for the emission from all such sources, or that some of their properties (such as the detailed behavior of electron densities and magnetic fields with increasing core distance) are tightly constrained.

Variability

To the chagrin of people who have devoted their careers to studying variability, I believe that its most important characteristic is that it's *no big deal*. I mean this in a very serious way, namely, that variability causes only modest variations in the global shape of the spectrum (see also Bregman 1986). This is well illustrated in the compendium of measurements by Kidger and Beckman (1985) which have been adapted for Figure 5. The lesson from this is that the overall shapes of the spectra are of fundamental importance, and that any variability is governed by the same underlying structure.

Having acknowledged the importance of the spectral shape, there is still information to be gained by studying the details of variability.

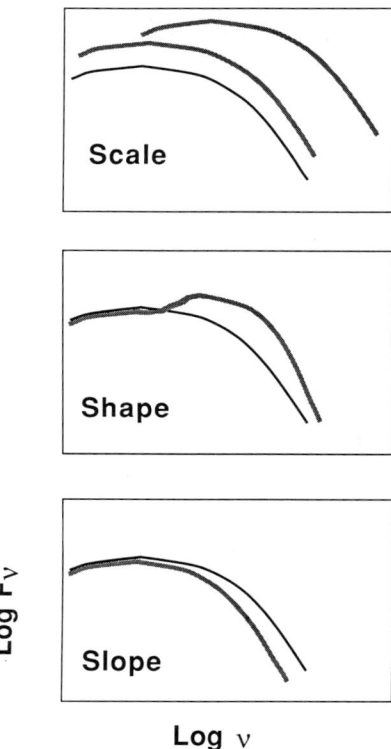

Figure 7 Schematic of types of variability seen in compact nuclei, as suggested by Jones et al. (1981).

The prototypical outburst (but you have to look hard for one!) starts at high frequencies, and is delayed, time-stretched, and attenuated at lower frequencies. It is somewhat depressing (except for the authors) to see that we still refer to the van der Laan (1966) and Pauliny-Toth and Kellermann (1968) expanding source models to explain the basic nature of these outbursts. More recent work, such as that of Marscher and Gear (1985) and Gear et al. (1986), has looked at extended injections of fresh particles and shock-driven particle accelerations (which they prefer) to explain the details of the spectral evolution. It would be interesting to know if a long-term average of diffusive shock acceleration events, as described by Smith (1986), would converge to the observed spectral shapes.

There are other sources or outbursts in which the fluctuations appear simultaneous over a large range of frequencies (e.g., Ennis, Neugebauer, and Werner 1982). Isolated correlations are also sometimes seen

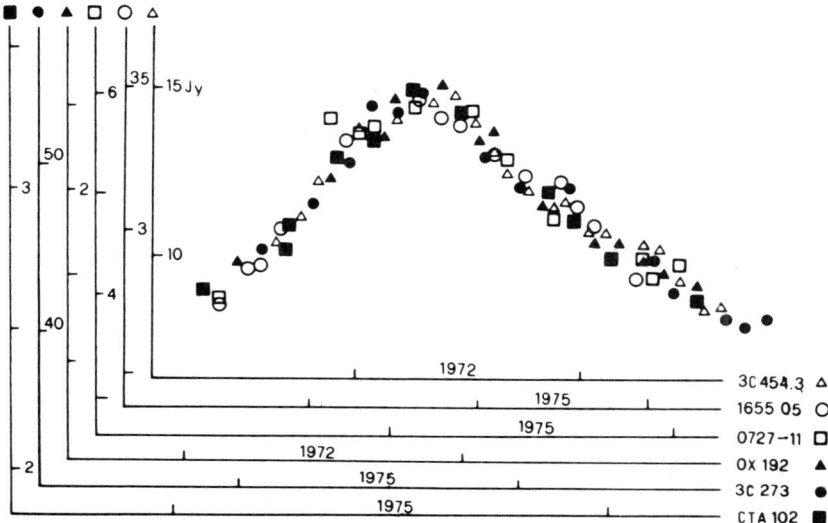

Figure 8 Burst profiles from various sources, scaled in amplitude and time scale. Taken from Legg (1984).

between variability at optical and radio wavelengths (Figure 6). This work is logistically very difficult, but helps provide an extreme example for outburst models. I am unaware of theoretical work on such variations. An insightful but sadly under-referenced synthesis of the types of broadband variability was given by Jones *et al.* (1981), and is illustrated in Figure 7.

One piece of work on variability which caught my attention recently is the remarkable constancy of the shape of outburst light curves (Figure 8), following $t^n e^{-t/\tau}$, documented by Legg (1984). Is anybody working on explaining this very pretty result?

Another big advance in variability studies is coming from the daily monitoring programs on the Green Bank interferometer by the Navy (Fiedler *et al.* 1987*b*), and from observations of short time-scale "flickering" (Simonetti, Cordes, and Heeschen 1985). The groups at the Naval Research Laboratory and the U.S. Naval Observatory observe intrinsic variations showing a time delay at lower frequencies, evidence for interstellar scintillation, and a new exciting class of "dropouts" which they interpret as occultations by a new class of interstellar cloud (Fiedler *et al.* 1987*a*). The entire study of low-frequency variability is opening up new tools for studying the interstellar medium, and I happily defer to Roberto Fanti for a fuller discussion (this Workshop, page 200).

Polarization

The time-averaged and fluctuating polarization properties of blazars and other active nuclei have provided a number of interesting insights into their structures. Radio polarizations are typically 1–10%, with fluctuations on time scales comparable (but sometimes shorter than) the total intensities. A whole zoo of variations has been observed, including large fluctuations in total intensity, fractional polarization, and polarization position angle which occur separately and in all possible combinations (Aller, Aller, and Hodge 1981; O'Dea et al. 1986)

One "characteristic" type of variation shows large jumps in position angle (70°–90°). This could be due to a change in opacity (Jones and O'Dell 1977; O'Dea et al. 1982) or to the changing relative contributions of two orthogonally polarized components (Komesaroff et al. 1984). We can look forward to the VLBI polarization studies that will follow up on this interesting behavior.

Rotation measure studies (Wardle 1977; Rudnick and Jones 1983) have demonstrated the extremely low populations of non-relativistic electrons in the immediate neighborhood of active nuclei, or, alternatively, unrealistically low fields, $B \leq 4 \times 10^{-7} \mu G$, in the broad-line regions. I feel (naturally!) that these results should be playing a more prominent role in constraining our models.

The underlying stochastic nature of blazars has been elucidated by polarization variations. Moore et al. (1982) and Moore, Schmidt, and West (1987) have described the variations in BL Lac's optical polarization as a "random walk" in (Q, U) space, resulting from the combination of about a hundred transient components. The multi-frequency, multi-epoch radio observations by Rudnick et al. (1985) have been interpreted as illustrating two types of polarization contributions (Jones et al. 1985). One is a steady ~ 2–3% contribution, with a magnetic field angle parallel to the VLBI axis; see Rusk and Seaquist (1985) for an even cleaner result. The second contribution, also ~ 2–3% in strength, appears random in both frequency and time. The combination of these two contributions usually yields a "characteristic" polarization angle for active sources, with variations most easily isolated in (Q, U) plots. One suggestion by Jones et al. is that observed polarization angle "rotation" events may simply be the statistical result of examining a large number of random walks. Such an explanation would be considerably more difficult for the amazing optical rotation and counter-rotation in BL Lac observed by Moore, Schmidt, and West (1987).

An important development in the interpretation of polarization observations is coming from consideration of turbulence in the flow (e.g., Jones 1986), perhaps coupled with weak shocks. The calculations show

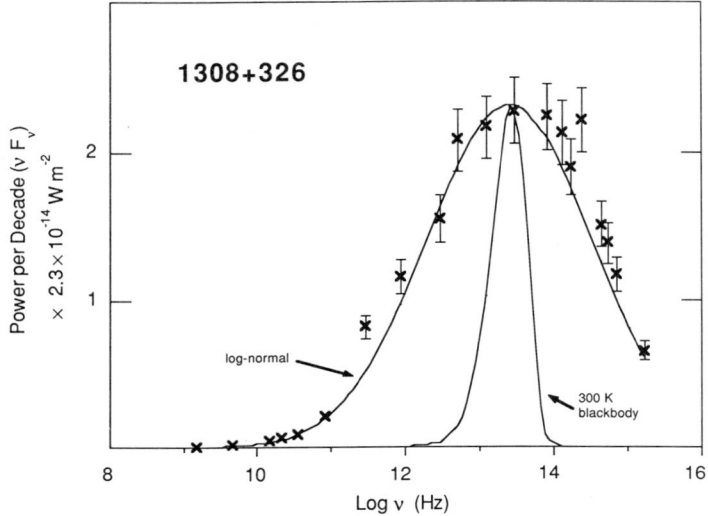

Figure 9 Another look at the "universal" (?) log-normal shape of blazar broadband spectra. Adapted from Jones, Rudnick, and Landau (1986).

that even weak flow enhancements, when Doppler boosted, may be the source of VLBI components. Hughes, Aller, and Aller (1985) are having very good success using shocks to model the polarization and total intensity variations in BL Lac (Aller, this Workshop, page 273). This line of investigation is worth keeping an eye on.

4. So What's Wrong?

After describing all of the information we have gained from spectral, variability and polarization studies, whence my pessimism? Let me share with you some

Nagging Doubts...

(a) It is easy to generate inhomogeneous models which give spectra that are smooth and gently curved, like the parabolae discussed above. But why do these inhomogeneous sources then appear to be dominated by only a few "components"? This *cosmic conspiracy* (Cotton et al. 1980) has received no attention at this Workshop, but it has not gone away.

(b) If the radio fluxes are Doppler boosted, how do they manage to connect smoothly to the infrared and optical emission, which presumably arise in spatially and dynamically distinct regions? Similarly, why do the optical luminosities, emission-line equivalent widths, and optical polarization behavior seem to exclude all possible fractional anisotropic contributions? (e.g., Moore and Stockman 1984).

(c) Is there really a simple universal form to the broadband spectra (see Figure 9 for another look), with perhaps only one parameter, such as luminosity, responsible for their differences? How do the sources contrive to be so similar?

(d) How can we get global, simultaneous variations in some sources, especially from radio to optical, when the emissions come from such different scales?

(e) Might the above difficulties be related to the sudden appearance or disappearance of VLBI components, or to the stationary–superluminal combinations such as 4C 39.25 (Shaffer et al. 1987)?

... and a Different Perspective?

At this point, I abandon all semblance of rational "scientific" argument, and share with you my "gut feel" about what may be going on. My strongest feeling is that there is some *unifying, regulating* process which is responsible for the simple broadband spectral shapes, the relations between sources, and the cosmic conspiracy. I believe that this process will *not* be found by starting from the detailed physics, but by appealing instead to more general fundamental laws. In this thinking, I am further influenced by the important role which seems to be played by stochastic phenomena. For example, I would not be surprised to find that all the things that we call "components" in VLBI are simply due to the "low pass filter" of our synthesized beams acting on a completely chaotic structure.

To what general principles might we appeal? The notion of *equilibrium* (e.g., hydrostatic and thermal) has served astrophysicists well in the past. Other successes have come from looking at *saturation* phenomena, such as the inverse Compton limit to brightness temperatures, or the Eddington limit. *Equipartition* and *entropy* have also led to understanding of diverse phenomena. *Noise* and *chaos* have played little role in astrophysics to date, but might prove very useful as statistical generators of log-normal spectra, variability, or pseudo-components.

I believe that we need to step back, and ask if any of these overall regulating mechanisms might be operable in compact radio sources. But whatever the answer we get, we'll sure have fun getting there.

I am pleased to acknowledge continuing contributions from Tom Jones and Jeff Pedelty, as well as Rob Landau, for insisting that we look at things the way they are, and not how we want them to be. None of these people, however, should be held accountable for the opinions expressed herein. At Minnesota, my work on active nuclei is supported by the National Science Foundation under grant AST 83-11973.

References

Aller, H. D., Aller, M. F., and Hodge, P. E. 1981, *Astron. J.*, **86**, 325.
Balonek, T. J. 1986, in *Continuum Emission in Active Galactic Nuclei*, ed. M. L. Sitko (Tucson: Kitt Peak National Observatory, National Optical Astronomy Observatories), p. 161.
Bregman, J. N. 1986, in *Continuum Emission in Active Galactic Nuclei*, ed. M. L. Sitko (Tucson: Kitt Peak National Observatory, National Optical Astronomy Observatories), p. 102.
Bregman, J. N., Glassgold, A. E., Huggins, P. J., Neugebauer, G., Soifer, B. T., Matthews, K., Elias, J., Webb, J., Pollock, J. T., Pica, A. J., Leacock, R. J., Smith, A. G., Aller, H. D., Aller, M. F., Hodge, P. E., Dent, W. A., Balonek, T. J., Barvainis, R. E., Roellig, T. P. L., Wiśniewski, W. Z., Rieke, G. H., Lebofsky, M. J., Wills, B. J., Wills, D., Ku, W. H.-M., Bregman, J. D., Witteborn, F. C., Lester, D. F., Impey, C. D., and Hackwell, J. A. 1986, *Astrophys. J.*, **301**, 708.
Cotton, W. D., Wittels, J. J., Shapiro, I. I., Marcaide, J., Owen, F. N., Spangler, S. R., Rius, A., Angulo, C., Clark, T. A., and Knight, C. A. 1980, *Astrophys. J. (Letters)*, **238**, L123.
Dyson, J. E. (ed.) 1985, *Active Galactic Nuclei* (Manchester: Manchester University Press).
Ennis, D. J., Neugebauer, G., and Werner, M. 1982, *Astrophys. J.*, **262**, 460.
Fiedler, R. L., Dennison, B., Johnston, K. J., and Hewish, A. 1987a, *Nature*, in press.
Fiedler, R. L., Waltman, E. B., Spencer, J. H., Johnston, K. J., Angerhoffer, P. E., Florkowski, D. R., Josties, F. J., Klepczynski, F. J., McCarthy, D. D., and Matsakis, D. N. 1987b, *Astrophys. J. Suppl.*, submitted.
Gear, W. K., Brown, L. M. J., Robson, E. I., Ade, P. A. R., Griffin, M. J., Smith, M. G., Nolt, I. G., Radostitz, J. V., Veeder, G., and Lebofsky, L. 1986, *Astrophys. J.*, **304**, 295.
Giuricin, G., Mardirossian, F., Mezzetti, M., and Ramella, M. (ed.) 1986, *Structure and Evolution of Active Galactic Nuclei* (Dordrecht: Reidel).
Harvey, P. M., Wilking, B. A., and Joy, M. 1982, *Astrophys. J. (Letters)*, **254**, L29.
Hughes, P. A., Aller, H. D., and Aller, M. F. 1985, *Astrophys. J.*, **298**, 301.
Jones, T. W. 1986, *Can. J. Phys.*, **64**, 463.
Jones, T. W. (ed.) 1986, "The Minnesota Lectures on Active Galactic Nuclei", *Publ. Astron. Soc. Pacific*, **98**, 129.
Jones, T. W., and O'Dell, S. L. 1977, *Astrophys. J.*, **214**, 522.
Jones, T. W., Rudnick, L., Aller, H. D., Aller, M. F., Hodge, P. E., and Fiedler, R. L. 1985, *Astrophys. J.*, **290**, 627.
Jones, T. W., Rudnick, L., and Landau, R. 1986, in *Continuum Emission in Active Galactic Nuclei*, ed. M. L. Sitko (Tucson: Kitt Peak National Observatory, National Optical Astronomy Observatories), p. 122.
Jones, T. W., Rudnick, L., Owen, F. N., Puschell, J. J., Ennis, D. J., and Werner, M. W. 1981, *Astrophys. J.*, **243**, 97.
Kidger, M. R., and Beckman, J. E. 1985, in *Structure and Evolution of Active Galactic Nuclei*, ed. G. Giuricin, F. Mardirossian, M. Mezzetti, and M. Ramella (Dordrecht: Reidel), p. 591.
Komesaroff, M. M., Roberts, J. A., Milne, D. K., Rayner, P. T., Cooke, D. J. 1984, *Monthly Notices Roy. Astron. Soc.*, **208**, 409.
Landau, R., Golisch, B., Jones, T. J., Jones, T. W., Pedelty, J., Rudnick, L., Sitko, M. L., Kenney, J., Roellig, T., Salonen, E., Urpo, S., Schmidt, G., Neugebauer, G., Matthews, K., Elias, J. H., Impey, C., Clegg, P., and Harris, S. 1986, *Astrophys. J.*, **308**, 78.
Landau, R., Jones, T. W., Epstein, E. E., Neugebauer, G., Soifer, B. T., Werner, M. W., Puschell, J. J., and Balonek, T. J. 1983, *Astrophys. J.*, **268**, 68.
Ledden, J. E., and O'Dell, S. L. 1985, *Astrophys. J.*, **298**, 630.

Legg, T. H. 1984, in *IAU Symposium 110, VLBI and Compact Radio Sources*, ed. R. Fanti, K. Kellermann, and G. Setti (Dordrecht: Reidel), p. 183.
Marscher, A. P., and Gear, W. K. 1985, *Astrophys. J.*, **298**, 114.
Miller, J. S. (ed.) 1985, *Astrophysics of Active Galaxies and Quasi-Stellar Objects*, (Mill Valley, CA: University Science Books).
Moore, R. L., McGraw, J. T., Angel, J. R. P., Duerr, R., Lebofsky, M. J., Rieke, G. H., Wiśniewski, W. Z., Axon, D. J., Bailey, J., Hough, J. M., Thompson, I., Breger, M., Schulz, H., Clayton, G. C., Martin, P. G., Miller, J. S., Schmidt, G. D., Africano, J., and Miller, H. R. 1982, *Astrophys. J.*, **260**, 415.
Moore, R. L., Schmidt, G. D., and West, S. C. 1987, *Astrophys. J.*, **314**, 176.
Moore, R. L., and Stockman, H. S. 1984, *Astrophys. J.*, **279**, 465.
Neugebauer, G., Miley, G. K., Soifer, B. T., and Clegg, P. E. 1986, *Astrophys. J.*, **308**, 815.
O'Dea, C. P., Balonek, T. J., Dent, W. A., and Kapitzky, J. E. 1982, in *Low Frequency Variability of Extragalactic Radio Sources*, ed. W. D. Cotton and S. R. Spangler (Green Bank: National Radio Astronomy Observatory), p. 115.
O'Dea, C. P., Dent, W. A., Kinzel, W. M., and Balonek, T. J. 1986, *Astron. J.*, **92**, 1262.
Oke, J. B., Neugebauer, G., and Becklin, E. E. 1970, *Astrophys. J.*, **179**, 1.
Pauliny-Toth, I. I. K., and Kellermann, K. I. 1968, *Astrophys. J. (Letters)*, **152**, L169.
Phillips, R. B., and Mutel, R. L. 1982, *Astrophys. J. (Letters)*, **257**, L19.
Phinney, E. S. 1985, in *Astrophysics of Active Galaxies and Quasi-Stellar Objects*, ed. J. S. Miller (Mill Valley: University Science Books), p. 453.
Rieke, G. H., and Lebofsky, M. J. 1979, *Ann. Rev. Astron. Astrophys.*, **17**, 477.
Rudnick, L., and Jones, T. W. 1982, *Astrophys. J.*, **255**, 39.
Rudnick, L., and Jones, T. W. 1983, *Astron. J.*, **88**, 518.
Rudnick, L., Jones, T. W., Aller, H. D., Aller, M. F., Hodge, P. E., Owen, F. N., Fiedler, R. L., Puschell, J. J., and Bignell, R. C. 1985, *Astrophys. J. Suppl.*, **57**, 693.
Rudnick, L., Jones, T. W., and Fiedler, R. 1986, *Astron. J.*, **91**, 1011.
Rudnick, L., Owen, F. N., Jones, T. W., Puschell, J. J., and Stein, W. A. 1978, *Astrophys. J. (Letters)*, **225**, L5.
Rudnick, L., Sitko, M. L., and Stein, W. A. 1984, *Astron. J.*, **89**, 753.
Rusk, R., and Seaquist, E. R. 1985, *Astron. J.*, **90**, 30.
Saikia, D. J., Swarup, G., and Kodali, P. D. 1985, *Monthly Notices Roy. Astron. Soc.*, **216**, 385.
Shaffer, D. B., Marscher, A. P., Marcaide, J., and Romney, J. D. 1987, *Astrophys. J. (Letters)*, **314**, L1.
Simonetti, J. H., Cordes, J. M., and Heeschen, D. S. 1985, *Astrophys. J.*, **296**, 46.
Sitko, M. L. (ed.) 1986, *Continuum Emission in Active Galactic Nuclei* (Tucson: Kitt Peak National Observatory, National Optical Astronomy Observatories).
Smith, D. R. 1986, in *Continuum Emission in Active Galactic Nuclei*, ed. M. L. Sitko (Tucson: Kitt Peak National Observatory, National Optical Astronomy Observatories), p. 111.
Stocke, J. T., Liebert, J., Schmidt, G., Gioia, I. M., Maccacaro, T., Schild, R. E., Maccagni, D., and Arp, H. C. 1985, *Astrophys. J.*, **298**, 619.
Ulvestad, J. S., and Antonucci, R. R. J. 1986, *Astron. J.*, **92**, 6.
Urry, C. M. 1986, in *Continuum Emission in Active Galactic Nuclei*, ed. M. L. Sitko (Tucson: Kitt Peak National Observatory, National Optical Astronomy Observatories), p. 91.
van der Laan, H. 1966, *Nature*, **211**, 1131.
Wardle, J. F. C. 1977, *Nature*, **269**, 563.
Wiita, P. J. 1985, *Physics Reports*, **123**, 117.
Worrall, D. M. 1986, in *Continuum Emission in Active Galactic Nuclei*, ed. M. L. Sitko (Tucson: Kitt Peak National Observatory, National Optical Astronomy Observatories), p. 97.

Infrared, Optical, UV, and X-ray Properties of Superluminal Radio Sources

CHRIS IMPEY

Steward Observatory, University of Arizona

1. Blazars: Faster than Light?

Most superluminal radio sources show high polarization and rapid variability at optical and infrared wavelengths. These properties are typical of blazars: the class of sources that includes BL Lac objects, highly polarized quasars (HPQs), and optically violently variable quasars (OVVs) (Angel and Stockman 1980). For the definition of this class, the strength of emission lines is of secondary importance for several reasons: (a) The lines emerge from a region $\sim 10^4$ times larger than the variable core. (b) There are sources which can look like either BL Lac objects or strong emission-line quasars, depending on the relative strength of lines and continuum. (c) The continuum properties of blazars consistently indicate the dominance of an extremely compact, active, synchrotron component. Blazars show degrees of linear polarization of up to 46%, and variability of up to a factor of 500 and on time scales as short as a day. In contrast, most quasars are not highly polarized and do not vary with large amplitudes or on short time scales (Figure 1). However, blazars do not form the tail of the distribution of quasar polarization and variability. Figure 2 shows that polarization, variability amplitude, and variability time scale are related: if a sample was selected by high polarization only, the *same sources* would also have the largest amplitude and shortest time scales of variability.

Current lists of probable and possible superluminal radio sources are drawn together in Tables 1, 2, and 3. Where available, ranges of optical polarization and position angle are included (however, many sources have only one measurement). It is striking that 13 out of 21 of the probable and 9 out of 10 of the possible superluminal sources are highly polarized in the optical ($p > 2.5\%$). At least 70% of the superluminal sources are identified with blazars. This is a lower limit, since the variable polarization in blazars can fall below 2–3%.

Even stubborn sources eventually reveal blazar properties. Until recently, 3C 273 was a prominent exception to the connection between superluminals and blazars. It is one of the classic superluminal sources

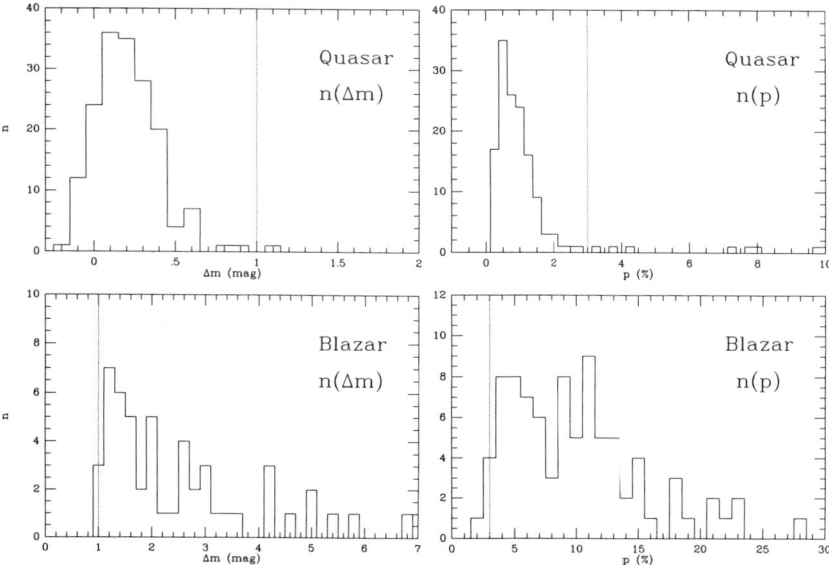

Figure 1 Histograms of optical variability amplitudes Δm and linear polarization p for quasars and blazars; note the different scales on the x-axes.

with a good kinematic model, but it has always shown low polarization ($\sim 0.5\%$) and quiescent optical and infrared fluxes. However, high accuracy polarimetry shows a "mini-blazar" component with all the standard properties: polarization $\sim 25\%$, night-to-night variability, and polarization position-angle rotations (Impey, Malkan, and Tapia, in preparation). This blazar component contributes less than 5% of the optical flux of 3C 273, so without high accuracy polarimetry it had escaped detection.

It is possible that *every* compact radio source will eventually be found to have both apparent superluminal motion and blazar properties. This becomes evident from studying Marshall Cohen's list of the strongest sources in the sky at 10 GHz, which takes account of historical variability (Cohen 1986). In the absence of an all-sky survey at such a high radio frequency, this is not a complete sample of strong compact radio sources, but it should be representative. Over 80% of the sources in this list are probably superluminal and the fraction is increasing. As the sensitivity of VLBI experiments increases, even relatively weak cores are found to be superluminal (e.g., Zensus and Porcas 1984). On the other hand, the detection of blazar properties is intimately connected with the presence of a compact radio source. Optical searches for blazars have been unsuccessful (Impey and Brand 1982; Borra and Corriveau 1984). However, in a recent optical polarization study of a complete sample

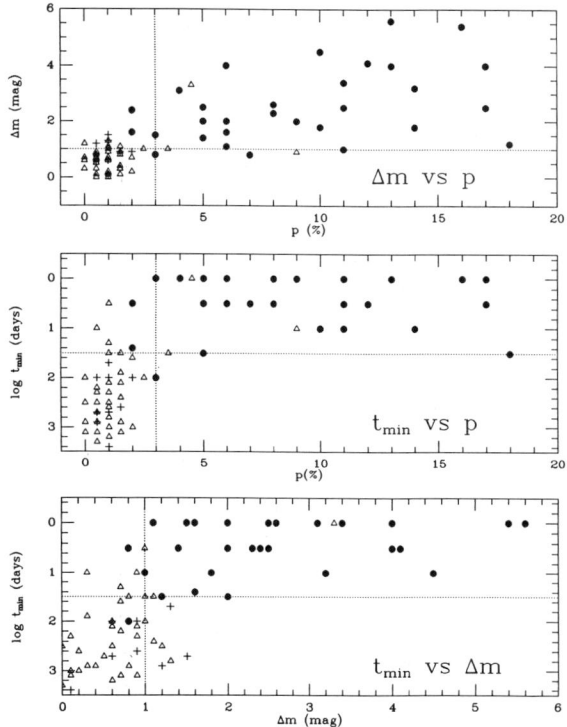

Figure 2 The relationship between high polarization, rapid optical variability, and large amplitude variability. Note that the correlations are poor because of variability in each parameter in each quasar. *Filled circles:* blazars and quasars; *open triangles:* optically selected QSOs. *References:* for polarization, Stockman *et al.* (1984) for QSOs, Angel and Stockman (1980) and Impey and Tapia (in preparation) for blazars; for variability, Bònoli *et al.* (1979) for QSOs, Angel and Stockman (1980) and Moore and Stockman (1984) for blazars.

of radio quasars ($S_{5 \text{ GHz}} > 1$ Jy), at least 40% of the sources show a polarization degree $p > 3\%$, and that fraction will increase as repeat measurements are made (Impey and Tapia, in preparation). The radio sources with high optical polarization are those with the flattest spectra from 2.7 to 5 GHz, once again indicating an association of compact radio sources with blazar activity.

Severe problems arise in the interpretation of blazar properties if the emission is assumed to be isotropic and nonrelativistic: the infrared luminosities often exceed the Eddington limit, the variability time scales are too short for standard models of accretion onto a gravitational engine (Impey *et al.* 1982), and the X-ray flux densities are often orders of magnitude below the prediction of the synchrotron-self-Compton model

Table 1 Superluminal Radio Sources

Source	Name	z	$S_{10\text{ GHz}}$ (Jy)	θ_{VLBI} (deg)	v_{app}/c	p_{opt} (%)	θ_{opt} (deg)
0212+735		2.367	2.3	103	3.9	8	97
0235+164		0.851	7.4	20	~ 30	6–43	15–175
0333+321	NRAO 140	1.258	2.9	134	4.8	1.0	~ 150
0430+052	3C 120	0.033	16.8	63	2.1–4.1	2.6	101
0723+679	3C 179	0.846	0.3	57	4.8	1.5	60
0735+178		0.424	2.7	45	2.8	3–35	0–175
0850+581	4C 58.17	1.322	1.2	154	3.9	...	...
0851+202	OJ 287	0.306	3.5	90	3.3	1–32	0–180
0906+430	3C 216	0.669	1.4	155	2.4	3–21	70–160
0923+392	4C 39.25	0.699	11.9	270	3.5	< 1	~ 102
1040+123	3C 245	1.029	1.0	105	3.1	< 0.5	...
1137+660	3C 263	0.656	0.6	110	1.3	< 1	~ 97
1150+812		1.25	1.1	178	4.1	...	...
1226+023	3C 273	0.158	60.0	50	5.1–8.0	0.5	45–65
1253−055	3C 279	0.538	20.3	50	2.0	4–19	10–180
1641+399	3C 345	0.595	16.3	45–105	1.4–9.5	2–16	10–170
1642+690	4C 69.21	0.751	3.1	196	7.9	17	8
1721+343	4C 34.47	0.206	~ 0.8	343	3.1	1.1	145
1901+319	3C 395	0.635	~ 1.7	118	15.5	...	...
1928+738	4C 73.18	0.302	3.0	162	7.0	0.8–1.2	145–163
1951+498		0.466	~ 0.2	90	1.3	...	...
2007+777		...	1.2	263	> 1	15	87
2200+420	BL Lac	0.070	16.4	3	2.4	2–23	0–180
2230+114	CTA 102	1.037	2.6	146	18.4	1–11	100–170
2251+158	3C 454.3	0.859	27.4	130	8.8	0–16	0–170

Table 2 Possible Superluminal Radio Sources

Source	Name	z	$S_{10\text{ GHz}}$ (Jy)	θ_{VLBI} (deg)	v_{app}/c	p_{opt} (%)	θ_{opt} (deg)
0153+744		2.340	1.2	155	≤ 1.3	...	...
0224+671	4C 67.05	...	~ 1.1	1	...	...	...
0355+508	NRAO 150	...	...	...	...	...	...
0415+379	3C 111	0.049	...	...	...	...	...
0454+844		...	1.5	146	...	19	50
0538+498	3C 147	0.545	1.5	~ 30	...	1.4	~ 155
0605−085	OH−010	0.870	2.4	46	...	10	127
0615+820		0.71˙	0.9	181	1.1	...	...
0716+714		...	2.3	16	...	14–29	10–163
0836+710		2.16˙	1.7	220	6.2	...	...
1038+528	OL 564	0.678	...	342	~ 2	...	...
1039+811		1.26˙	0.8	292	≤ 2.5	...	...
1510−089		0.361	3.0	305	...	1–8	28–139
1749+701		...	1.4	315	...	4–12	86–125
1803+784		...	2.6	277	...	35	96
1807+698	3C 371	0.051	2.8	262	...	0–12	65–100
1845+797	3C 390.3	0.057	2.0	139	...	1–4	149–165
2223−052	3C 446	1.404	9.7	112	...	4–17	10–160

Table 3 Subluminal Radio Sources

Source	Name	z	$S_{10\text{ GHz}}$ (Jy)	θ_{VLBI} (deg)	v_{app}/c	p_{opt} (%)	θ_{opt} (deg)
0108+388	OC 314	0.323	0.9	241	≤ 1	...	...
0316+413	3C 84	0.017	65.0	12	0.19	1–6	100–160
0710+439	OI 417	0.517	1.1	345	≤ 1	...	...
0711+356	OI 318	1.620	0.7	158	< 1	...	...
1228+127	M 87	0.004	33.2	331	< 0.06	< 0.5	...
1322−427	Cen A	0.002	61.0	51	...	9	147
1607+268	CTD 93	...	0.7	...	< 1	...	...
1637+826	NGC 6251	0.023	~0.4	300	< 0.4	...	...
1957+405	Cyg A	0.057	...	284	...	...	...
2021+614	OW 637	0.223	2.0	213	< 0.6	...	...
2134+004	PHL 61	1.936	13.0	62	< 0.4	1.3	120

Table 4 Average Spectral Indices α in Seven Wavebands

Spectral Region (log ν)		Blazar	Quasar	QSO	Ref.
cm	9.00– 9.70	$+0.20 \pm 0.33$ (27)	-0.20 ± 0.57 (32)	$+0.05 \pm 0.30$ (12)	1
mm	9.70–10.48	-0.02 ± 0.46 (27)	-0.49 ± 0.46 (32)	...	1
IRAS	12.48–13.48	-0.96 ± 0.21 (24)	-0.86 ± 0.44 (19)	-0.62 ± 0.66 (17)	2
IR	13.48–14.48	-1.06 ± 0.06 (22)	-1.25 ± 0.28 (22)	-0.90 ± 0.11 (13)	3
Opt	14.48–14.78	-1.51 ± 0.10 (25)	-0.60 ± 0.05 (85)	-0.55 ± 0.49 (12)	4
UV	14.78–15.30	-1.56 ± 0.11 (25)	-0.99 ± 0.11 (21)	-0.73 ± 0.34 (7)	5
Xray	16.70–18.20	-1.20 ± 0.51 (12)	-0.59 ± 0.06 (17)	-1.00 ± 0.16 (13)	6

Note that slightly differing numbers of sources (noted in parentheses) are averaged in different wavelength ranges. To calculate the zero points between spectral zones and especially to bridge the gaps in the submillimeter and hard UV, simultaneous energy distributions from the above authors have been used. No correction to the rest frame by $1/(1+z)$ was made as a number of sources do not have redshifts. However, spectral shapes over the broad range of ten decades are not affected by this omission.

References: 1. Kühr et al. 1981; Condon et al. 1981; 2. Neugebauer et al. 1986; Impey and Neugebauer, in preparation; 3. Capps, Sitko, and Stein 1982; Neugebauer et al. 1979; Impey, unpublished; 4. Ghisellini et al. 1986; Richstone and Schmidt 1980; 5. Ghisellini et al. 1986; Kinney et al. 1986; Bechtold et al. 1984; 6. Madejski 1985; Wilkes and Elvis 1986.

(Madejski 1985; Urry 1984). Furthermore, all of the parameters of the incoherent synchrotron model are stretched by the most extreme blazars: the magnetic fields required to prevent excessive inverse-Compton X-radiation are high ($B \sim 10^{2-3}$ G); electron acceleration time scales are short ($t_{\text{loss}} \sim 1$–10 s); efficient reacceleration must be achieved (10^{3-4} reaccelerations within the emitting volume); and submillimeter brightness temperatures are high ($T_b \sim 10^{11-13}$ K). Each of these conflicts is eased or removed if there is relativistic motion near the line of sight (Doppler factor $\delta \sim 2$–10). While this evidence for relativistic motion from optical and infrared studies is model-dependent and indirect, it is consistent with the conclusions from VLBI observations.

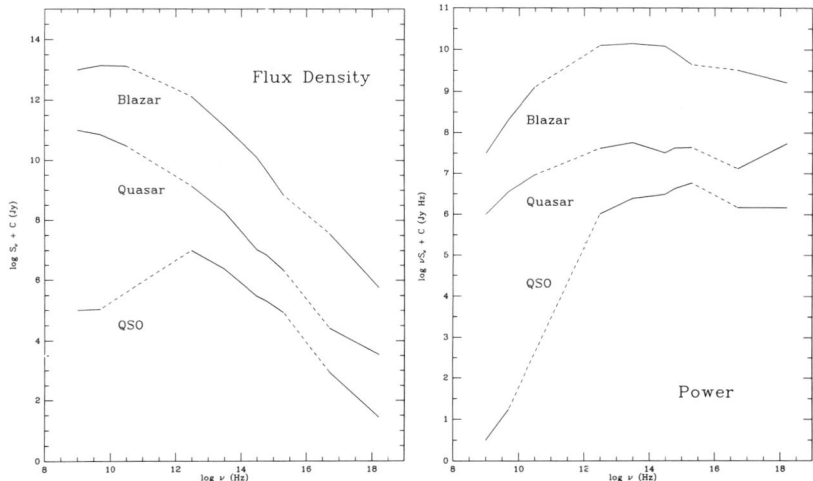

Figure 3 *Left:* spectral energy distributions for blazars, radio quasars, and optically selected QSOs. Mean spectral indices are shown. *Right:* the same presented in terms of flux per unit logarithmic bandwidth (νS_ν) for an indication of the energy budget of the different types of source.

2. Energy Distributions: 10^8 to 10^{18} Hz

Sufficient multifrequency data are available to compare the energy distributions of different types of active galactic nucleus. For ease of interpretation, it is preferable to omit low-luminosity objects where the contribution from starlight and dust may be substantial. One approach to the study of energy distributions is to combine coordinated, simultaneous observations at many frequencies into a snapshot of an individual source. This observationally demanding program has been carried out for around a dozen quasars and BL Lac objects (Bregman 1986; Worrall 1986). An alternative approach is to look at the mean energy distribution for different classes of object, recognizing that selection effects or incompleteness may affect each class. Luckily, there is remarkably good coverage of the superluminal radio sources from 10^8 to 10^{18} Hz, with detection rates of ~ 80% from the Einstein, IUE, and IRAS satellites. Is it coincidence that the superluminal sources have the appropriate energy distributions to *just* permit detection with different detector technologies over ten decades in wavelength?

Using published and unpublished data, mean spectral indices α ($S_\nu \propto \nu^\alpha$) in six spectral regions from 10^8 to 10^{18} Hz can be calculated (Table 4). The regions are divided according to the different detectors involved: cm, mm, IRAS, infrared, optical, IUE, and X-ray.

Figure 3 shows the energy distributions for blazars, quasars, and optically selected QSOs plotted in terms of flux density S_ν (a) and flux

Table 5 Comparison of Continuum Spectra of Different Classes

SL/Blazar	Quasar	QSO
weak UV bump	moderate UV bump	strong UV bump
1 component: mm–UV	1 component: mm–IR	1 component: IR–X-ray
power: 70% in IR	power : equal in IR, UV, X-ray	power: 50% in UV

per unit logarithmic bandwidth νS_ν (b). The blazar class of 25 sources includes 80% of the known and suspected superluminal sources. The quasars (22 sources) are radio selected, but show low-level optical polarization and variability. A compact radio component is generally far less prominent in these objects than in blazars. Finally, good multifrequency data exist for about a dozen optically selected QSOs. There are significant differences betweeen the three classes (differences at radio frequencies are of course expected due to selection biases): (a) The blazars are characterized by one smooth, continuously curving component from 10^9 to 10^{15} Hz. A parabola is as good a fit to these spectra as a series of power laws (Landau et al. 1986). The broad UV "bump", often associated with a hot thermal component, is weak or absent (major exceptions are 3C 273 and 3C 446). The X-rays are almost always much stronger than the extrapolation of the UV spectrum would indicate. About 70% of the total energy is emitted in the infrared between 1 and 100 μm. (b) The average radio quasar has a smoothly curving energy distribution from 10^9 to 10^{14} Hz, but the curvature is weaker than in the superluminal sources and the scatter in each spectral index is larger. The broad UV bump is considerably stronger in quasars than in blazars. The X-rays are also stronger, but have a spectrum which indicates that their origin is smoothly connected with the infrared component. The energy budget indicates roughly equal amounts of energy at IR, UV, and X-ray frequencies. (c) Optically selected QSOs show evidence for one power-law component from the infrared to the X-ray region, with a dramatic fall-off at long wavelengths. The broad UV bump is very strong, and on average contributes $\sim$ 50% of the energy output. These differences are summarized in Table 5 and illustrated in Figure 4 which shows mean spectral index plotted against frequency. It is striking how homogeneous the energy distributions of the blazars are over such an enormous energy range.

3. Unified Schemes

The simplest interpretation of the superluminal effect is based on relativistic motion near the line of sight. If the apparent speed is known, and the Doppler boosting factor is inferred, for example, from X-ray studies, a simple kinematic model for the emitting region is completely determined. Since the observed flux density $S_{\rm obs}$ is a strong function

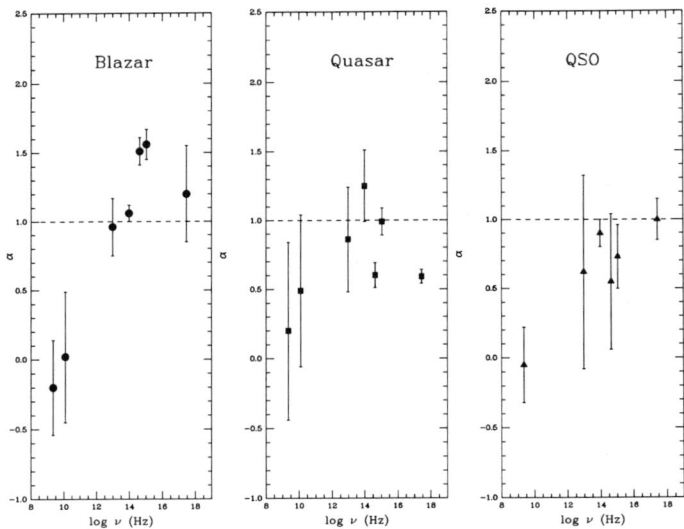

Figure 4 Spectral indices for the three classes of source are plotted as a function of frequency. Error bars are taken from Table 4, and scaled by $\sqrt{n}$ to a number of observations $n = 25$, to allow rough intercomparison between different frequency ranges.

of the Doppler factor δ, $S_{\text{obs}} = \delta^{3-\alpha} S_{\text{int}}$ (where S_{int} is the intrinsic flux density, and α is the spectral index), the effects of Doppler boosting can be dramatic. Phinney (1985) has pointed out that only 10^{-6} of the volume of a subrelativistic jet needs to be moving with Lorentz factor $\gamma \sim 12$ near the line of sight to outshine the entire jet. This kinematic model has several implications: (a) The relativistic flux enhancement is so strong that Doppler favoritism must be important. Beamed sources may not be rare in surveys selected by a beamed parameter, such as high-frequency radio flux. This is confirmed by the large fraction of superluminal sources (and blazars) in complete samples selected at 5 GHz. (b) The beamed sources may represent the intrinsically weakest sources in the parent population of misdirected beams. Urry and Shafer (1984) found that the luminosity functions of the beamed and misdirected sources can be very different. (c) Wills and Browne (1986) have shown that there is an orientation effect in the width of quasar emission lines, in that the upper envelope of the Hβ line widths increases as the ratio R of core-to-extended radio flux density increases. This result is consistent with different viewing angles onto a geometrically thin disc confining the gas motion near an active core.

A principal goal of the "unified schemes" (e.g., Orr and Browne 1982) is to identify the large population of sources with misaligned jets. The general criticism of these schemes is that the dependence of the

Table 6 A Simple Kinematic Model

γ	β	θ_c (deg)	$\delta_{max}/\delta_{min}$	x
1.2	0.5528	72.5	2.2	0.349
2	0.8660	54.7	7.5	0.211
5	0.9798	35.3	49.5	0.092
10	0.9950	25.2	199.5	0.048
20	0.9987	18.0	799.5	0.024
50	0.9998	11.4	4999.5	0.010

γ: Lorentz factor. β: dimensionless velocity: $\beta = (1-\gamma^{-2})^{1/2}$. θ_c: the maximum angle to the line of sight at which the Doppler factor $\delta = \gamma^{-1}(1-\beta\cos\theta)^{-1}$ exceeds 1. $\delta_{max}/\delta_{min}$: ratio of maximum to minimum Doppler factor for sources of given γ, $= 1/(1-\beta)$. x: fraction of sources in a randomly oriented sample for which $\delta > 1$; $x = (1 - \cos\theta_c)/2$.

model parameters on R or any other orientation-dependent parameter is usually too weak, given the strong Doppler favoritism expected. Another problem is the uncomfortably large angular sizes, after deprojection, of the large-scale structures in some superluminal sources (de Bruyn and Schilizzi 1984). However, we also know from the study of individual sources that a simple ballistic model must be an incomplete description. In 3C 345, which is presumably exceptional only in the high quality of its data, we see misalignments, intrinsic curvature, acceleration of a single component, and acceleration between successive components (Biretta, Moore, and Cohen 1986; Biretta and Cohen, this Workshop, page 40). Furthermore, there is evidence for intrinsically one-sided emission in at least 3C 273 (Davis 1986) and NGC 6251 (Jones, this Workshop, page 162). Therefore, models with a range in component Lorentz factor and ejection angle are needed. One should also consider mixtures of beamed and isotropic components (e.g., 3C 273); for example, assume a simple kinematic model with Lorentz factor $\gamma = (1 - \beta^2)^{-1/2}$, $\beta = v/c$, and Doppler factor $\delta = [\gamma(1-\beta\cos\theta)]^{-1}$ (where θ is the angle of the motion to the line of sight). Then given γ, we can calculate β, the angle θ_c within which $\delta > 1$, the ratio $\delta_{max}/\delta_{min} = (1+\beta)/(1-\beta)$, and the fraction x of sources with $\delta > 1$ in a randomly oriented sample (Table 6).

For a mixture of beamed and isotropic components, $S_{tot} = S_b + S_i$, and a ratio $f = S_b/S_i$, $\theta_c = \arccos[\beta^{-1}(1 - \gamma^{-1}f^{1/(2+\alpha)})]$. In Figure 5, x is plotted as a function of γ and f. As the beamed component is more heavily diluted by the isotropic component, the fraction of sources with $\delta > 1$ decreases. There is also a minimum Lorentz factor required before any net flux enhancement is seen. For $f < 0.03$, $\gamma > 3$ is required, and a maximum of 2% of the source population will show beaming effects, *regardless* of the distribution of γ. It is clear that the effects of Doppler favoritism can be muted with such a two-component model, and there is good evidence that this is the case in the classic superluminal sources.

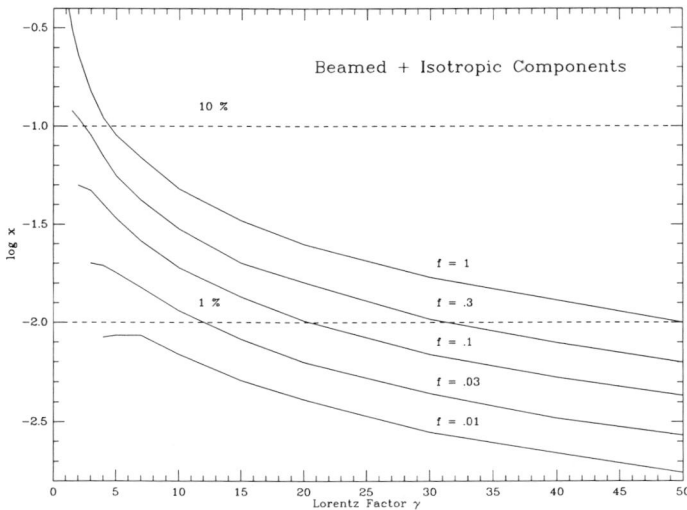

Figure 5 The fraction x of randomly oriented sources showing Doppler-enhanced flux (i.e., $\delta > 1$) versus the Lorentz factor γ, for different ratios f of beamed to isotropically emitted flux.

4. Emission Lines

Of the superluminal sources in Tables 1 and 2, roughly 75% have emission lines typical of normal quasars, while 25% have relatively weak lines (line-to-continuum ratio < 0.1). There are seven BL Lac objects in the list (0235+164, 0454+844, 0716+714, OJ 287, 1749+701, 1803+784, BL Lac), and three "transition" objects which can look like either a BL Lac object or an HPQ depending on the continuum strength (3C 279, 3C 371, 3C 446). Genuinely lineless objects are rare; 90% of the sources in Tables 1–3 have known redshifts, and for most BL Lac objects emission lines were found when they were observed with sufficiently good signal-to-noise ratio. Unfortunately, it is difficult to go beyond this broad classification of line strength. The published spectroscopy on strong radio sources has been acquired using many detector and spectrograph combinations, and the data cover a wide range of signal-to-noise ratio, and are often presented without absolute calibration. This situation is gradually being remedied (Lawrence et al., this Workshop, page 260). In principle, the distribution of line-to-continuum ratios is a good test of the degree of beaming of the optical continuum, since the broad-line region is dispersed at $\sim 10^{18}$ cm from the core with a covering factor of $\sim 10\%$ and should be immune from the effects of relativistic enhancement. Bregman et al. (1986) have argued that the variations of ultraviolet line and continuum intensity in 3C 446 rule out

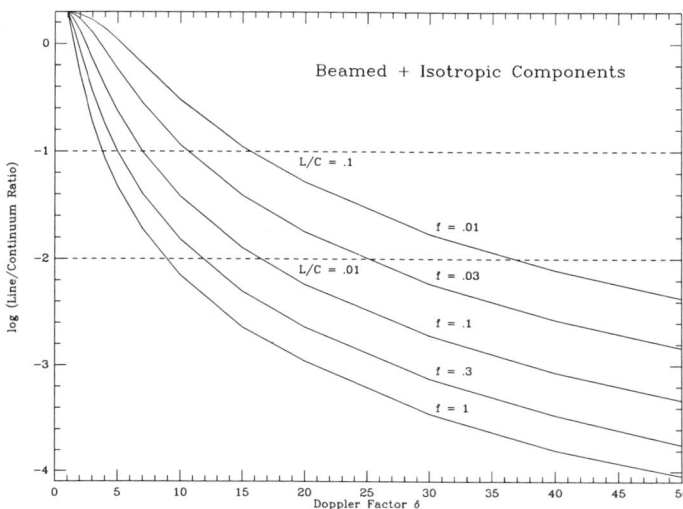

Figure 6 The observed line-to-continuum ratio is plotted as a function of the Doppler boosting factor for various values of f. The line-to-continuum ratio with no Doppler boosting is taken to be two. The dotted lines indicate the region occupied by BL Lac objects.

a large degree of beaming (more than a factor of two) of the ultraviolet continuum. Kinney et al. (1986) have made the same claim for a small set of quasars in which line and continuum fluxes are correlated with very small scatter. However, their sample does not include strong flat-spectrum radio sources like the superluminal objects.

It is straightforward to make the logical modification to the simple beaming picture and include an isotropic component. The dependence of the line-to-continuum ratio L/C on Doppler factor δ and ratio f of beamed to isotropic flux is shown in Figure 6, assuming that $L/C = 2$ for a quasar unaffected by relativistic motion, i.e., optically selected. (The same curves apply to the decrease of line equivalent width from a typical quasar value of 100 Å.) As the dilution becomes greater, L/C has a more gradual fall-off with increasing δ. This dilution would have a substantial effect on a spectroscopic classification based on line strength. The band between the dotted lines in Figure 6 is the region where an object would be labeled weak-lined or BL Lac (equivalent widths 0.5–5 Å). If the dilution is modest ($f \sim 1$), a δ of 5 implying a $\gamma \sim 2$–3 within θ_c is sufficient to yield a BL Lac spectrum. But if f is as low as 0.01, the implied γ for a BL Lac spectrum is 10-15, with only a small fraction of 3–5% of the parent population showing the effect. In the unified models for relativistic motion, a realistic scheme may have to include distributions of γ, ejection angle, and f.

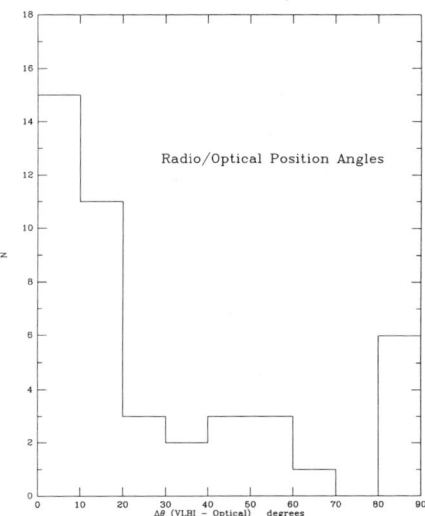

Figure 7 Histogram of differences between the *median* position angle of optical polarization and the position angle of smallest VLBI structure (seven BL Lac objects have been omitted because they show no preferred position angle).

The range of the ratio f in strong radio sources can be constrained with spectroscopic and optical continuum monitoring of complete samples. There is already an indication that low polarization sources have a larger ratio of L_{opt}/L_{rad} than high polarization sources. This difference may well be ascribed to an isotropic component which dominates in the low polarization sources. The distribution of γ will emerge from VLBI studies of complete radio samples (e.g., Pearson and Readhead 1984) with the bias that large values of γ near the line of sight may be under-represented owing to the small proper motions. In fact, the study of the proper motions of VLBI components allows another test of the beaming hypothesis (Blandford and Rees 1978; Angel and Stockman 1980). Sources with small f will show low polarization, but for higher values of f, Doppler boosting permits the beamed, polarized component to dominate the flux density. Within the emission cone ($\theta < \theta_c$), the observed proper motion of the components will be large and will increase for sources with larger γ. However, for sources with very small angles to the line of sight ($\theta \sim 0°$), the motion will be small. The statistics are still very limited, but the numbers given in Table 7 suggest that this is a useful framework for the discussion of future data. The high polarization objects represent beamed sources with systematically higher values of β_{app}. However, *within* the set of beamed sources are the BL Lacs, where the lines have been swamped by beamed continuum. First indications

Table 7 Optical Emission Lines and Superluminal Motion

Source Type	β_{app}	Interpretation
Quasar (low pol., strong lines)	0/11 with $\beta_{app} > 8$	Isotropic compt. dominates (large θ or low γ)
Blazar (high pol., strong/weak lines)	7/11 with $\beta_{app} > 8$	Beamed compt. dominates (small θ, high γ)
BL Lac (high pol., weak lines)	3/4 with $\beta_{app} < 4$	Beamed compt. dominates (very small θ, high γ)

are that they do indeed show smaller proper motions than strong-lined polarized sources, consistent with a pole-on orientation.

5. Polarization: Alignment in the Core?

The beauty of the VLBI maps is their direct indication of the geometry in the radio source near the active core (subject to deprojection uncertainties). Continuum studies at other wavelengths can offer no such structural information; the best resolution of ground-based optical data is a thousand times worse than that of the best high-frequency VLBI maps†.

Most superluminal sources show high optical polarization. Is there a geometric connection between the VLBI structural axis and the plane of polarization? Such a connection has been indicated by Ulrich (1984), Biermann et al. (1981), and most recently by Rusk and Seaquist (1985) who added recent VLBI data and presented a good discussion of the radio and optical data. The distribution of the difference between polarization position angle and VLBI structural axis is strongly peaked at zero. Charles Lawrence, Santiago Tapia, and I have been trying to confirm this result systematically by measuring optical polarization for the complete sample of Pearson and Readhead (1984). Figure 7 shows the histogram of position-angle differences for all radio sources with VLBI position angles and optical polarimetry. The size of the sample previously considered is doubled by the addition of data from the literature and preliminary results from the Pearson-Readhead sample. The peak at zero position-angle difference has been strengthened by the increased sample size, but there is now also the hint of a second peak at 90°. For a source to be included in the histogram it had to have (a) a VLBI position angle for the milliarcsecond structure, and (b) a single polarization measurement with $p/\sigma(p) > 3$, or a preferred position angle in numerous

---

† Most superluminal sources are variable at optical and infrared wavelengths, and the light travel times for the fastest variations imply that the size of the emitting volume is 1–5×10^{15} cm (the intrinsic source size must be reduced by δ if there is relativistic motion towards the observer). Since it is the core that has the most violent flux variations in VLBI data, the implication is that the optical and infrared flux is emerging from even closer to the core than the VLBI components.

polarization measures. Objects with no preferred polarization position angle—those having less than two-thirds of the measurements within a 40° range—were excluded. The strong alignment in Figure 7 is remarkable, given that the average error in the VLBI position angle was 3°–5° and that of the polarization position angle was 10°–15°. In future, a similar comparison may be possible with the VLBI polarimetry that is now becoming available (Roberts and Wardle, this Workshop, page 193). Most of the width of the peak at zero can be attributed to observational errors. In addition, if the radio and optical emitting volumes are not coincident, intrinsic curvature could smear out $\Delta\theta$; the low polarization objects with $p/\sigma(p) > 3$ could suffer from contamination by low levels (0.2–0.4%) of interstellar polarization; and, if there is relativistic ejection near the line of sight, aberration effects could systematically affect $\Delta\theta$. However, this effect has been minimized by excluding seven classic BL Lac objects which do not have preferred position angles. Overall, there must be a near perfect alignment between VLBI structure axis and polarization position angle.

An interpretation of this alignment depends on the assumed emission mechanism for the polarized flux. If the polarization is imprinted on the continuum by scattering outside the core, there are at least two possible mechanisms. (a) The easiest to rule out is scattering by dust grains. Any simple dust cloud model has extreme difficulty in generating the required infrared luminosities (10^{46}–10^{48} erg s^{-1}) within the small volume defined by variability and light-travel-time arguments. Also, producing high degrees of polarization (20–45%) requires a severe asymmetry of the dust distribution, and that further increases the energy problems. Also, the wavelength dependence of polarization in the optical and infrared is too weak for small-grain scattering. (b) Another possibility is electron scattering by nonspherical clouds. If the radio axis is aligned by rotation, then it is likely that any electron-scattering cloud would be elongated perpendicular to the radio axis. Therefore a value of $\Delta\theta = 0°$ would represent an optically thin scattering cloud, and $\Delta\theta = 90°$ would represent the optically thick case. Electron scattering is a possible choice only for the low polarization sources. For blazars it is virtually certain that the emission mechanism from millimeter to ultraviolet wavelengths is the optically thin incoherent synchrotron process. Consequently, $\Delta\theta = 0°$ implies that the projected component of the magnetic field is perpendicular to the VLBI structure axis.

There are two useful ways to subdivide the distribution of $\Delta\theta$. One is to divide by redshift as in Figure 8a, and it is clear that the strong peak at $\Delta\theta = 0°$ consists mostly of high luminosity sources ($z > 0.5$). The distribution for $z < 0.5$ is flat, and includes low luminosity radio galaxies such as 3C 120 and 3C 390.3 where dust emission is likely to be impor-

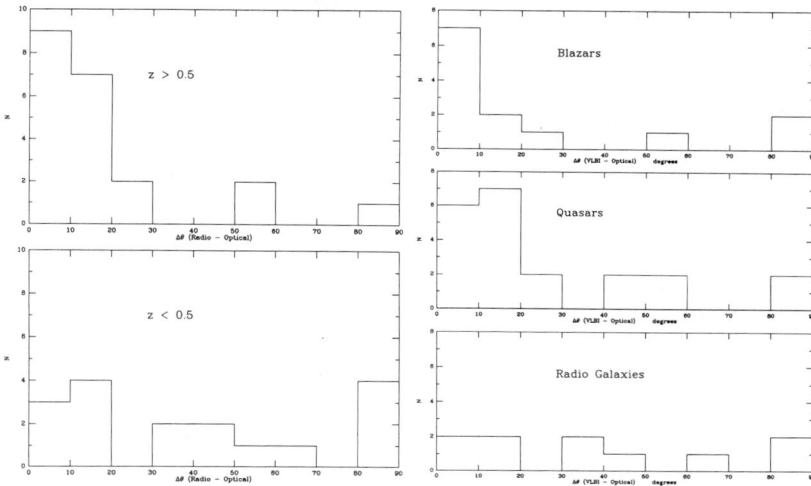

Figure 8 *Left:* the histogram of Figure 7 subdivided by redshift. Most of the sources with $\Delta\theta > 30°$ have redshifts less than 0.1. *Right:* the same histogram subdivided according to optical characteristics into blazars, low polarization quasars, and radio galaxies.

tant. Antonucci (1984) pointed out that thermally dominated Seyfert-2 galaxies cluster around $\Delta\theta = 90°$. For low luminosity sources, disentangling the effects of different emission mechanisms will be difficult. Figure 8*b* shows another breakdown of the sample. The lowest panel includes all objects with $z < 0.06$, i.e., radio galaxies where starlight is a prominent contributor to the optical continuum. These objects show no clear relationship between radio axis and polarization axis. The top two panels include objects of higher luminosity ($L_{\mathrm{opt}} > 10^{44}$ W) divided into low-polarization quasars and high-polarization blazars. The key result is that *both* classes show an equally strong alignment of VLBI structure axis and polarization position angle. This may be the first good evidence that the low (1–2%) polarization in most radio quasars is due to the synchrotron process, suggesting that we can associate both the low quasar polarization and the high blazar polarization with collimation and ejection of VLBI components.

6. Cautionary Remarks

A selection effect that is often neglected is the role of the optical counterpart of the strong radio source. Obviously, the type of galaxy within which a superluminal source is embedded affects issues such as the shape of the gravitational potential, the long-term fuel supply, and the quantities of gas and dust in the nucleus. There is often a complicated interaction between the active galactic nucleus and the host galaxy (Balick and Heckman 1982). One-sided radio emission is generally interpreted as either a beaming effect or intrinsic one-sided emission. However, the possibility that dense thermal material contains and quenches the radio emission must also be considered. Only one in 10^9 elliptical galaxies evolves into a strong 3C source, so the evolutionary path of those galaxies may be highly unusual. The space density of superluminal sources is so low ($\sim 10^{-11}$ Mpc^{-3}) that we do not have any nearby examples to study the effects of interaction. At lower luminosities (similar to 3C 273), there is a relationship between the radio properties of the nucleus and the type of host galaxy (Malkan 1984; Boroson and Oke 1984; Boroson, Persson, and Oke 1985). Therefore, the identification of sources as quasars or galaxies is important in statistical VLBI studies. There is a range of radio luminosities (10^{26-27} W Hz^{-1}, or $z \sim 0.3$–0.4) where the prominence of the nonthermal optical continuum can account for the difference between quasars and radio galaxies. It is also likely that the optical continuum is weakly beamed (Browne and Wright 1985). Therefore, complete radio samples may have orientation biases built into them by the type of identification chosen.

7. Prospects

Superluminal radio sources give us a unique insight into the kinematics of regions at the heart of active galactic nuclei. Studies at other wavelengths are an important part of the picture. Superluminal sources are distinctive enough that their non-radio properties can be used to identify them when VLBI is not available (for example, in the southern hemisphere). Any radio source with strong optical polarization and variability and a smoothly curved synchrotron spectrum from 10^{11} to 10^{15} Hz is a good bet to be a superluminal. The key to a statistical understanding of relativistic ejection will be studies of complete samples, selected either by compact or by extended emission. Multifrequency studies can add to the weight of data on the smooth energy distributions for superluminal sources. The narrow range of fundamental synchrotron parameters, such as turnover frequency and optically thin spectral index, for sources covering factors of thousands in luminosity must be important, but the factors that regulate sources to produce such repeatable energy spectra are unknown. The luminosity function of complete samples can be used to constrain a parent population, because the beamed and unbeamed population will have different slopes and space densities. Monitoring of the optical-infrared continua will reveal the relative amount of beamed and isotropic components. Homogeneous emission-line data will discriminate between the various unified schemes, for instance, via the dependence of line strength on parameters such as R. With larger samples, the distribution of $\beta_{\rm app}$ may show whether the beam is closest to the line of sight in BL Lac objects. Finally, the remarkably precise polarization connection should be explored more thoroughly, particularly with higher signal-to-noise ratio data on the low polarization objects and VLBI polarimetry. These alignments may be a key to understanding the collimation mechanism in the cores of active galactic nuclei.

I thank Charlie Lawrence, Santiago Tapia, Matt Malkan, and Gerry Neugebauer for permission to quote joint data in advance of publication. Discussions with Marshall Cohen, Ian Browne, and Tony Zensus have guided my thoughts on the properties of the superluminal sources. Part of this work was carried out under a Caltech Weingart Fellowship.

References

Angel, J. R. P., and Stockman, H. S. 1980, *Ann. Rev. Astron. Astrophys.*, **18**, 321.
Antonucci, R. R. J. 1984, *Astrophys. J.*, **278**, 499.
Balick, B., and Heckman, T. M. 1982, *Ann. Rev. Astron. Astrophys.*, **20**, 431.
Bechtold, J., Green, R. F., Weymann, R. J., Schmidt, M., Estabrook, F. B., Sherman, R. D., Wahlquist, H. D., and Heckman, T. M. 1984, *Astrophys. J.*, **281**, 76.
Biermann, P., Duerbeck, H., Eckart, A., Fricke, K., Johnston, K. J., Kühr, H., Liebert, J., Pauliny-Toth, I. I. K., Schleicher, H., Stockman, H., Strittmatter, P. A., and Witzel, A. 1981, *Astrophys. J. (Letters)*, **247**, L53.
Biretta, J. A., Moore, R. L., and Cohen, M. H. 1986, *Astrophys. J.*, **308**, 93.
Blandford, R. D., and Rees, M. J. 1978, in *Pittsburgh Conference on BL Lac Objects*, ed. A. M. Wolfe (Pittsburgh: University of Pittsburgh), p. 328.
Bònoli, F., Braccesi, A., Federici, L., Zitelli, V., and Formiggini, L. 1979, *Astron. Astrophys. Suppl.*, **35**, 391.
Boroson, T. A., and Oke, J. B. 1984, *Astrophys. J.*, **281**, 535.
Boroson, T. A., Persson, S. E., and Oke, J. B. 1985, *Astrophys. J.*, **293**, 120.
Borra, E. F., and Corriveau, G. 1984, *Astrophys. J.*, **276**, 449.
Bregman, J. N. 1986, in *Continuum Emission in Active Galactic Nuclei*, ed. M. L. Sitko (Tucson: Kitt Peak National Observatory, National Optical Astronomy Observatories), p. 102.
Bregman, J. N., Glassgold, A. E., Huggins, P. J., and Kinney, A. L. 1986, *Astrophys. J.*, **301**, 698.
Browne, I. W. A., and Wright, A. E. 1985, *Monthly Notices Roy. Astron. Soc.*, **213**, 97.
Capps, R. W., Sitko, M. L., and Stein, W. A. 1982, *Astrophys. J.*, **255**, 413.
Cohen, M. H. 1986, in *Highlights of Modern Astrophysics*, ed. S. L. Shapiro and S. A. Teukolsky (New York: Wiley), p. 299.
Condon, J. J., O'Dell, S. L., Puschell, J. J., and Stein, W. A. 1981, *Astrophys. J.*, **246**, 624.
Davis, R. J. 1986, in *IAU Symposium 119, Quasars*, ed. G. Swarup and V. K. Kapahi (Dordrecht: Reidel), p. 211.
de Bruyn, A. G., and Schilizzi, R. T. 1984, in *IAU Symposium 110, VLBI and Compact Radio Sources*, ed. R. Fanti, K. Kellermann, and G. Setti (Dordrecht: Reidel), p. 165.
Ghisellini, G., Maraschi, L., Tanzi, E. G., and Treves, A. 1986, *Astrophys. J.*, **310**, 317.
Impey, C. D., and Brand, P. W. J. L. 1982, *Monthly Notices Roy. Astron. Soc.*, **201**, 849.
Impey, C. D., Brand, P. W. J. L., Wolstencroft, R. D., and Williams, P. M. 1982, *Monthly Notices Roy. Astron. Soc.*, **200**, 19.
Kinney, A. L., Huggins, P. J., Bregman, J. N., and Glassgold, A. E. 1986, *Astrophys. J.*, **291**, 128.

Kühr, H., Witzel, A., Pauliny-Toth, I. I. K., and Nauber, U. 1981, *Astron. Astrophys. Suppl.*, **45**, 367.
Landau, R., Golisch, B., Jones, T. J., Jones, T. W., Pedelty, J., Rudnick, L., Sitko, M. L., Kenney, J., Roellig, T., Salonen, E., Urpo, S., Schmidt, G., Neugebauer, G., Matthews, K., Elias, J. H., Impey, C., Clegg, P., and Harris, S. 1986, *Astrophys. J.*, **308**, 78.
Madejski, G. M. 1985, Ph. D. thesis, Harvard University.
Malkan, M. A. 1984, *Astrophys. J.*, **287**, 555.
Moore, R. L., and Stockman, H. S. 1984, *Astrophys. J.*, **279**, 465.
Neugebauer, G., Miley, G., Soifer, B. T., and Clegg, P. E. 1986, *Astrophys. J.*, **308**, 815.
Neugebauer, G., Oke, J. B., Becklin, E. E., and Matthews, K. 1979, *Astrophys. J.*, **230**, 79.
Orr, M. J. L., and Browne, I. W. A. 1982, *Monthly Notices Roy. Astron. Soc.*, **200**, 1067.
Pearson, T. J., and Readhead, A. C. S. 1984, in *IAU Symposium 110, VLBI and Compact Radio Sources*, ed. R. Fanti, K. Kellermann, and G. Setti (Dordrecht: Reidel), p. 15.
Phinney, E. S. 1985, in *Astrophysics of Active Galaxies and Quasi-Stellar Objects*, ed. J. S. Miller (Mill Valley: University Science Books), p. 453.
Richstone, D. O., and Schmidt, M. 1980, *Astrophys. J.*, **235**, 361.
Rusk, R., and Seaquist, E. R. 1985, *Astron. J.*, **90**, 30.
Stockman, H. S., Moore, R. L., and Angel, J. R. P. 1984, *Astrophys. J.*, **279**, 485.
Ulrich, M.-H. 1984, in *IAU Symposium 110, VLBI and Compact Radio Sources*, ed. R. Fanti, K. Kellermann, and G. Setti (Dordrecht: Reidel), p. 73.
Urry, C. M. 1984, Ph. D. thesis, The Johns Hopkins University.
Urry, C. M., and Shafer, R. A. 1984, *Astrophys. J.*, **280**, 569.
Wilkes, B., and Elvis, M. 1986, in *Continuum Emission in Active Galactic Nuclei*, ed. M. L. Sitko (Tucson: Kitt Peak National Observatory, National Optical Astronomy Observatories), p. 56.
Wills, B. J., and Browne, I. W. A. 1986, *Astrophys. J.*, **302**, 56.
Worrall, D. M. 1986, in *Continuum Emission in Active Galactic Nuclei*, ed. M. L. Sitko (Tucson: Kitt Peak National Observatory, National Optical Astronomy Observatories), p. 97.
Zensus, J. A., and Porcas, R. W. 1984, in *IAU Symposium 110, VLBI and Compact Radio Sources*, ed. R. Fanti, K. Kellermann, and G. Setti (Dordrecht: Reidel), p. 163.

Superluminal Radio Sources: What Does X-ray Emission Tell Us?

DIANA M. WORRALL

Harvard-Smithsonian Center for Astrophysics

1. Introduction

X-ray emission has provided one of the strongest indicators of relativistic beaming in compact extragalactic radio sources. The argument that a static homogeneous synchrotron source would produce (by the inverse Compton mechanism) several orders of magnitude more X-ray luminosity than is observed has been used as evidence for relativistic beaming in more than 30 sources (e.g., Marscher et al. 1979; Madejski and Schwartz 1983; Madau, Ghisellini, and Persic 1987). X-ray observations always set a lower limit to the amount of beaming for a radio source of known size, but confirmation that the observed X-rays have a self-Compton origin is still required, as recently emphasized by Marscher (1986).

Superluminal motion provides an independent measure of the amount of beaming, which can be compared with the lower limit obtained from X-ray measurements (e.g., Cohen and Unwin 1984). Results have been obtained for NRAO 140, where the X-ray prediction that the source should be superluminal was subsequently observationally confirmed (Marscher and Broderick 1982), and for 3C 345, where the relativistic beaming factor implied by the observed superluminal motion is comparable to the minimum required by the X-rays (Unwin et al. 1983). However, similar arguments applied to 3C 273 have led to an unexplained excess of X-ray emission (Unwin et al. 1985).

Sources with superluminal motion include flat radio spectrum (FRS) QSOs, highly polarized QSOs (HPQs) including the optically violently variable sources (OVVs), and BL Lac objects. The approach I will follow here is to contrast the statistical relationships between X-ray, optical, and radio luminosities amongst different categories of active galactic nuclei (AGN) to try to discover where there may be energy mechanisms in common.

2. Correlations of X-ray, Optical, and Radio Emission

Figure 1 shows the correlation of monochromatic 2 keV X-ray luminosity with monochromatic 2500 Å optical luminosity and with 5 GHz radio luminosity for AGN of various classes. All luminosities are given in $(2h)^{-2}$ erg s^{-1} Hz^{-1} assuming a Friedmann cosmology with $H_0 = 100h$ km s^{-1} Mpc^{-1} and $q_0 = 0$. The sources of the data and values assumed for the spectral indices in the X-ray, optical, and radio bands (α_x, α_o, and α_r, respectively; $S \propto \nu^{-\alpha}$), for the K-corrections, are as follows:

(a) The FRS QSO sample includes all but 13 of the 61 sources for which data are given by Worrall et al. (1987). The 13 deleted QSOs form part of the HPQ sample described below. Luminosity K-corrections for the 48 remaining FRS QSOs assume $\alpha_x = 0.5$, $\alpha_o = 0.5$, and measured values of α_r. The value for α_x is the average found by Wilkes and Elvis (1987) for fits to radio-loud QSOs observed with the Einstein Observatory IPC detector. Five objects not in the FRS list of Worrall et al. (1987) have been added to the figure (but not the sample) since they have reported superluminal motion. These are 3C 120 (usually classified as a Seyfert galaxy), NRAO 140, 1928+738, and the lobe-dominated radio QSOs 3C 245 and 3C 263. Only core contributions to the radio luminosities of the lobe-dominated sources are used, and data are given in Worrall et al. (1987). The X-ray luminosities for 3C 120 and NRAO 140 have been extracted from IPC images in the identical manner used for the FRS QSOs. The X-ray luminosity for 1928+738 has been calculated from the 1 keV flux density given by Eckart et al. (1985). Optical and radio data are given by Véron-Cetty and Véron (1985), Walker et al. (1984), Marscher and Broderick (1981), and Kühr et al. (1979).

(b) The HPQs use data given by Ledden and O'Dell (1985). Only those HPQs observed at X-ray energies by the Einstein Observatory are included, giving a total of 25 objects. Luminosity K-corrections assume $\alpha_x = 1.0$, $\alpha_o = 1.4$, and $\alpha_r = 0.0$. The value for α_o is the average found by Ghisellini et al. (1986) for a combined sample of BL Lac objects and HPQs. In the absence of a statistical study of the X-ray spectra of HPQs, the value for BL Lac objects has been adopted for α_x (see below).

(c) Data for the BL Lac objects have also been taken from Ledden and O'Dell (1985), selecting only objects with measured redshift. (The superluminal source 0735+178 has been added to the figure, but it is not included in the sample because its redshift is from an absorption system.) Objects which were discovered through their X-radiation are not used, since they would bias an analysis of X-ray emission as a function of radio and optical emission. Luminosity K-corrections assume the same spectral indices as for the HPQs. It should be noted that

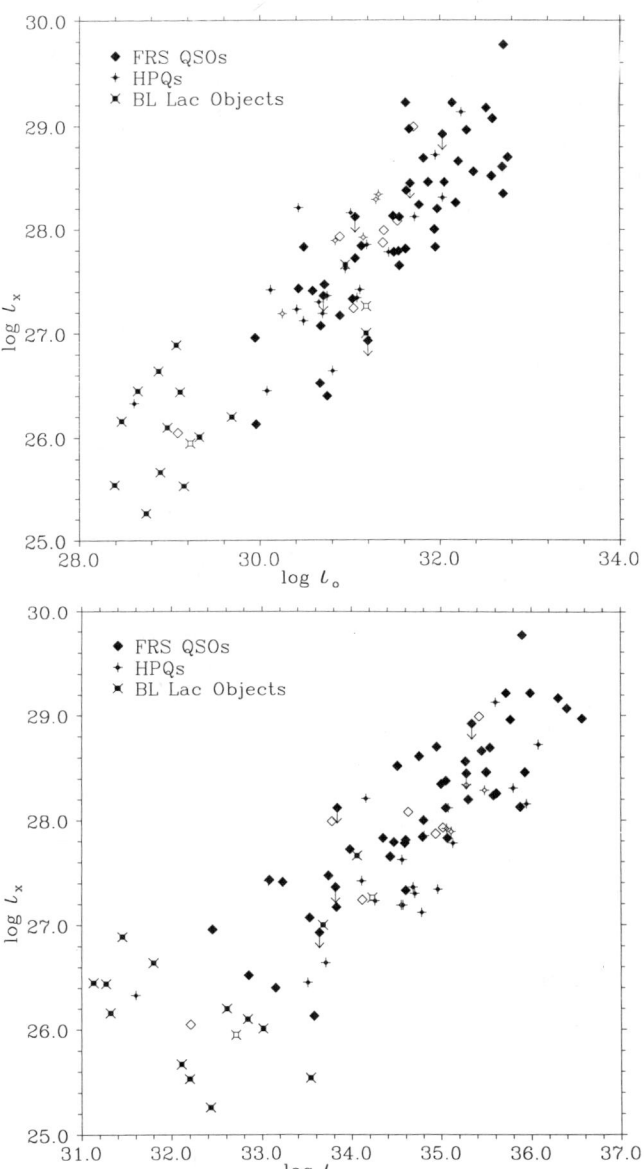

Figure 1 Correlations of monochromatic 2 keV X-ray luminosity with (a) 2500 Å optical luminosity and (b) 5 GHz radio luminosity. Units are $(2h)^{-2}$ erg s^{-1} Hz^{-1}. Open symbols show superluminal radio sources. In order of decreasing X-ray luminosity, the superluminal QSOs are NRAO 140, 3C 273, 3C 263, 4C 39.25, 3C 245, 1928+738, and 3C 120; the superluminal HPQs are CTA 102, 3C 454.3, 3C 345, 3C 279, and 3C 216; the superluminal BL Lac objects are 0735+178 and BL Lacertae.

the value for α_x gives an unacceptable fit to several individual BL Lac objects (Madejski 1985).

The following is inferred from Figure 1:

(a) All the objects show the same X-ray/optical correlation.

(b) The X-ray/radio correlation is worse than the X-ray/optical correlation, particularly in the low-luminosity region dominated by BL Lac objects. The dispersion of the X-ray/optical correlation for radio-quiet objects increases for objects with 2500 Å luminosities $\lesssim 10^{30}(2h)^{-2}$ erg s^{-1} Hz^{-1}; this has been attributed to the larger and more uncertain intrinsic X-ray absorptions at low luminosities (Zamorani 1986), but since the increased dispersion here is found only in the X-ray/radio correlation, some other cause is indicated. Variability will cause dispersion, since the luminosity determinations are not from simultaneous measurements. Worrall et al. (1987) calculate that, for the FRS data, the correlations between X-ray luminosity and a combined function of optical and radio luminosity are only significantly affected if the differential variability in the different bands is more than a factor ~ 2. In all three wave bands, BL Lac objects are typically much more variable than this on time scales of order a month. However, such variability should not cause a larger dispersion in the X-ray/radio correlation than in the X-ray/optical correlation since the 5 GHz variability is generally less than the optical variability.

(c) The superluminal sources, although few in number, show similar luminosity ratios to the other sources. This tends to rule out a model in which the emission in only one or two of the three wave bands is relativistically boosted unless *all* objects in Figure 1 are beamed towards the observer at small angles to the line of sight. In this case, they should have less luminous counterparts beamed away from the observer, that might appear as steep radio spectrum QSOs (Orr and Browne 1982) or radio-quiet objects (Scheuer and Readhead 1979). If the jets of the objects in Figure 1 are distributed evenly over all angles to the line of sight, and are beamed in all three wave bands, there is the problem of understanding why the ratio of emission-line strength to continuum radiation is not enhanced for sources with jets directed away from the line of sight (e.g., Kinney et al. 1986).

Although the selection effects in the sample of objects in Figure 1 are poorly defined, none of the objects was discovered by X-rays, so a regression of X-ray luminosity onto radio or optical luminosity is permitted if the X-ray upper limits are considered as well as the detections. Objects have been included regardless of whether they were discovered in the radio or the optical band. This was also the case for the analysis of the FRS QSOs by Worrall et al. (1987), although Ledden and O'Dell (1985) separated the radio-selected objects into a subclass in their analysis of BL Lac objects and HPQs. An attempt has been made to derive the dependence of l_x on l_o and l_r using the fitting method of Avni et al. (1980), which treats both X-ray detections and upper limits. In this method the distribution of residuals around the best-fit function is assumed to be Gaussian, and the standard deviation of the distribution, σ_G, is treated as a free parameter. Models of the form $l_{x27} = f(l_{o31}, l_{r35})$ are constructed, where l_{x27}, l_{o31}, and l_{r35} are monochromatic X-ray, optical and radio luminosities in units of 10^{27}, 10^{31}, and 10^{35} $(2h)^{-2}$ erg s^{-1} Hz^{-1}. Best fit values are found for σ_G and the parameters of the function f. The relative goodness of fit of a particular data set to various models is indicated by the smallness of $S = -2\ln L$, where L is the likelihood function, and S is distributed like χ^2.

The first question to be addressed is, are the HPQs more like the BL Lac objects or the FRS QSOs? The three samples have been fitted separately to the model $f(l_{o31}, l_{r35}) = A l_{o31}^\alpha l_{r35}^\beta$. The sum of the values of S for separate fits to the HPQ and FRS samples is smaller by 4.5 than the value of S for a fit to the HPQ and FRS samples combined. There is a probability of 3×10^{-2} that this improvement would be obtained randomly. However, the sum of the values of S for separate fits to the HPQ and BL Lac samples is smaller by 11.7 than the value of S for a fit to the HPQ and BL Lac samples combined. There is a probability of only 6×10^{-4} that this improvement would be obtained randomly, and the conclusion is that the HPQ sample is significantly more similar to the FRS sample than to the BL Lac sample. Although BL Lac objects and HPQs certainly exhibit similarities at radio and optical wavelengths and have been given the collective name "blazars" (Angel and Stockman 1980), these results suggest that blazars should not be treated as a coherent group of objects when their multifrequency spectra are analyzed. Ledden and O'Dell (1985) compare BL Lac objects discovered optically or by X-rays with a blazar sample of HPQs and only radio-discovered BL Lac objects. Too few BL Lac objects have measured redshifts for this approach to be used here, and the HPQ sample will be combined with the FRS QSOs into an expanded FRS sample.

The second and more complex question to be addressed is, to what extent can it be concluded that the X-rays from objects under consideration are produced by the self-Compton mechanism? This question is examined for the BL Lac objects and FRS QSOs separately.

(a) Figure 2 shows the 90% confidence combined error contours for parameters α and β. The BL Lac objects give a best fit for negative β, and are consistent with $\beta = 0$, whilst requiring positive α. This is what might have been expected from examination of Figure 1, and supports the conclusions of Stocke et al. (1985) and Maraschi et al. (1986) that the observed X-rays might be dominated by an isotropic emission

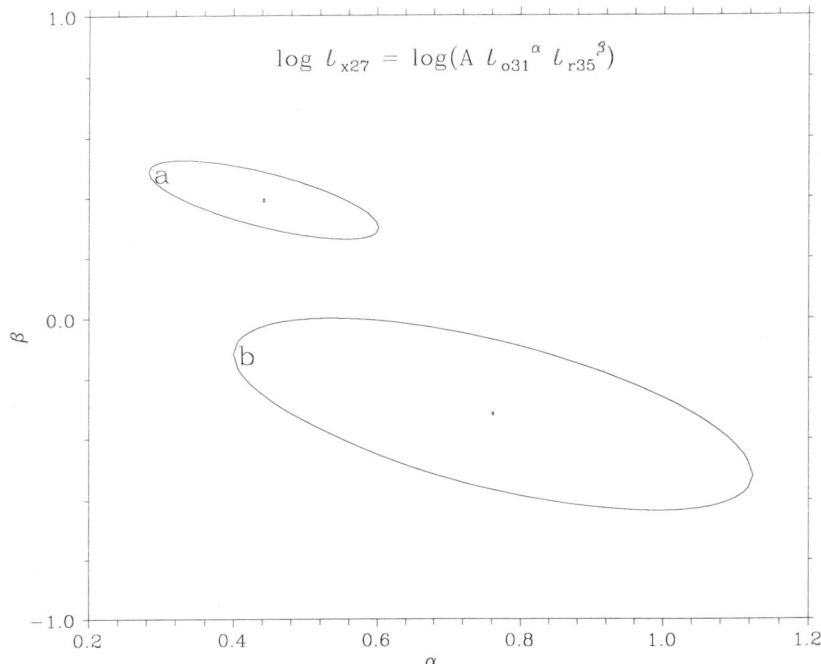

Figure 2 Contours of 90% confidence for two parameters, α and β, for a fit to the function $l_{x27} = Al_{o31}^{\alpha} l_{r35}^{\beta}$ of (a) radio-loud QSOs (FRS+HPQs) and (b) BL Lac objects.

component which is not directly related to the relativistically boosted radio emission. For BL Lac objects, therefore, the X-rays can be used in conjunction with radio data to set only a lower limit to the amount of relativistic beaming. In the few cases where a flattening of X-ray spectra to higher energies is observed (see Urry 1986; Madejski 1985), this may indicate that the self-Compton emission is emerging from under a steeper spectrum component. However, it is then surprising that neither of the two best examples of this, Mrk 421 and 2155−304 (Urry 1986), is a radio-discovered BL Lac object.

(b) Unlike BL Lac objects, FRS QSOs show a strong correlation between X-ray luminosity and radio luminosity (Ku, Helfand, and Lucy 1980; Zamorani et al. 1981; Owen, Helfand, and Spangler 1981; Ledden and O'Dell 1983; Kembhavi, Feigelson, and Singh 1986; Worrall et al. 1987; Browne and Murphy 1987), but radio-quiet QSOs *also* emit X-rays. Worrall et al. (1987) found that when their FRS sample was fitted to a joint function of optical and radio luminosity, $f = Al_{o31}^{\alpha} + Bl_{r35}^{\beta}$, the results were consistent with a radio-loud QSO consisting of an average radio-quiet QSO plus a separate radio-emitting region which provides some extra X-ray emission. The radio-emitting region then provides the dominant X-ray contribution in 80% of the sources considered. The same function has been fit to the FRS sample here. The best fit is $l_{x27} = 2.1\, l_{o31}^{0.68} + 4.5\, l_{r35}^{1.0}$ and, within 95% confidence errors, the term in l_o now gives more X-ray emission than is observed from an average radio-quiet QSO. Thus it is possible that some of the optical luminosity is related to the level of radio emission and may be relativistically boosted. In this case, the adopted function is not appropriate. A further free parameter must be introduced into the function unless the optical and radio emission are assumed to be related in some known way to another observable quantity (Browne and Murphy 1987). It is reasonable to assume that the X-ray luminosities of at least those FRS QSOs for which the radio term is larger than the optical term in the above expression might be dominated by self-Compton emission, e.g., 50% of the objects in the sample. As in the case of BL Lac objects, more measurements of X-ray spectral indices may help to clarify the situation. Wilkes and Elvis (1987) have found that radio-loud QSOs tend to have flatter X-ray spectral indices ($\alpha_x < 1.0$) than radio-quiet QSOs. They suggest that this may be because self-Compton emission dominates the X-radiation in radio-loud objects. All superluminal QSOs with reported X-ray spectral indices are among the group with $\alpha_x < 1.0$; 3C 120 (Halpern 1985), 3C 273 (Worrall et al. 1979), 3C 263, 3C 279 (Wilkes and Elvis 1987), 3C 345 (Bregman et al. 1986), and NRAO 140 (Marscher 1986).

3. Summary

The X-ray versus optical and X-ray versus radio correlations of radio-loud QSOs and superluminal radio sources are similar. This argues against a model in which the emission in only one or two of the three wave bands is relativistically boosted, unless *all* these objects exhibit beaming at small angles relative to the observer's line of sight. A regression analysis shows that highly polarized QSOs and optically violently variable QSOs are more similar to other flat-spectrum, radio-loud QSOs than to BL Lac objects, and it is reasonable to assume that self-Compton emission dominates the X-ray emission from at least half of the sources in this class. The X-ray versus radio correlation for BL Lac objects is poor, and there is support for the hypothesis that their X-ray emission is dominated by an isotropic component which is not directly related to relativistically boosted radio emission.

Support for this work was provided by NASA contract NAS 8-30751 and Smithsonian Institutional funds.

References

Angel, J. R. P., and Stockman, H. S. 1980, *Ann. Rev. Astron. Astrophys.*, **18**, 321.
Avni, Y., Sołtan, A., Tananbaum, H., and Zamorani, G. 1980, *Astrophys. J.*, **238**, 800.
Bregman, J. N., Glassgold, A. E., Huggins, P. J., Neugebauer, G., Soifer, B. T., Matthews, K., Elias, J., Webb, J., Pollock, J. T., Pica, A. J., Leacock, R. J., Smith, A. G., Aller, H. D., Aller, M. F., Hodge, P. E., Dent, W. A., Balonek, T. J., Barvainis, R. E., Roellig, T. P. L., Wiśniewski, W. Z., Rieke, G. H., Lebofsky, M. J., Wills, B. J., Wills, D., Ku, W. H.-M., Bregman, J. D., Witteborn, F. C., Lester, D. F., Impey, C. D., and Hackwell, J. A. 1986, *Astrophys. J.*, **301**, 708.
Browne, I. W. A., and Murphy, D. W. 1987, *Monthly Notices Roy. Astron. Soc.*, in press.
Cohen, M. H., and Unwin, S. C. 1984, in *IAU Symposium 110, VLBI and Compact Radio Sources*, ed. R. Fanti, K. Kellermann, and G. Setti (Dordrecht: Reidel), p. 95.
Eckart, A., Witzel, A., Biermann, P., Pearson, T. J., Readhead, A. C. S., and Johnston, K. J. 1985, *Astrophys. J. (Letters)*, **296**, L23.
Ghisellini, G., Maraschi, L., Tanzi, E. G., and Treves, A. 1986, *Astrophys. J.*, **310**, 317.
Halpern, J. P. 1985, *Astrophys. J.*, **290**, 130.
Kembhavi, A., Feigelson, E. D., and Singh, K. P. 1986, *Monthly Notices Roy. Astron. Soc.*, **220**, 51.
Kinney, A. L., Huggins, P. J., Bregman, J. N., and Glassgold, A. E. 1986, *Astrophys. J.*, **291**, 128.
Ku, W. H.-M., Helfand, D. J., and Lucy, L. B. 1980, *Nature*, **288**, 323.
Kühr, H., Nauber, U., Pauliny-Toth, I. I. K., and Witzel, A. 1979, Preprint No. 55 (Bonn: MPI für Radioastronomie).
Ledden, J. E., and O'Dell, S. L. 1983, *Astrophys. J.*, **270**, 434.
Ledden, J. E., and O'Dell, S. L. 1985, *Astrophys. J.*, **298**, 630.
Madau, P., Ghisellini, G., and Persic, M. 1987, *Monthly Notices Roy. Astron. Soc.*, **224**, 257.
Madejski, G. M. 1985, Ph. D. thesis, Harvard University.
Madejski, G. M., and Schwartz, D. A. 1983, *Astrophys. J.*, **275**, 467.
Maraschi, L., Ghisellini, G., Tanzi, E. G., and Treves, A. 1986, *Astrophys. J.*, **310**, 325.
Marscher, A. P. 1986, in *Continuum Emission in Active Galactic Nuclei*, ed. M. L. Sitko (Tucson: Kitt Peak National Observatory, National Optical Astronomy Observatories), p. 143.
Marscher, A. P., and Broderick, J. J. 1981, *Astrophys. J.*, **249**, 406.
Marscher, A. P., and Broderick, J. J. 1982, *Astrophys. J. (Letters)*, **255**, L11.
Marscher, A. P., Marshall, F. E., Mushotzky, R. F., Dent, W. A., Balonek, T. J., and Hartman, M. F. 1979, *Astrophys. J.*, **233**, 498.
Orr, M. J. L., and Browne, I. W. A. 1982, *Monthly Notices Roy. Astron. Soc.*, **200**, 1067.
Owen, F. N., Helfand, D. J., and Spangler, S. R. 1981, *Astrophys. J. (Letters)*, **250**, L55.
Scheuer, P. A. G., and Readhead, A. C. S. 1979, *Nature*, **277**, 182.
Stocke, J. T., Liebert, J., Schmidt, G., Gioia, I. M., Maccacaro, T., Schild, R. E., Maccagni, D., and Arp, H. C. 1985, *Astrophys. J.*, **298**, 619.
Unwin, S. C., Cohen, M. H., Biretta, J. A., Pearson, T. J., Seielstad, G. A., Walker, R. C., Simon, R. S., and Linfield, R. P. 1985, *Astrophys. J.*, **289**, 109.
Unwin, S. C., Cohen, M. H., Pearson, T. J., Seielstad, G. A., Simon, R. S., Linfield, R. P., and Walker, R. C. 1983, *Astrophys. J.*, **271**, 536.
Urry, C. M. 1986, in *Continuum Emission in Active Galactic Nuclei*, ed. M. L. Sitko (Tucson: Kitt Peak National Observatory, National Optical Astronomy Observatories), p. 91.
Véron-Cetty, M.-P., and Véron, P. 1985, *A Catalogue of Quasars and Active Nuclei (2nd Edition)*, European Southern Observatory Scientific Report No. 4. (Garching: European Southern Observatory).

Walker, R. C., Benson, J. M., Seielstad, G. A., and Unwin, S. C. 1984, in *IAU Symposium 110, VLBI and Compact Radio Sources*, ed. R. Fanti, K. Kellermann, and G. Setti (Dordrecht: Reidel), p. 121.
Wilkes, B. J., and Elvis, M. 1987, *Astrophys. J.*, submitted.
Worrall, D. M., Giommi, P., Tananbaum, H., and Zamorani, G. 1987, *Astrophys. J.*, **313**, 596.
Worrall, D. M., Mushotzky, R. F., Boldt, E. A., Holt, S. S., and Serlemitsos, P. J. 1979, *Astrophys. J.*, **232**, 683.
Zamorani, G. 1986, in *IAU Symposium 119, Quasars*, ed. G. Swarup and V. K. Kapahi (Dordrecht: Reidel), p. 223.
Zamorani, G., Henry, J. P., Maccacaro, T., Tananbaum, H., Sołtan, A., Avni, Y., Liebert, J., Stocke, J., Strittmatter, P. A., Weymann, R. J., Smith, M. G., and Condon, J. J. 1981, *Astrophys. J.*, **245**, 357.

Optical Spectra of Superluminal Sources

CHARLES R. LAWRENCE, ANTHONY C. S. READHEAD,
TIMOTHY J. PEARSON, AND STEPHEN C. UNWIN

Owens Valley Radio Observatory and Palomar Observatory,
California Institute of Technology

1. Introduction

The lack of a tight relationship between the radio and optical brightness of extragalactic radio sources means that even complete samples of the strongest radio sources contain objects difficult to observe spectroscopically. Nevertheless, big telescopes and CCD spectrographs now make spectroscopic completeness a reasonable goal.

In 1984, we began a program to obtain spectra at least good enough to determine redshifts for all of the 65 sources in the Pearson-Readhead VLBI sample (Pearson and Readhead 1984; Pearson, Readhead, and Barthel, this Workshop, page 94). At that time, redshifts were unknown for about one-third of the sample, and three objects, 0108+388, 1624+416, and 2342+821, were unidentified. At present, redshifts are known for 54 of the 65, and identifications are complete (0108+388 and 1624+416 are resolved with $m_r = 22.0$, 2342+821 is unresolved with $m_r = 20.5$). The redshift distribution is given in Figure 1. With the possible exceptions of 1749+701 and 0454+788, two objects with discouragingly featureless spectra, there is hope that by the end of 1987 redshifts will be known for all 65 sources. We are using the double spectrograph on the Palomar 200 inch telescope, with typical wavelength coverage from 3800–9800 Å. Total integration times range from ten minutes to several hours. Eight sources in the sample have been identified as superluminal with varying degrees of certainty (Pearson, Readhead, and Barthel, this Workshop, page 94). Their optical spectra, shown in Figure 2, are the subject of this paper.

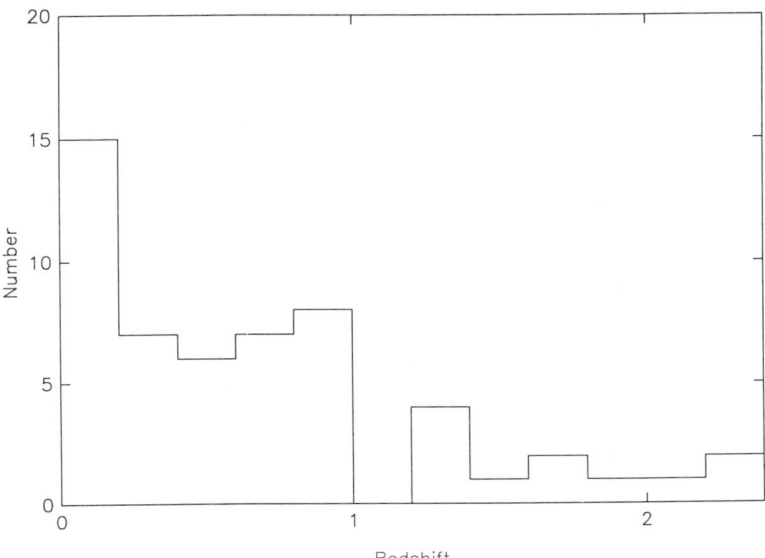

Figure 1   Redshift distribution of the Pearson-Readhead sample.

2. Radio Source Identifications

Before turning to the spectra, we digress for a remark on name-calling in radio astronomy. The importance of optical counterparts in the study of radio sources needs no comment, but consider how the popular labels *galaxy*, *quasar*, and *BL Lac object* are assigned.

First, an image of the optical counterpart is examined, often on the Palomar Observatory Sky Survey prints. If the image looks fuzzy, the object is called a galaxy; if sharp, a stellar object (perhaps with a color). In the latter case, a spectrum is required to determine the appropriate label. If the spectrum has emission lines, the object is called a quasar, otherwise it is called a BL Lac object. That is usually all there is to it.

Unfortunately, the distinction between quasars and BL Lac objects depends on data quality. For example, in 1984 eight members of the Pearson-Readhead sample appeared in the literature as "BL Lac objects" or "possible BL Lac objects". So far, we have found lines, sometimes easily, in five of the eight. The distinction between galaxies and the others also depends on data quality, because we now expect high-dynamic-range images of quasars and BL Lac objects to show fuzz. However, if we agree that one should not need a CCD to spot the fuzz of a *real* galaxy, this distinction becomes tolerably clear.

The moral: never buy a BL Lac object without an extended warranty.

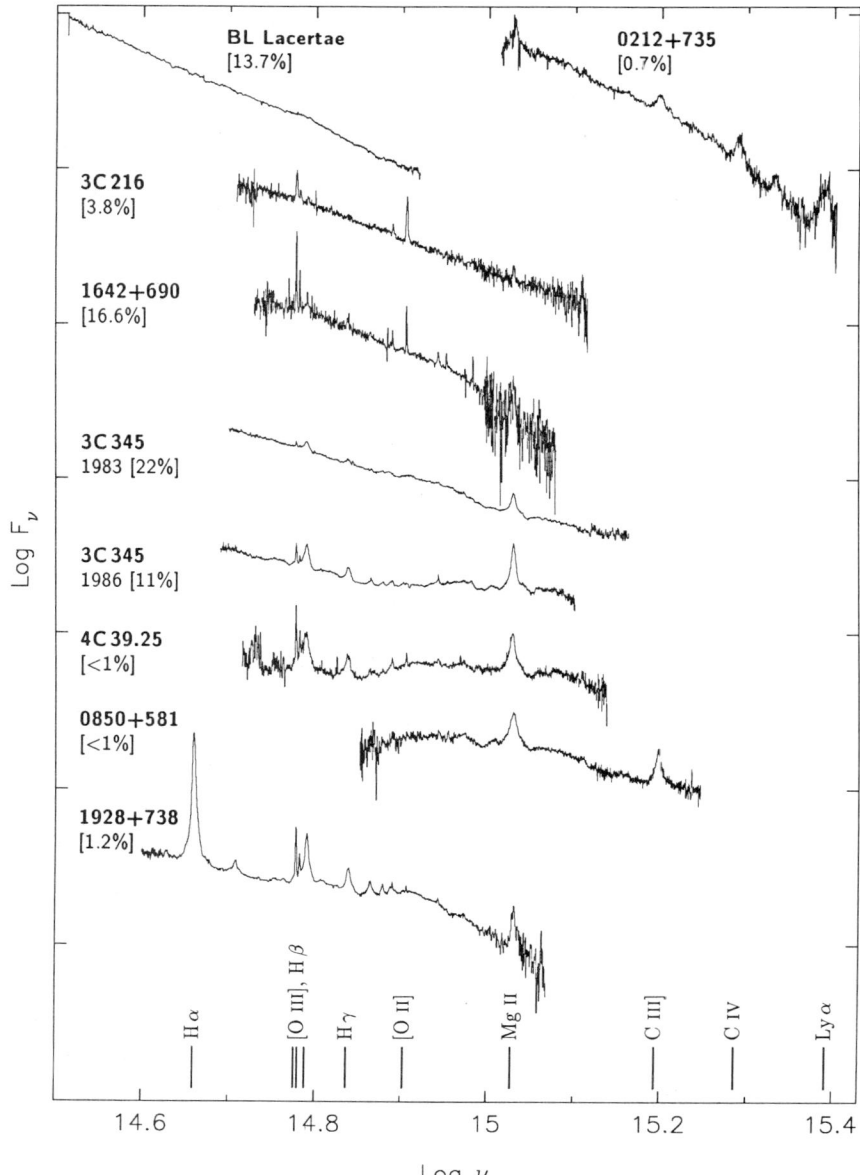

Figure 2 Spectra of five confirmed and three possible superluminal sources in the Pearson-Readhead sample. Broad-band optical polarizations are in square brackets. The 1986 polarization of 3C 345 was measured in 1986 July, 10 days after the spectrum was taken; the 1983 polarization was interpolated to the date of the spectrum from Smith et al. (1986). Other polarizations were measured in 1986 July or December.

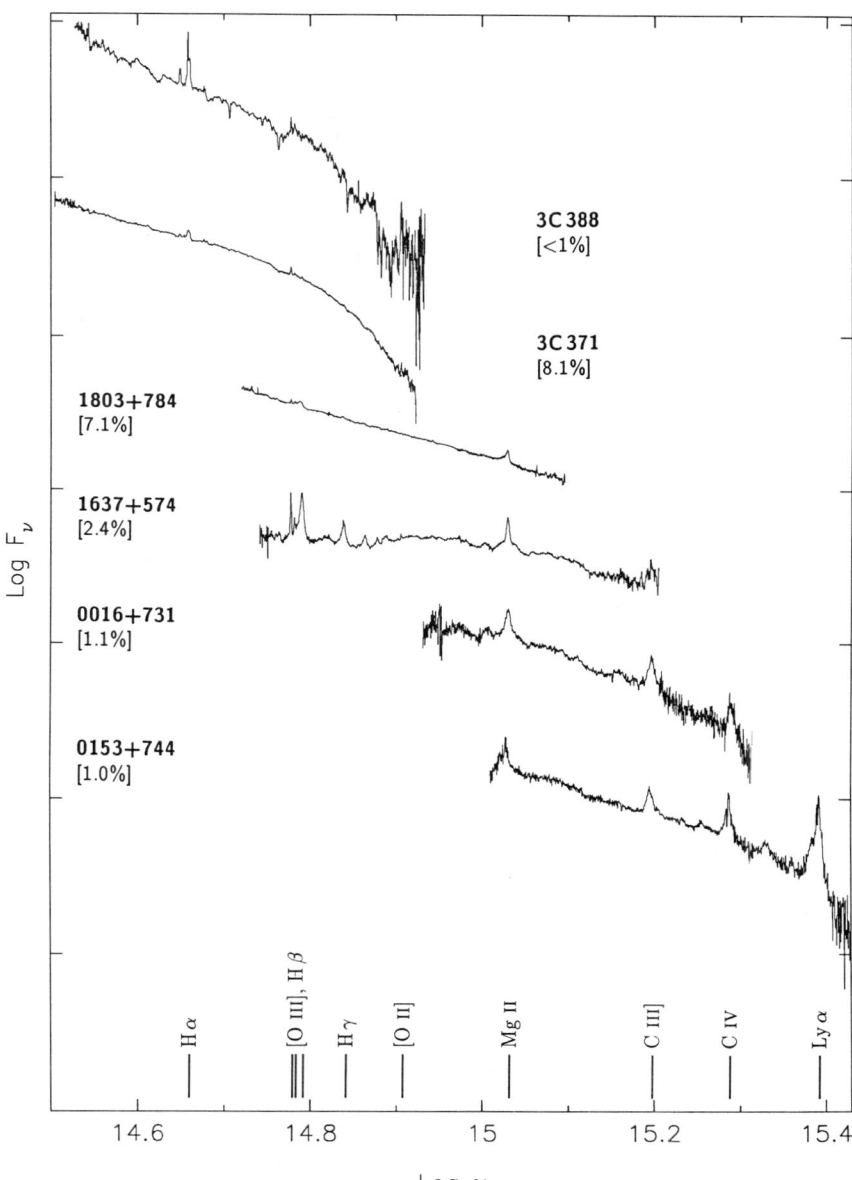

Figure 3 Spectra of sources from the Pearson-Readhead sample for which there is as yet no evidence for superluminal motion. Only the polarization measurement of 1803+784 was contemporaneous with the spectrum.

3. Spectra

The question we would like to address is the following: can superluminal sources be recognized by their optical spectra? Figure 2 shows spectra of the five confirmed and three possible superluminal sources in the Pearson-Readhead sample. Figure 3 shows spectra of other sources in the sample chosen for comparison. (3C 388 was chosen because its spectrum has many stellar features, the rest because their spectra are similar in one way or another to those in Figure 2. It should be emphasized that the spectra in Figure 3 were not randomly chosen. In fact, some of the sources could turn out to be superluminal.) The spectra were taken through slits of $1''$ or $2''$, and calibrated with observations of the F-subdwarf stars measured by Oke and Gunn (1983).

Narrow-slit spectra are subject to many frequency-dependent systematic errors. The most serious, a result of atmospheric dispersion, can distort the overall shape of the spectrum. Orienting the slit in the direction of the dispersion greatly reduces the error, and we do this routinely. Nevertheless, since the orientation is fixed during a typical 50 minute exposure, the potential for systematic errors remains. Suffice it to say that not every feature in the spectra shown is real, and in the poor spectrum of BL Lac, almost *no* feature is real. Spectral pundits will recognize the problems, but they are not important for the present purpose.

The most obvious features of the spectra in Figures 2 and 3 are emission lines. In low redshift objects (e.g., 1928+738) the most prominent lines are usually the Balmer lines, the [O III] $\lambda\lambda 4959, 5007$ doublet, and Mg II $\lambda 2798$. At higher redshifts, Ly α, C IV $\lambda 1549$, and C III] $\lambda 1909$ are the most prominent. Although emission-line widths in active nuclei range from hundreds to tens of thousands of kilometers per second, and may cover quite a range even in a single object, the generic adjectives *broad* and *narrow* are often applied to lines. Examination of the H β–[O III] trio near 5000 Å in many of the objects shows immediately the motivation for these terms.

The next most obvious feature is the "continuum", that is, the stuff between the obvious lines. This can have several origins. In the case of 3C 388, the continuum is starlight. In the other "red" objects, the continuum is probably synchrotron emission. In some objects, a forest of thousands of Fe II lines, both in emission and in absorption of the Balmer continuum, gives the appearance of a bumpy continuum surrounding the Mg II line at 2798 Å (see Wills, Netzer, and Wills 1985). Other iron features are seen sometimes near the H β–[O III] trio. Emission from an optically thick accretion disk may underly the Fe II–Balmer continuum bump, and extend it to $10^{15.3}$ Hz. None of our spectra extends much past this frequency, making it hard to identify this feature; however, Bregman et al. (1986) report emission in 3C 345 consistent with a blackbody between 10 000 K and 40 000 K, and Smith et al. (1986) required such emission to model the wavelength-dependent polarization of the source.

Aside from the effect of iron absorption on the Balmer continuum just described, absorption features are not prominent in any of the superluminal spectra. However, absorption is seen in some of the comparison spectra. 0153+744 has what appear to be strong absorption features in the Ly α, C IV, and Mg II lines. The prominent absorption features in the galaxy 3C 388 are due to stars. BL Lac has stellar absorption lines, but they are small compared to systematic errors in the spectrum shown in Figure 2. 1803+784 has a fairly convincing feature at $\lambda 3834$ Å, where the K line of Ca II appears.

Now back to the question of whether superluminal sources can be recognized by their optical spectra. Figure 2 shows marked differences in the spectra. BL Lac and 1928+738 represent the extremes in the relative strength of lines and continuum (quantified by equivalent width $= \int F_{\text{line}}/F_{\text{cont}}\, d\lambda$). BL Lac is also extreme in continuum slope, with 4C 39.25 and 0850+581 at the other extreme in this case. Moreover, comparison of the spectra in Figure 2 with those in Figure 3 reveals some similarities. 0850+581 and 4C 39.25 are much like 1637+574 and 0016+731, 1803+784 (a "BL Lac object") is much like 3C 345 (1983), and 0212+735 (another "BL Lac object") is much like 0153+744.

The answer to the question, then, is "no". Fortunately, there is more to be learned than that. All spectra in Figure 2 have a power-law component and significant optical polarization except 4C 39.25 and 0850+581. However, the two spectra of 3C 345 suggest the possibility that 4C 39.25 and 0850+581 may have power-law components at lower rest frequencies than we observed. Specifically, when the power-law continuum that dominated the 1983 spectrum of 3C 345 faded, it left behind a spectrum above $10^{14.85}$ Hz much like those of 0850+581, 4C 39.25, and 1637+574. Below $10^{14.85}$ Hz, the red continuum still dominates, while above $10^{14.85}$ Hz, the lowest rest frequency we observed in 0850+581, the synchrotron component of 3C 345 is hard to detect. The apparent turn-up at the end of the spectrum of 4C 39.25 may be the fading end of such a power law, although the warning about slit spectra should be kept in mind. Spectra of 0850+581 and 4C 39.25 out to a few microns would be quite interesting. At the moment, our observations cannot rule out the possibility that *all* superluminal sources in the Pearson-Readhead sample have significant power-law components in their spectra. Of course, some of the sources in Figure 3 also have strongly polarized continua. While none of these is known to be superluminal, only 3C 388, a radio galaxy with no strong core at all, would be a big surprise if it later turned out to be superluminal. To summarize, then, it is not possible to identify a superluminal source by its spectrum, but the superluminal–blazar connection discussed by Chris Impey (this Workshop, page 233) is strongly supported by the Pearson-Readhead sources.

Only two of the more than 50 sources observed to date, 3C 216 and 1642+690, have spectra dominated by narrow emission lines but show no sign of stellar features in the continuum. Narrow-line quasars (e.g., 1E 0449−184 and Q 2016+112) are unusual. Narrow-line *blazars* may be more unusual, and the fact that the only two in the Pearson-Readhead sample are superluminal is intriguing.

As previously mentioned, the spectrum of 1803+784 is remarkably similar to the 1986 spectrum of 3C 345. Both show a nice Balmer series, Mg II, and [O III]. The H β line in 3C 345 has two components, of FWHM 3400 and 7800 km s^{-1}; that of 1803+784 has components of FWHM 3600 and 7800 km s^{-1}. The broad components in each case are about half the peak intensity of the narrow components. Such broad lines in blazars fit uneasily into the disk-orientation picture of Wills and Browne (1986).

The spectrum of 1803+784 in Figure 3 was taken in 1986 July. Thirteen months earlier the source was $1^m.2$ brighter, and the emission lines were almost lost in the (systematic) noise. Ignoring the small difference in redshift, 1803+784 (1985), 1803+784 (1986), 3C 345 (1983), and 3C 345 (1986) form a steady sequence of decreasing continuum strength relative to the emission lines. BL Lac may be a BL Lac, but 1803+784, which used to be a BL Lac, looks more like a 3C 345!

References

Bregman, J. N., Glassgold, A. E., Huggins, P. J., Neugebauer, G., Soifer, B. T., Matthews, K., Elias, J., Webb, J., Pollock, J. T., Pica, A. J., Leacock, R. J., Smith, A. G., Aller, H. D., Aller, M. F., Hodge, P. E., Dent, W. A., Balonek, T. J., Barvainis, R. E., Roellig, T. P. L., Wiśniewski, W. Z., Rieke, G. H., Lebofsky, M. J., Wills, B. J., Wills, D., Ku, W. H.-M., Bregman, J. D., Witteborn, F. C., Lester, D. F., Impey, C. D., and Hackwell, J. A. 1986, *Astrophys. J.*, **301**, 708.
Oke, J. B., and Gunn, J. E. 1983, *Astrophys. J.*, **266**, 713.
Pearson, T. J., and Readhead, A. C. S. 1984, in *IAU Symposium 110, VLBI and Compact Radio Sources*, ed. R. Fanti, K. Kellermann, and G. Setti (Dordrecht: Reidel), p. 15.
Smith, P. S., Balonek, T. J., Heckert, P. A., and Elston, R. 1986, *Astrophys. J.*, **305**, 484.
Wills, B. J., and Browne, I. W. A. 1986, *Astrophys. J.*, **302**, 56.
Wills, B. J., Netzer, H., and Wills, D. 1985, *Astrophys. J.*, **288**, 94.

Emission-Line Profile Changes in 3C 390.3

J. BEVERLEY OKE

Palomar Observatory, California Institute of Technology

1. Introduction

A program has been under way for many years to monitor the spectral characteristics of selected bright Seyfert galaxies and quasars. Until 1981 the program was carried out with the Multichannel Spectrometer on the 200-inch telescope. These observations yielded absolute spectral energy distributions and line intensities. In 1981, with the completion of the CCD Double Spectrograph, it was possible to expand the program to monitor the continuum energy distribution, the broad-line and narrow-line intensities, and the detailed broad-line profiles. The Seyfert galaxies include 3C 382 and 3C 390.3 which are objects with very broad lines, and 3C 120 which has relatively narrow lines. The quasars include 3C 279, 3C 345, and 3C 446. In the case of the three Seyfert galaxies, IUE observations of Ly α and C IV λ1550 have also been obtained every six months for many years.

In this talk I will discuss primarily the results obtained for 3C 390.3. We have over thirty spectra taken during the last six years. Multichannel observations go back to 1969 (Yee and Oke 1981).

2. Continuum Changes in 3C 390.3

The light curve for 3C 390.3 from 1969 to late 1979 has been published by Lloyd (1984). Our own measurements of the continuum flux agree with Lloyd's observations. The object varied wildly between B magnitudes 14.8 and 16.4 from 1967 to 1973 and then remained moderately constant from 1974 to 1979 at $B \approx 16.3$ mag. Our more recent observations (Figure 1) show that 3C 390.3 remained near this minimum until late 1984. Since then it has brightened by about 0.3–0.4 mag. During the interval discussed below the magnitude remained relatively constant. The continuum flux near H α is shown in the upper part of Figure 1.

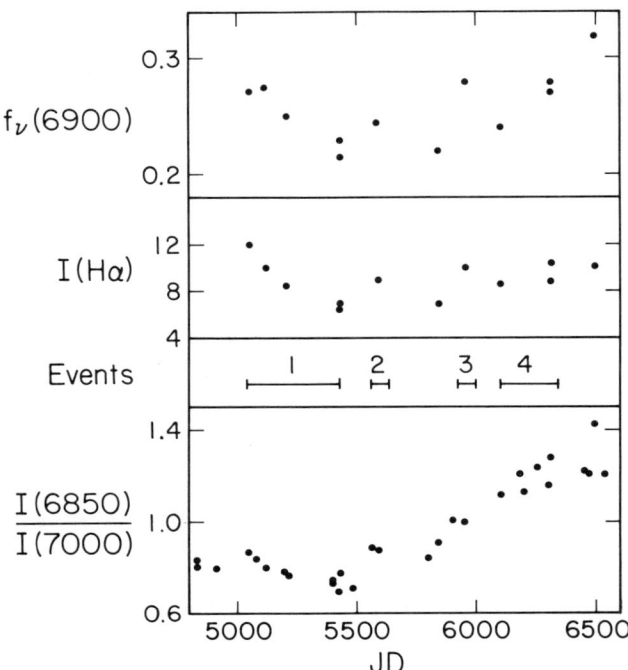

Figure 1 The continuum flux of 3C 390.3 at 6940 Å in units of 10^{-25} erg s^{-1} cm^{-2} Hz^{-1}, the Hα total intensity in units of 10^{-13} erg s^{-1} cm^{-2}, the four events discussed in the text, and the ratio of intensities of the flux in Hα at 6850 and 7000 Å, as a function of the Modified Julian Date (JD $-$ 2 440 000).

3. Changes in the Broad-Line Profiles

Four specific events where the broad lines changed have been identified. The time periods in which they occurred are shown in Figure 1. In two of these, events 2 and 3, the Hα line intensity changed but the overall line profile shape remained the same. In the other two events the line intensity changed and the profile also changed shape. The observed time scales for all of these events were between 150 and 200 days. Since monitoring of 3C 390.3 has not been done much more frequently than this (Figure 1), the time scale could be significantly less than 150 days.

Event 1 occurred between early 1982 and early 1983. Figure 2a shows the Hα line profile and absolute intensity in March 1982, August 1982, and April 1983. The profiles include several narrow-line features. These are redshifted [O I] at 6659 and 6726 Å, [N II] at 6921 and 6958 Å, Hα at 6936 Å, and [S II] at 7099 and 7114 Å. These narrow lines come from a different region than the broad lines and are not considered here. However, they make it very difficult to say anything about the broad-line

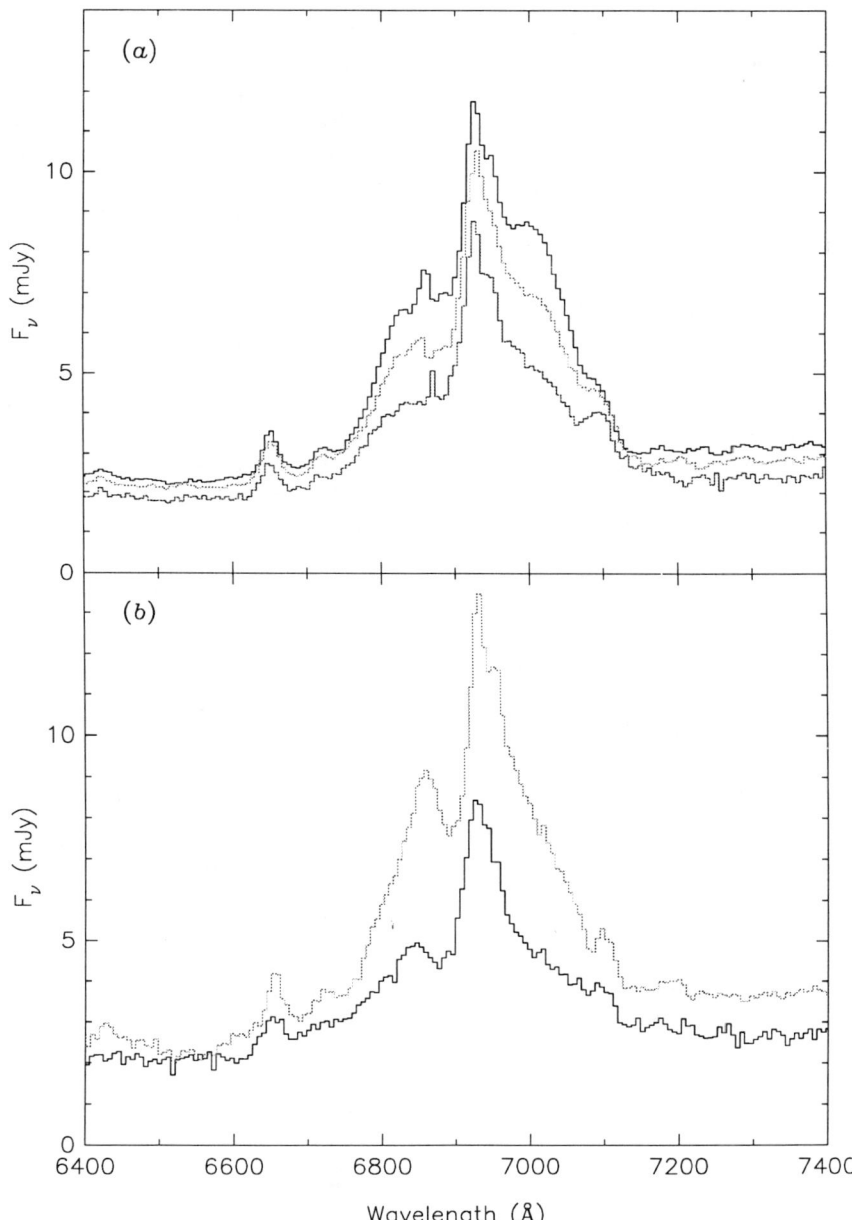

Figure 2 Hα profiles of 3C 390.3. The vertical and horizontal scales are the same in all cases. (a) Profiles measured on 1981 March 22 (*top*), 1981 August 28 (*center*), and 1983 April 9 (*bottom*). (b) Profiles measured on 1985 February 8 (*lower*) and 1985 August 31 (*upper*).

H α profile near the center of the line. The whole of the broad H α profile has collapsed by about a factor two during the interval. Furthermore, the two broad peaks at 6850 and 7000 Å, which were initially prominent, have nearly disappeared in the lowest plotted profile.

Event 4 reflects an increase in the H α line intensity between February 1985 and August 1985. The two profiles are shown in Figure 2b. Not only does the broad line increase at all wavelengths, but the increase in the red is greater than in the blue. It should be noted that although the line is of comparable intensity in August 1985 and early 1982 (Figure 2) the profiles are in fact quite different.

A study of the H α line profile and intensity changes allows a number of general statements to be made.

(a) Changes in the broad-line region occur on time scales of six months or less. Additional observations made under non-photometric weather conditions suggest that these time scales are longer than a month or two. Because of light-travel-time effects the time scales for variations indicate that the maximum radius of the broad-line region is 0.1 to 0.5 light years.

(b) Changes of the overall H α line intensity accompany changes in the continuum (Figure 1).

(c) H α broad-line intensity changes occur over the whole line profile, corresponding to velocities of up to ± 9500 km s^{-1}. In particular, the events noted above show the blue wing and red wing changing at the same time.

(d) The line changes in the blue and the red wings are not simply proportional to each other. Figure 1 shows the ratio of the broad-line intensities at 6850 and 7000 Å. More observations are shown here than in the other plots since non-photometric data are included. This ratio has shown a fairly smooth change with time. In 1981 to 1984 the ratio was nearly constant at 0.8. In mid-1984 it began to increase to the present value of 1.2.

(e) We also mention here that 3C 390.3 has a very broad H α profile while 3C 120 has a profile which is a factor three narrower.

4. Models for the Broad-Line Region

Accretion Disk

If the broad-line clouds form an accretion disk, then the line broadening is caused by the rotation of the disk about a central massive source. The rate at which changes occur reflects the size of the disk where the H α emission occurs. Changes in overall intensity reflect changes in the continuum flux which occurred several months earlier. It is not yet clear from the observations that such a delay is present or absent. Changes on the blue and on the red side of the line should occur simultaneously since the light travel time is similar for the approaching and receding gas clouds. The observed simultaneous changes are the strongest argument for an accretion disk model. The change in the ratio of intensity of the blue to the red side of the line implies a somewhat asymmetrical disk. Since the rotation period for the disk is 20 to 100 years, this is not surprising. The accretion disk model implies that 3C 390.3 is being seen edge-on while 3C 120 with its narrow lines is seen pole-on. The superluminal character of 3C 120 also indicates that we see it nearly pole-on.

Bipolar Flow Along the Axis of Rotation

In such a model the line broadening is produced by the velocity of ejection (or infall) along the polar directions. 3C 390.3 should then be seen pole-on while 3C 120 is seen edge-on. This model appears to be ruled out by the superluminal behavior of 3C 120, unless the radio clouds behave completely differently from the gas clouds producing the broad lines. In this model the blue and red wings of the line are produced respectively by material flowing outwards or inwards towards the center of the object. In 3C 390.3 one should see a time delay between changes in the blue and red side of the line due to the light travel time involved. As noted in point (c) above no time delay is observed. The slow change of the intensity ratio of the blue to red side of the line would be expected since it would represent a difference in the fluxes in the two polar directions at different times, or it could be a consequence of absorption of radiation coming from the far side. Overall, observations suggest that this model is not correct.

Equatorial Inflow

Another possible model suggested by comments in point (b) above is to have inflow along the equator of the system. This implies that 3C 390.3 is seen edge-on and 3C 120 pole-on, which agrees with the superluminal observations. Such a model still has the serious problem that the red side of the line is formed by material on the observer's side of the object while the blue side of the line comes from material on the far side of the object. Owing to light travel time effects this does not allow the blue and the red sides of the line to change at the same time.

5. Summary

The broad-line H α profile measurements of 3C 390.3 show changes which favor an accretion-disk model for the broad-line region. Models with either bipolar or equatorial flows disagree with the observations. Analysis of the already existing observations of the broad lines in 3C 120 should help to further delineate among the various models. It is important to obtain more line-profile observations and they should be made more frequently than in the past. At present, we know that the changes occur in less than six months. A very few observations suggest that changes do not occur in less than one month.

References

Lloyd, C. 1984, *Monthly Notices Roy. Astron. Soc.*, **209**, 697.
Yee, H. K. C., and Oke, J. B. 1981, *Astrophys. J.*, **248**, 472.

Evidence for Shocks in Relativistic Jets

HUGH D. ALLER, PHILIP A. HUGHES
AND MARGO F. ALLER
Radio Astronomy Observatory, University of Michigan

Blandford and Königl (1979) proposed that the superluminal components observed with VLBI could be produced by shocks propagating down a preexisting relativistic jet. For the past several years, we have been obtaining and analyzing data that strongly support this hypothesis. Recently we have been making quantitative comparisons of the predictions of models invoking shocks with multifrequency total flux density and linear polarization data obtained in the University of Michigan variability program.

The basic idea of these models is illustrated in Figure 1. A preexisting relativistic jet is assumed to be observed at a small angle to the line of sight. This orientation of the relativistic flow can account for the apparent superluminal motions, the high apparent brightness of the components, and the high degrees of polarization observed in some sources. If a shock forms in the jet flow (e.g., because a higher velocity flow overtakes a slower one), the magnetic field will be compressed, and in the case of an axial compression, the magnetic field will be preferentially aligned perpendicular to the jet axis. This alignment, which will occur even if the magnetic field is originally highly turbulent, can lead to quite high degrees of linear polarization for even moderate amounts of compression (Hughes, Aller, and Aller 1985). We have studied both the total flux density and the linear polarization spectra of bursts to test the validity of this type of model.

The series of outbursts in BL Lac that started in 1980 (Mutel and Phillips, this Workshop, page 60) can be well described by a shock model. Jones (1982) showed that a good fit could be made to the total flux density data (both the integrated flux densities and the brightness profiles obtained from VLBI) by a model invoking weak shocks propagating down a preexisting jet. At centimeter wavelengths, two of the bursts in BL Lac have reached degrees of polarization of 10% in integrated flux density with the electric vector of the polarized emission parallel to the orientation of the VLBI structure (Aller, Aller, and Hughes 1985). The multifrequency time evolution of the polarization in these

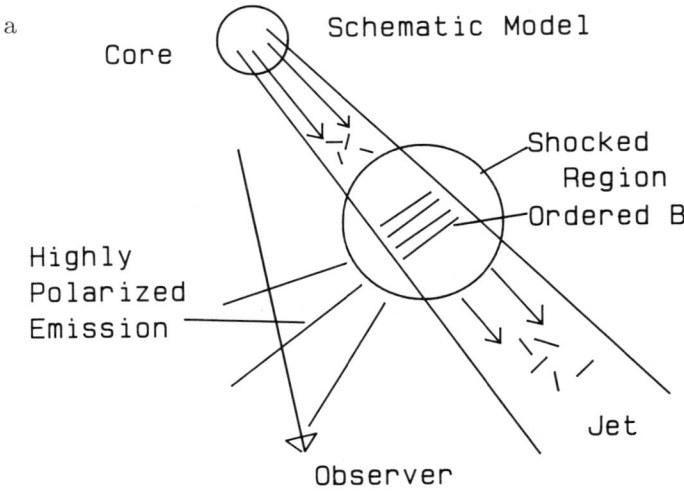
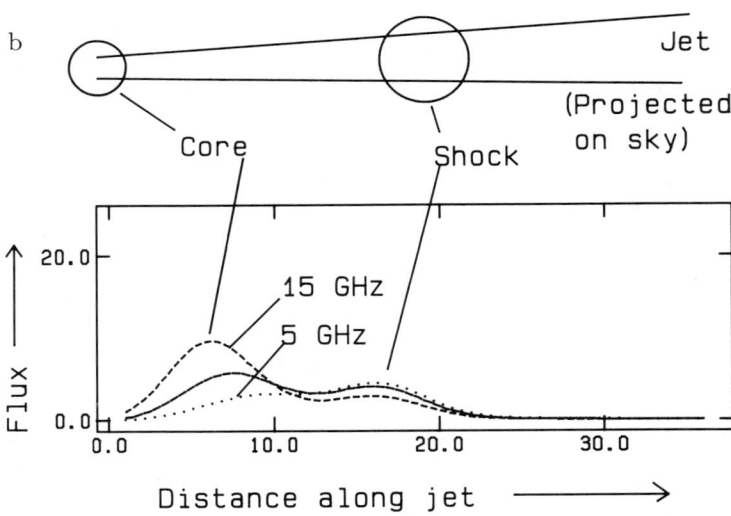

Figure 1 Schematic illustration of the shock model. (*a*) The orientation of the jet and observer in space. (*b*) The appearance of the source on the sky.

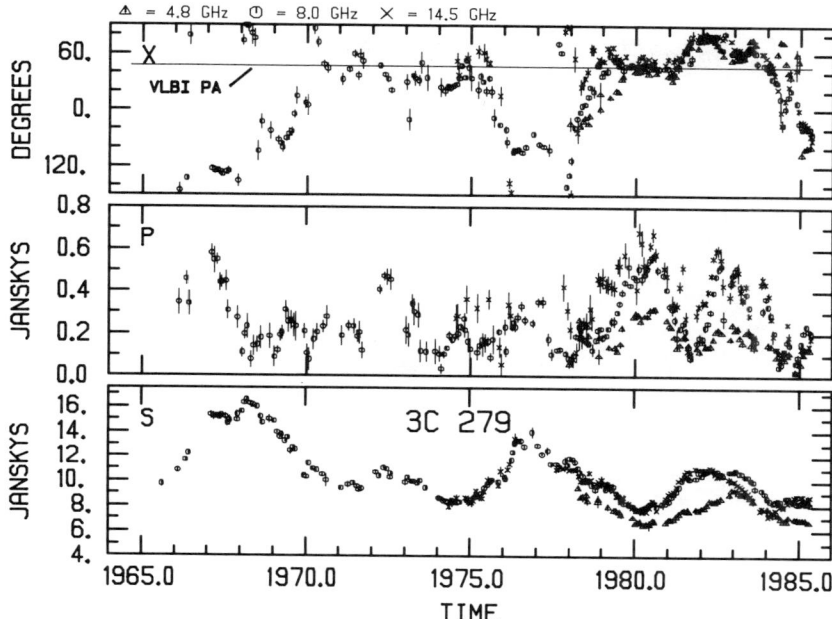

Figure 2 Monthly averages of the total flux density, degree of polarization, and polarization position angle of 3C 279.

bursts has been fit remarkably well by a model invoking weak shocks similar to that employed by Jones (Hughes, Aller, and Aller 1985). One interesting aspect of these model fits is that the polarization and flux density behavior require only quite weak shocks with Mach numbers in the range of 1.2 to 1.5 (Hughes, Aller, and Aller 1986).

Another source which has exhibited outbursts with relatively high degrees of polarization is 3C 279 (Figure 2). Over the long term, the polarization electric vector has alternated between being parallel and perpendicular to the VLBI structure (Unwin, this Workshop, page 34). Since 1980, this source has exhibited a more complex behavior than BL Lac; the individual outbursts (evident in the polarization data shown in the upper panel of Figure 3) are very closely spaced in time. Using the outbursts in polarized flux density to estimate the parameters and times of shock events, we have predicted the evolution of the total flux density. The curves shown in Figure 3 are the shock model fits to the polarization data at 4.8, 8.0, and 14.5 GHz; and as shown in the lower panel of the figure, the model correctly predicts that the total flux density curves vary smoothly during the time period. It is unclear whether more VLBI intensity maps made at intervals of less than a year would have resolved these events into separate source components.

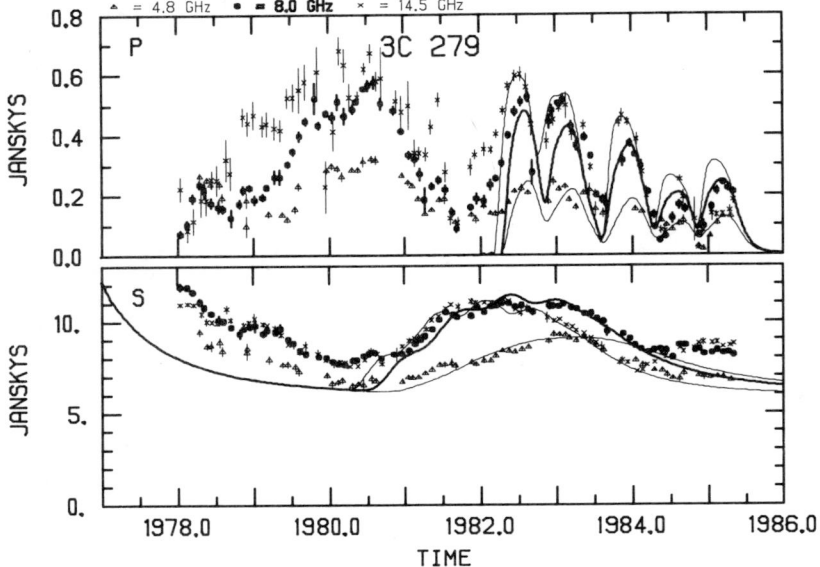

Figure 3 The polarized flux density and total flux density of 3C 279 at three frequencies, with the theoretical curves of a shock model superimposed.

For a closer examination of complex sources such as 3C 279, we have been developing computer programs to investigate arbitrary source and magnetic field geometries. This involves solving in detail the transfer of polarized radiation through a relativistic plasma (Jones and O'Dell 1977) and accounting for highly turbulent magnetic fields. Some of the results for a model with weak shocks propagating through a preexisting jet composed of many cells with randomly oriented magnetic fields are illustrated in Figures 4 and 5. The assumed model parameters closely approximate the physical conditions believed to exist in these sources (Jones *et al.* 1985). The predicted integrated polarized flux densities and total flux densities versus time for a single shock are illustrated in Figure 4. The spectral evolution and polarization behavior shown are typical of what is observed in many sources. Figure 5 shows predictions of the brightness profiles along a jet that might be observed by VLBI. The major cause of the decay of the components as they move down the jet is adiabatic expansion of the emitting region. Figure 5*b* shows that in a series of shock events the apparent expansion speed can change in an apparently arbitrary manner, and the core is not always the brightest component. Also note that because of opacity effects, the apparent distance between components can be frequency dependent. The complex behavior generated by even such simple models

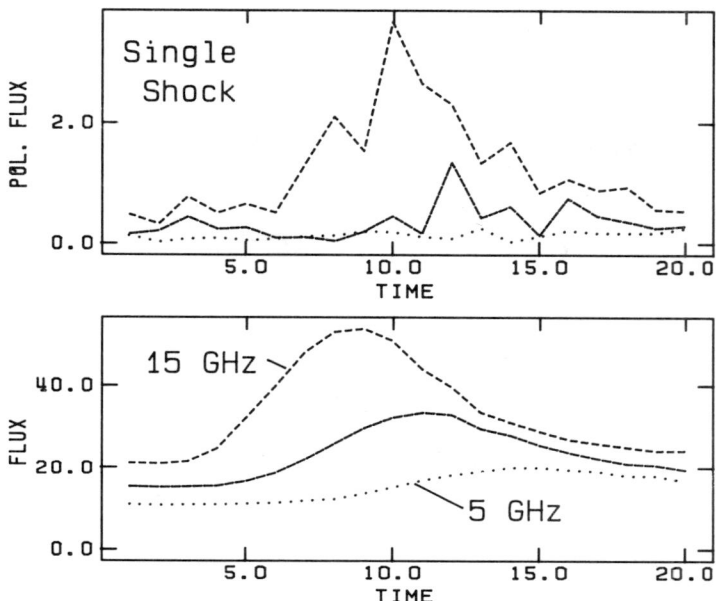

Figure 4 Integrated total flux density and polarized flux density predicted by a single shock model.

as these emphasizes that care is needed to interpret VLBI maps properly.

The ability of the shock mechanism to reproduce so many characteristics of both the observed time variations and source morphologies in such a natural manner makes it tempting to argue that all VLBI components and temporal "outbursts" are the result of shocks or similar disturbances propagating down preexisting jet flows. This hypothesis has several important implications.

(*a*) The component velocity observed by VLBI will not be the particle flow velocity but rather the velocity of the shock front; this velocity could be considerably less than the flow speed or even zero. Thus the Doppler boosting factor of the radiation emitted by the component may be quite different from that implied by the Lorentz factor required to produce the apparent component velocity.

(*b*) Only weak shocks are required, so that most of the jet energy continues to flow (unseen by us) down the jet channel toward some outer source component.

(*c*) Extragalactic jets should be regarded as quasi-steady-state structures: we see highly variable output from them because of the disturbances propagating down the jets, but the total output of the central

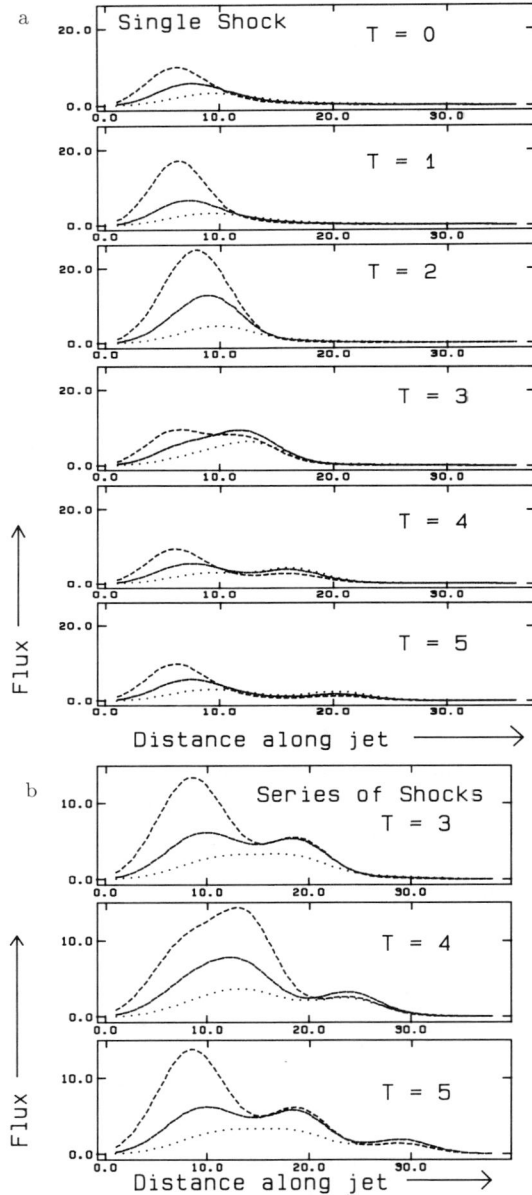

Figure 5 Predicted brightness profiles for shocks propagating down a preexisting jet at three frequencies versus time: (a) a single shock, (b) a series of three shocks. The brightness distributions have been convolved with an "instrumental beam" which is three units wide.

collimated power source need not vary significantly on short time scales.

(d) Because we are observing only a propagating disturbance, we may see only a small portion of the physical jet flow at any one time.

This work was supported in part by National Science Foundation grant AST 85-1093.

References

Aller, H. D., Aller, M. F., and Hughes, P. A. 1985, *Astrophys. J.*, **298**, 296.
Blandford, R. D., and Königl, A. 1979, *Astrophys. J.*, **232**, 34.
Hughes, P. A., Aller, H. D., and Aller, M. F. 1985, *Astrophys. J.*, **298**, 301.
Hughes, P. A., Aller, H. D., and Aller, M. F. 1986, *Can. J. Phys.*, **64**, 466.
Jones, T. W. 1982, *Bull. Am. Astron. Soc.*, **14**, 963.
Jones, T. W., and O'Dell, S. L. 1977, *Astrophys. J.*, **215**, 236.
Jones, T. W., Rudnick, L., Aller, H. D., Aller, M. F., Hodge, P. E., and Fiedler, R. L. 1985, *Astrophys. J.*, **290**, 627.

Synchro-Compton Emission from Superluminal Sources

ALAN P. MARSCHER

Department of Astronomy, Boston University

1. Introduction

Successful models for superluminal motion must reproduce the observed morphologies and rapid structural changes of the compact radio components in superluminal sources. But these properties represent only a subset of the observed characteristics of these sources. We must also require that the models be able to explain the spectrum, polarization, time variability, and other aspects of the emission from superluminal sources. This is where many attractive phenomenologies fail. For example, Hodge and Aller (1979) have shown that the dipole magnetic field model of Sanders (1974) and Milgrom and Bahcall (1978) predicts much higher circular polarization than is observed. Also, the scattering of radio waves required by Lynden-Bell's (1977) light-echo model is difficult to achieve.

The basic emission mechanism which appears to operate in superluminal sources is incoherent, random-pitch-angle synchrotron radiation in a non-uniform magnetic field. The nearly random pitch angles essentially cancel out the otherwise strong circular polarization of single-particle synchrotron radiation. The linear polarization, however, can still be quite high, and is regulated by the randomness of the local magnetic field directions. These properties correspond quite well to the observed polarization characteristics of compact radio sources, which have low circular polarizations (usually less than 1%) and moderate linear polarizations (1–10%, often higher in regions resolved by polarization VLBI; Roberts and Wardle, this Workshop, page 193).

A by-product of synchrotron radiation in compact regions is (inverse) Compton scattering ("self-Compton" emission). In this process, the radio and infrared synchrotron photons are scattered to higher frequencies, principally in the X-ray and gamma-ray range. Quasars and active galactic nuclei are in fact observed to be luminous X-ray sources. Furthermore, the X-ray flux is correlated with the radio flux density arising in compact regions. Thus, one question which begs to be addressed is whether most of the X-rays from sources with compact radio emission have a self-Compton origin. Since it is possible to derive the physical parameters of compact radio sources using VLBI observations and total flux density measurements, one would think that this question could be readily answered. What is discovered instead is that in many cases the parameters of the compact radio components are such that one would expect much brighter self-Compton X-ray emission than is observed. Rather than being a depressing defeat of previously accepted synchrotron theory, this discrepancy actually provides us with important constraints on the geometries and motions of the emission regions.

In this review, I will critique the customary application of synchro-Compton theory to real compact radio sources. The procedure typically used is to make the most simplifying assumptions regarding geometry. These assumptions can result in misleading conclusions if the observations provide insufficient constraints. Hence, part of the review will be a sort of *caveat observator* for studies of compact radio sources. It ends up that extremely bizarre-looking geometries are expected for superluminal components viewed from the observer's frame of reference.

After this, I will explore the paltry evidence for and against a self-Compton origin of the X-rays in radio-loud quasars and active galactic nuclei. It appears that the evidence simultaneously supports and rejects the self-Compton hypothesis.

Finally, I will present some ideas on the phenomenology of superluminal motion and try to understand why it appears to be explained by beaming models until one tries to find a clear case of a non-superluminal —and then discovers that the latter are rather rare. Even one of our two best documented non-superluminals, 4C 39.25, has recently undergone a superluminal event (Shaffer and Marscher, this Workshop, page 67). While the few detected non-superluminals may not eliminate existing beaming models, they do strain them to their limits.

I feel obliged to point out that most of the original ideas on compact radio source emission were generated by Terrell (1966), Kellermann and Pauliny-Toth (1969), Jones, O'Dell, and Stein (1974a, 1974b), and Burbidge, Jones, and O'Dell (1974). The basic relativistic jet model is from Blandford and Rees (1978) and Blandford and Königl (1979). Although I will not cite these papers after every concept which arose from them, it should be recognized that I had little to do with the origin of the thoughts on synchro-Compton theory presented here. My contribution has been a more practical one of trying to find the best methods for combining the theory with those observations which are feasible given the currently available techniques.

2. Basic Synchro-Compton Theory of Ideal Sources

Let us embark on a temporary trip to a land where all sources are uniform and spherical. The relativistic electrons which reside in this fantasy world have power-law energy distributions of the form

$$N(E) = N_\circ E^{-(2\alpha+1)}$$

between some limits $E_1 \leq E \leq E_2$, with $N(E) = 0$ outside these limits. At high frequencies (but still much lower than the synchrotron critical frequency ν_2 corresponding to electrons of energy E_2), the source is optically thin and the flux density is a power-law function of frequency: $S_\nu \propto \nu^{-\alpha}$. At some frequency ν_m the spectrum peaks and the flux density decreases toward lower frequencies owing to synchrotron self-absorption, with an asymptotic spectrum $S_\nu \propto \nu^{\frac{5}{2}}(\nu \ll \nu_m)$. Let S_m be the flux density at ν_m obtained by extrapolating the straight-line optically thin slope. [By doing this, we save having to multiply by e^{τ_m}, where $\tau_m(\alpha)$ is the optical depth of a line of sight passing through the center of the sphere. Since this term is different for different α's, I prefer to leave it out. Also, S_m is about as easy to estimate as the actual flux density at ν_m, since the latter usually involves an interpolation between observed frequencies.] The value of the spectral index can be determined, in principle at least, from the slope of the spectrum at optically thin frequencies. The other observables are S_m, ν_m, the redshift z of the associated optical object, the angular diameter θ, and the apparent speed relative to the speed of light $\beta_{\rm app}$, which, if interpreted properly, could yield information about the amount of relativistic beaming involved. The relativistic beaming effects are included through the Doppler factor

$$\delta = [\gamma(1 - \beta\cos\theta)]^{-1},$$

where β is the speed (in units of c) of the bulk motion of the emitting region, $\gamma = (1-\beta^2)^{-\frac{1}{2}}$, and θ is the angle which the bulk velocity vector makes to the line of sight, with $\theta = 0$ corresponding to motion directly at us. The luminosity distance D can be obtained from the redshift in the usual way if we adopt standard Friedmann cosmology.

Under the above assumptions and definitions, we can write down the following expressions for S_m and the optical depth τ_m at frequency ν_m (Marscher 1983 and references therein):

$$S_m = \tfrac{\pi}{6} c_1(\alpha) N_\circ B^{1+\alpha} \nu_m^{-\alpha} \theta^3 D(1+z)^{-(1+\alpha)} \delta^{3+\alpha} \qquad (1)$$

and

$$\tau_m(\alpha) = c_2(\alpha) N_\circ B^{\frac{3+2\alpha}{2}} \nu_m^{-\frac{5+2\alpha}{2}} \theta D(1+z)^{-\frac{9+2\alpha}{2}} \delta^{\frac{5+2\alpha}{2}}, \qquad (2)$$

Table 1 Functions of the spectral index α

α	τ_m	c_1	c_2	c_3	b
0.25	0.25	1.01×10^{-18}	2.29×10^{12}	2.7×10^{-22}	1.8
0.50	0.48	3.54×10^{-14}	1.17×10^{17}	3.2×10^{-19}	3.2
0.75	0.69	1.44×10^{-9}	6.42×10^{21}	3.9×10^{-16}	3.6
1.00	0.88	6.3×10^{-5}	3.5×10^{26}	5.0×10^{-13}	3.8

α	n	t	d	e	f
0.25	7.9	0.14	130	0.45	$0.95[(\nu_2/\nu_1)^{1/4} - 1]$
0.50	0.27	0.16	43	0.70	$0.27 \ln(\nu_2/\nu_1)$
0.75	0.012	0.15	18	0.95	$0.47[1 - (\nu_1/\nu_2)^{1/4}]$
1.00	5.9×10^{-4}	0.20	9.1	1.2	$0.23[1 - (\nu_1/\nu_2)^{1/2}]$

where $c_1(\alpha)$, $c_2(\alpha)$, and $\tau_m(\alpha)$ are tabulated in Table 1. (I recommend a logarithmic interpolation for values of α between those listed in the table.) The derivation of the dependence on z and δ is somewhat tricky. The factors most often overlooked are: (a) the line-of-sight depth through the source is δR instead of R/γ owing to light-travel-time effects; and (b) the angular diameter in terms of the radius and luminosity distance is $\theta = \frac{2R}{D}(1+z)^2$.

Only two unknowns appear in equations (1) and (2): B and $N_\circ$. We therefore obtain the following relations between B and $N_\circ$ and the observables:

$$B = 10^{-5} b(\alpha) \theta^4 \nu_m^5 S_m^{-2} \left(\frac{\delta}{1+z} \right) \text{ G}, \quad (3)$$

and

$$N_\circ = n(\alpha) D_{\text{Gpc}}^{-1} \theta^{-(4\alpha+7)} \nu_m^{-(4\alpha+5)} S_m^{2\alpha+3} (1+z)^{2(\alpha+3)} \delta^{-2(\alpha+2)}, \quad (4)$$

where $b(\alpha)$ and $n(\alpha)$ are tabulated in Table 1. Here and in what follows we express D_{Gpc} in units of 10^9 pc, θ in milliarcseconds, ν_m in GHz, and S_m in Jy. The cgs units of $N_\circ$ depend on the value of α: $\text{erg}^{2\alpha} \text{cm}^{-3}$. The extra D_{Gpc}^{-1} factor in equation (4) originates from the linear dependence of S_m and τ_m on the extent of the source along the line of sight. Hence, for non-spherical geometry with line-of-sight depth x, we must multiply the right-hand side by R/x. Non-spherical geometries will also result in minor modifications to the values of the functions of α tabulated in Table 1.

Once $N_\circ$ is determined, it is possible to integrate the function $N(E)$ over energy to obtain the number density of relativistic electrons or the function $N(E)E$ over energy to obtain the energy density in relativistic electrons. The limits E_1 and E_2 are rarely known (especially true for E_1); for typical values of α (0.5 to 1) this makes little difference for the energy density but renders estimates of the number density highly uncertain. In doing the integrals, we relate E_1 and E_2 to the observed cutoffs ν_1 and ν_2 through the formula $\nu \approx 4.2 \times 10^{18} BE^2$, where random pitch angles have been assumed. Since the resulting number density is highly uncertain, I give only the expression for the energy density in relativistic electrons:

$$U_{\rm re} \approx f(\alpha,\nu_1,\nu_2) D_{\rm Gpc}^{-1} \theta^{-9} \nu_m^{-7} S_m^4 (1+z)^7 \delta^{-5} \quad {\rm erg\ cm^{-3}}, \qquad (5)$$

where $f(\alpha,\nu_1,\nu_2)$ is given in Table 1.

The self-Compton flux density is somewhat more complicated to derive. Gould (1979) does a good job with the basics for spherical geometry. Adapting his results to the present formulation, we can obtain

$$S_\nu^C(E_{\rm keV}) \approx d(\alpha) \theta^{-2(2\alpha+3)} \nu_m^{-(3\alpha+5)} S_m^{2(\alpha+2)}$$
$$\times (h\nu)_{\rm keV}^{-\alpha} \ln\left(\frac{\nu_2}{\nu_m}\right) \left(\frac{1+z}{\delta}\right)^{2(\alpha+2)} \quad \mu{\rm Jy}, \qquad (6)$$

where $(h\nu)_{\rm keV}$ is the energy of the scattered photon in keV, and $d(\alpha)$ is tabulated in Table 1. It is instructive to cast this expression in a different form which involves the brightness temperature at frequency ν_m, T_m:

$$S_\nu^C(E_{\rm keV}) \approx \{T_m[e(\alpha) \times 10^{12}{\rm K}]^{-1}\}^{3+2\alpha} S_m \nu_m^{1+\alpha} (h\nu) E_{\rm keV}^{-\alpha}$$
$$\times \ln\left(\frac{\nu_2}{\nu_m}\right) \left(\frac{1+z}{\delta}\right)^{2(\alpha+2)} \quad \mu{\rm Jy}. \qquad (7)$$

This form illustrates the strong dependence of the self-Compton flux density on the brightness temperature, which is, in principle, an observable quantity. Since for most sources X-rays are detected at or below a level of 1 μJy, the brightness temperature T_m is limited to a few times 10^{11} K unless δ significantly exceeds unity. Notice that expression (7) does not depend on $D_{\rm Gpc}$, and hence depends only weakly on the assumption of spherical symmetry.

Another instructive quantity is the ratio of self-Compton flux density at frequency ν^C to the synchrotron flux density at ν^S:

$$\frac{S_\nu^C}{S_\nu^S} \approx c_3(\alpha) RN_{\circ} \ln\left(\frac{\nu_2}{\nu_m}\right) \left(\frac{\nu^S}{\nu^C}\right)^\alpha. \qquad (8)$$

This relationship is useful when comparing the (probably time-variable) X-ray flux with the synchrotron flux density of the region of the source which is suspected to be producing the X-rays via the self-Compton process.

As is well known, the self-Compton spectrum has the same slope as the optically thin synchrotron spectrum, $-\alpha$, between the limits

$$5.5 \times 10^{-9} \left(\frac{E_1}{mc^2}\right)^2 \nu_m \ll (h\nu)_{\text{keV}} \ll 5.5 \times 10^{-9} \left(\frac{E_2}{mc^2}\right)^2 \nu_2, \quad (9)$$

where ν_m and ν_2 are measured in GHz. Near each of the limits, the spectral slope changes gradually. Near the lower limit there is a broad turnover, and near the upper limit there is a gradual high-energy cutoff.

Ideal sources are therefore straightforward to analyze. Unfortunately, they do not seem to exist. In the next section I will critically examine the assumptions made above and try to place the formalism into the context of observational reality.

3. Real Sources

The difficulties encountered when one attempts to apply the expressions of Section 2 to real sources begin with the compound nature of compact extragalactic radio sources. Most sources contain a number of components, each becoming self-absorbed at a different frequency. The composite spectrum is usually remarkably flat; this fact has been noted and given the overstated nickname of the "Cosmic Conspiracy" (Cotton et al. 1980). Indeed, there must be a physical reason behind this, the most likely one in my opinion being that the "components" are all bright spots in an underlying jet which coordinates the physical parameters of what appear to be distinct components. In any case, the multicomponent nature of compact sources complicates the type of analysis outlined in the previous section.

In order to dissect accurately the spectrum of a compact radio source into the individual component spectra, one needs VLBI observations at many frequencies with resolution sufficiently high that no components appear blended with others. The reality is that the best-studied sources might have been observed using VLBI at four frequencies with resolutions which differ by a factor of ten or so. Hence, some assumptions are required if one is to have any hope at all of dissecting the spectrum. The most common assumption is that each component has the spectrum of a uniform synchrotron source. This is usually relaxed for the core since current theory identifies it with the base of a nonuniform jet. Hence, the spectral index of the core at frequencies below the turnover is usually left as a free parameter. It is difficult to overcome the under-resolution problem at lower frequencies. Some investigators prefer "super-resolution" model fitting or maximum-entropy maps. I find these to be unreliable and prefer to supplement them with the information on the relative positions of the centroids of blended components,

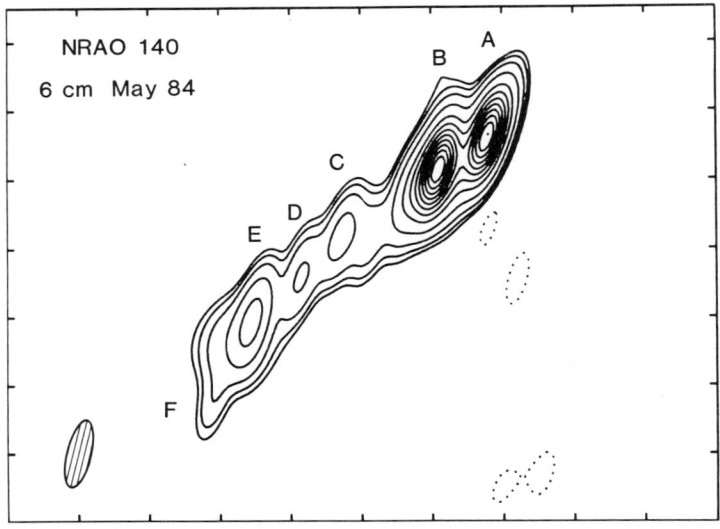

Figure 1 6 cm map of NRAO 140. Tick marks are 2 mas apart.

using higher frequency (and therefore higher resolution) maps to fix the individual component positions. The flux densities of the individual components at the lower frequency can then be adjusted to produce the observed positions of the brightness centroids of the blended components at that frequency. The results can then be checked against the super-resolved flux densities. This runs the risk of obtaining misleading results if there are spectral-index gradients. Except for the core, however, one does not expect strong spectral-index gradients, although this theoretical statement may not be of much comfort to observers.

For complicated sources containing more than three or four components, I find it essential to have features in the spectrum which indicate the approximate turnover frequencies of one or more components. In Figures 1 and 2, I show a 6 cm map of NRAO 140 and my dissection of the 1984 spectrum. The steepening in the total spectrum above 3 GHz indicates that component B peaks near that frequency. The assumption of uniform synchrotron source spectra for the non-core components (the core is identified with component A1, defined at 2.8 cm), the VLBI data, and the total flux density measurements then constrain the spectral dissection to something resembling Figure 2.

Since it was so helpful in the spectral dissection, how reliable is the assumption of uniformity? Nonuniformities manifest themselves as frequency-dependent angular sizes at frequencies below the turnover. Rarely, however, is a source resolved using Earth-based VLBI at frequencies below the turnover. Hence the angular size dependence on

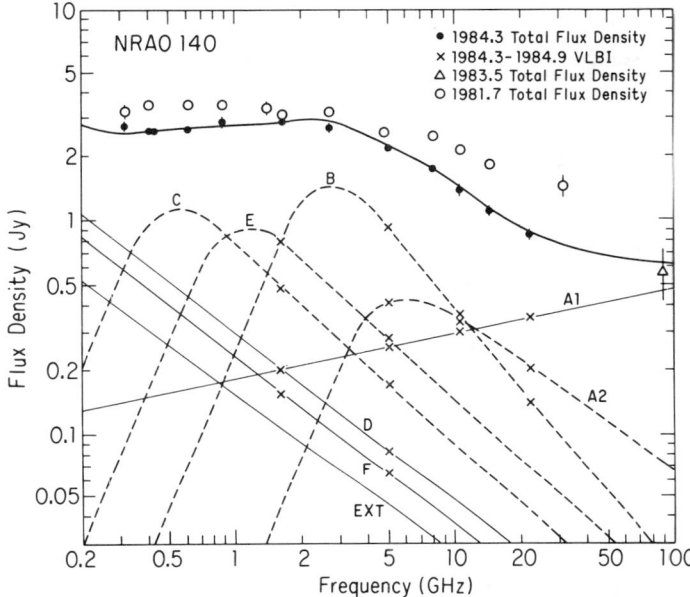

Figure 2 Spectral dissection of NRAO 140. From Marscher (in preparation), who gives references to the sources of the data.

frequency cannot ordinarily be determined. Nonuniformities caused by the existence of subcomponents can be all but ruled out if the component is over-resolved at higher frequencies; subcomponents should still be present on higher frequency maps, unless there are so many of them that they are individually undetectable. Except for the core, for which evidence suggests that the extent along the jet axis increases toward lower frequencies, I know of no cases where frequency-dependent sizes are indicated by the data. Still, only strong frequency dependences would be detectable. We need, therefore, to explore the theoretical likelihood, and the consequences with regard to analysis of the spectrum, of modestly frequency-dependent angular sizes.

In fact, the prevailing relativistic jet model predicts some frequency-dependent structure. In this context, it is revealing (and somewhat startling!) to draw an example of a feature in a relativistic jet. In Figure 3, a feature whose thickness along the jet axis is only $0.03r$ in the proper frame, where r is the distance from the convergent point of the conical jet, is shown in the observer's frame. Light-travel effects cause the far side of the feature to be situated quite a distance upstream of the near side, i.e., the observer sees the back side at an earlier stage in its evolution. When viewed in projection, this effect is nearly cancelled and the component might appear nearly circular. Still, the fact that the

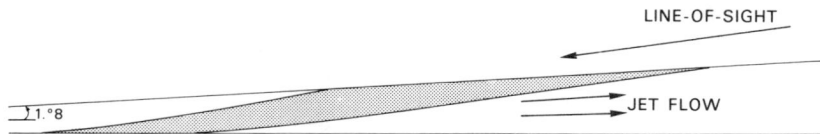

Figure 3 Example of a feature (*shaded*) in a relativistic jet in the observer's frame.

far side lies upstream where the densities and energies of the relativistic electrons are higher than at the near side means that the component is nonuniform, mainly along the line of sight. Inhomogeneities along the line of sight lead to a frequency-dependent depth along the line of sight, which in turn causes the self-absorption turnover to be broadened and the spectral index below the turnover to be less than 2.5. Therefore, the spectral dissection which assumed uniform components will not be correct in detail. Unfortunately, the limited resolution at lower frequencies prevents us from determining the shape of the spectrum below the turnover for most components. VLBI from space might someday help in this regard.

A nagging worry of mine (shared by many others in the field) with respect to synchrotron source analysis lies in the extremely strong dependence of the total energy requirements and expected self-Compton X-ray flux on the parameters ν_m and θ. Despite the likelihood of inhomogeneous components, it is often possible to constrain the value of the turnover frequency ν_m such that the uncertainty in this parameter is not so large as to render the analysis worthless. Of more concern is the angular diameter θ. First, the radiative transfer assumes that the components are spherical, yet we get better model fits to the VLBI data if the components are described by elliptical Gaussians. I handle this by noting that the visibility curve of a slice of a Gaussian of FWHM θ_G roughly coincides with that of a sphere of angular diameter $\theta_S \approx 1.8\theta_G$ (for a component less than 50% resolved). If the Gaussian is elliptical with major and minor FWHM axes θ_{G_a} and θ_{G_b}, I determine the value of θ_s of a sphere which subtends the same solid angle as the ellipse: $\theta_s = 1.8\sqrt{\theta_{G_a}\theta_{G_b}}$. This is the value of θ which I use in the expressions of Section 2. This procedure is justified by the dependence of the derived energy density of relativistic electrons, magnetic field, and self-Compton flux on the brightness temperature, which is a function of the solid angle which the component subtends on the sky, and not of the detailed geometry.

Another concern is that the value of θ used in the analysis should be that measured at the turnover frequency. However, since the Compton problem and energy requirements are most severe for compact components with turnovers at lower frequencies, the resolution is insufficient to determine the angular size near the turnover. Instead, the angular size is measured at a higher frequency and *assumed* to be the same at the turnover.

Frequency-dependent angular sizes at frequencies above the turnover could occur if the magnetic field or upper cutoff energy of the electrons decreases with distance from the center of the component. Nevertheless, synchrotron radiation is rather broadband, and an upper-frequency cutoff would also affect the flux density at the turnover in a manner which would partially compensate for the non-constant angular size. In addition, the optically thin spectrum would be steepened by a frequency-dependent angular size. In the least extreme case, suppose (unrealistically) that the depth along the line of sight is independent of frequency, and that the transverse dimensions vary as $\theta \propto \nu^{-\varsigma}$. The optically thin flux density would then vary as

$$S_\nu \propto \theta^2 \nu^{-\alpha} \propto \nu^{-(2\varsigma+\alpha)}.$$

For a value $\varsigma \geq 0.25$, the spectral index would be steepened by ≥ 0.5; I therefore think that $\varsigma = 0.25$ is a rough upper limit, since α is likely to be in the range 0.5 to 1 and the observed slope is not typically steeper than about 1. For NRAO 140 (Marscher and Broderick 1985), the angular size of the component with the Compton problem was measured at a frequency roughly 4 times ν_m. Hence, $\theta(\nu_m)$ could be pushed to

$$\sim 4^{0.25} \theta_{\text{measured}} \sim 1.4\, \theta_{\text{measured}}.$$

This is not a sufficient increase to negate the Compton or the total energy problem.

I therefore conclude that the Compton problem and total energy requirements are not substantially mitigated by considering source structures more complicated than the multiple, uniform-component model used by most investigators. Hence, I recommend keeping the assumed model simple unless forced to do otherwise by the observations. Still, Compton problems of order a factor of 1 to 30 might simply be consequences of our simplifying assumptions. I also want to emphasize that it is very difficult to analyze a complex radio source with accuracy sufficient to substantiate claims of Compton or total energy problems. I really find it difficult to take such claims seriously unless they are backed up by contemporaneous VLBI at three or more frequencies. Nevertheless, those few self-Compton calculations which are reliable provide lower limits to the Doppler factor which are distant-independent, and therefore quite useful.

As far as "predictions" of superluminal motion go, based on the existence of a Compton problem, the cases of 4C 39.25 and 3C 395 now make it clear that stationary structure can exist in sources which show evidence for relativistic motion. I therefore retract my previous prediction of superluminal motion in NRAO 140, even though it worked.

4. Are the X-rays Self-Compton?

As indicated in the introduction, the answer to the above question is a hearty "I don't know, and neither does anybody else". Still, it is worthwhile to examine the evidence for the self-Compton process and likewise to explore the seriousness of the evidence against it. I shall confine my comments to the X-ray emission observed in radio-loud quasars. This class of quasars has been found to have X-ray emission in excess of that found in radio-faint quasars, for a given optical luminosity. Hence, it is already established in those sources that the X-ray emission is related to that in the radio, in particular with the radio emission arising from compact components.

Owen, Helfand, and Spangler (1981) and Ledden and O'Dell (1985) have found a tight correlation between 90 GHz flux density and 0.5–3 keV X-ray flux in quasars which are bright at millimeter wavelengths. Strong millimeter emission is associated with the most compact radio components, and the brightness tends to be highly variable. The tight correlation is destroyed if either the millimeter or the X-ray emission of a typical source varies by a factor $\gtrsim 2$ *unless* the variations are synchronous and in the same sense at the two wavebands. Flux monitoring surveys at both wavebands are incomplete, so it is not yet clear whether this is a problem. Equation (8) shows that if the quantity RN_o is constant as the component evolves, the synchrotron and self-Compton emission will vary together. Another way for the variations to be directly related would be if the X-rays represent the extension of the synchrotron spectrum to higher frequencies. While the evidence for this seems good in the case of some BL Lac objects with steep soft X-ray spectra (Urry 1986), the majority of quasars have X-ray spectra flatter than would be expected under this scenario.

In the source for which I have obtained data, NRAO 140, the X-rays did indeed decline along with the core flux density (Marscher 1986). Furthermore, my best spectrum, obtained using EXOSAT, gives an X-ray spectral index of about 0.5 (i.e., $S_\nu \propto \nu^{-0.5}$). Although the optically thin spectral index of the core is well known in only a few sources (and NRAO 140 is not one of them), this value falls within the range of 0.5 to 1 observed in those few. Thus, the expected mimicking of the synchrotron slope is consistent with the data, although that is not saying much. As is well known, analysis of the synchrotron spectrum of a source often leads to *overprediction* of the self-Compton X-ray flux, so the best that can be said is that this analysis shows that the X-rays *could* be self-Compton in origin.

Actually, the evidence against the self-Compton process producing the X-rays is more formidable, although this may also be an illusion of overly simple assumptions. In 3C 273, the spectral index of the optically thin submillimeter emission is 0.7 in the quasi-steady and 1.2 in the variable component (Clegg et al. 1983; Robson et al. 1983; Marscher and Gear 1985). The X-ray spectral index, on the other hand, is reported to be closer to 0.4 (Worrall et al. 1979; Halpern 1982; Petre et al. 1984) at soft energies, at both high and low brightness. An even flatter slope has been reported at hard X-rays by Bezler et al. (1984) during a high brightness state. Many of my X-ray friends point out that the X-ray spectra are subject to systematic errors. If they stand up, however, either 1 keV lies just above the Compton-scattered turnover or the X-rays are not Compton-scattered descendants of the core's synchrotron photons.

Another strike against the self-Compton hypothesis is the discrepant time scales of variability in the millimeter to far-infrared and X-ray wavebands. Again concentrating on the bright X-ray quasar 3C 273, the X-rays have been reported to vary on a time scale of about a day, whereas the millimeter brightness changes over a time scale of weeks or longer. Since we have been conditioned from our youth to translate time scales to size scales (and from there to distance from the "central engine"), the immediate reaction is to declare the mass of the black hole by equating the size scale to a Schwarzschild radius. This temptation should be resisted at all costs.

In the context of the relativistic jet model, it is possible for geometries to arise in which the components are foreshortened along the line of sight. This is crucial since the line-of-sight dimension could be the only one which determines the time scale of variability. If we call this dimension x, the time scale can be as short as $(\frac{x}{c})(\frac{1+z}{\delta})$. In the relativistic shock model of Marscher and Gear (1985), electrons are accelerated in the shock front and subsequently lose energy by radiation and expansion as they convect away from the shock front. The highest energy electrons can lose energy quite rapidly, and hence radiate only in a thin shell behind the shock front. The high-frequency radiation therefore arises from a very thin region. If the observer is looking nearly down the jet axis, this translates to a small value of x and therefore a time scale of variability much shorter than one would expect given the *transverse* dimension of the component (which corresponds to the observed angular size). The possibility that the parameter x could be a function of frequency would lead to a frequency-dependent time scale of variability for a single event in a single component. For these reasons I really do not believe that time scales of variability by themselves necessarily void the hypothesis that the millimeter wave components produce the X-rays through the self-Compton process.

Actually relating time scales to size scales, i.e., determining the appropriate value of x, is quite tricky. If a thin component is not viewed face-on, x is roughly the line-of-sight displacement of the near edge from the far edge. The same is true if the front of the component is curved. As is illustrated in Figure 3, front-to-back time smearing must also be taken into account, although this is compensated to a large degree by length contraction in the direction of motion.

Another non-obvious consequence of paper-thin component geometries is that the synchrotron spectrum viewed by an electron inside the source is quite different from that detected by the observer. Most of the photons in the former case arrive from directions parallel to the front of the component. Therefore, the optical depth is proportional to the transverse dimension R, as opposed to x as seen from the observer's perspective. The turnover frequency as seen by an electron near the center of the component is then related to the observed turnover by

$$\nu'_m \approx \left(\frac{R}{x}\right)^{\frac{2}{2\alpha+5}} \nu_m,$$

where I have left out the $(1+z)$ and δ factors because I would need to take them right back out to make my point. The lower limit to relation (9) needs to have ν_m replaced by ν'_m, that is, the self-Compton low-frequency turnover moves to higher frequencies by the factor $(R/x)^{\frac{2}{2\alpha+5}}$. This is important since it is within about a decade above this frequency that the self-Compton X-ray shape remains flattened relative to the optically thin synchrotron slope.

Another point is that the frequency dependence of x steepens the observed optically thin slope of the synchrotron radiation. This steepening is not seen by an electron inside the component, so the observed optically thin synchrotron spectral index can be steeper (typically by 0.5) than the self-Compton slope.

I am afraid that what I have done is to destroy the predictive ability of the self-Compton hypothesis so that it cannot be disproven. Indeed, it cannot, given the currently available data. What is really needed is close flux density monitoring at millimeter and submillimeter wavelengths as well as in the X-ray; this must include accurate spectral indices. Although it may be nearly impossible to drive the last nail into the self-Compton coffin, totally uncorrelated millimeter–submillimeter and X-ray variations would render it a highly unlikely choice for the X-ray emission mechanism.

5. Alternatives to the Standard Model

The standard model for superluminal motion, in which a relativistic jet points almost directly at us, has fallen upon some hard times. Although most of us still admire it for bringing us the unification for which we had been searching, it has proven quite difficult to find the unbeamed counterparts to the supposedly beamed compact radio sources. Superluminal motion is found in nearly every compact source. Even the much heralded counterexample 4C 39.25 has now been planted firmly in the ranks of the superluminals. There are actually only a very few documented cases of non-superluminals and no one knows how many of these will remain after further observation. The source 4C 39.25 has superluminal motion, no component with the properties of a classical "core", and two-sided (although asymmetric) arcsecond-scale jet structure. Is this some intermediate state between highly beamed and unbeamed sources or can superluminal motion occur in weakly beamed sources?

In this section, I will examine some of the assumptions we tend to make when interpreting superluminal sources, and then have some fun perturbing the jet models in order to force them to produce superluminal motion for large angles between the jet axis and the line of sight.

Positions of the Knots and the Core

We must be careful to keep in mind that we measure superluminal motion by the change in position of the brightness centroid of a "knot" relative to that of the "core". If the knot itself has constant or symmetrically changing structure and the core is stationary, the measured speed will be accurate, to within the accuracy of the distance determination and angular separation measurements. However, it is possible for the knot to expand back toward the core; this is expected if the knot is actually a shock wave. In this case, the brightness centroid moves backward as viewed from the shock front, and the observer measures an apparent speed slower than that of the shock front. If the expansion ends (for example, when radiative losses cease to be important), acceleration up to the apparent speed of the shock front will occur. I would expect, however, that expansion will always occur, so for the shock-in-jet model the apparent superluminal motion requires a higher Lorentz factor than $\gamma = c/v_{\rm app}$, where $v_{\rm app}$ is the apparent speed of the centroid of the component.

It is well known that the gradients expected in density and magnetic field in a jet lead to a frequency-dependent apparent position of the base of the jet. If the base is identified as the core, the core-knot separation should be frequency dependent. This can also happen if the line-of-sight thickness of the knot x is frequency dependent. Such frequency dependence has been reported by Biretta, Moore, and Cohen (1986) for 3C 345, and is in the expected sense of smaller separation at longer wavelengths.

I also have a nagging doubt, perhaps unfounded, that what we call the "core" is actually the base of the jet. The observation by Bartel et al. (1986) that the core of 3C 345 did not move appreciably relative to an unrelated radio source over a ten-year period had convinced me that the Blandford-Königl picture was correct. However, the acceleration of a superluminal knot observed by Biretta, Moore, and Cohen (1986) leads to the possibility that components start out with zero velocity. Could the "core" simply be the newest knot, always starting out at the same spot as the last one? One possible scenario, borrowed from some of Peter Scheuer's remarks, is that the "core" is the point in a bent jet where the components are coming right at us. They are then brightest because they are beamed as strongly as possible, but have zero transverse velocity. Figure 4 shows the geometry. The strong beaming of the apparently stationary feature causes it to dominate over the smaller, less strongly beamed, upstream features, which would form a weak counterjet. If this scenario is correct (which I doubt), very-high-frequency VLBI might detect superluminal contraction in the counterjet. Independent of whether the geometry of Figure 4 is relevant, we must keep in mind that the repeated generation of superluminal knots implies that the "core" is often (perhaps always) a conglomerate of the real core (if there is one) and a new knot which has not yet been resolved. In some cases, the flux density of the knot might exceed that of the core, thereby reducing or eliminating apparent superluminal separation of other knots from this conglomerate "core".

Broad Beams

Another bug in my ear is the apparent lack of jets viewed directly down the beam. If the opening half-angle of a relativistic jet is the inverse tangent of the Mach number, as suggested by Blandford and Königl (1979), it is given by $\tan\psi \approx (\sqrt{3}\gamma)^{-1}$, where γ is the Lorentz factor of the jet. This is supposed to hold for relativistically hot gas inside the jet. For this angle, it is straightforward to calculate that over 40% of the sources in a flux-limited survey would be viewed down the beam, if γ is the same for each source. Since it is difficult to identify any class of source as those whose beams are pointing directly at us, either (a) the opening angles are much narrower than this, (b) all jets bend, and the section pointing at us is the "core", as in Figure 4, or (c) we need to allow wider viewing angles in our jet models for superluminal sources. Many of those who have been studying samples essentially free

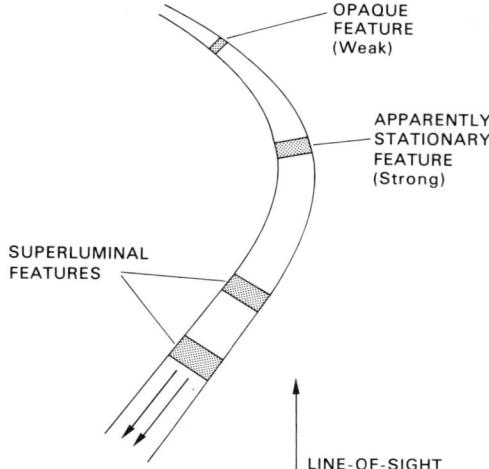

Figure 4 Stationary and superluminal features in a bent relativistic jet.

of orientation bias strongly favor option (c). I personally have favored option (a), but I must admit that the work of Zensus and Porcas (this Workshop, page 126) and others makes me feel quite uncomfortable. For this reason, I now describe some preliminary results on the gas dynamics of relativistic jets (Daly and Marscher, research in progress) and relate these results to radio source statistics.

As Königl (1980) has shown, jets containing relativistically hot gas behave in a fashion quite similar to those containing nonrelativistic gas, from a fluid-dynamical standpoint. There are, however, two important differences: the speed of sound of a relativistic gas is constant at $c/\sqrt{3}$, and a relativistic gas eventually becomes nonrelativistic as it cools. The latter property causes the gas to switch from a realm of constant sound speed and adiabatic index $\hat{\gamma} = 4/3$ to one of diminishing sound speed and $\hat{\gamma} = 5/3$. The constancy of the sound speed while the gas remains relativistically hot allows the gas to accelerate to relativistic speeds as it expands, since the gas keeps trying to expand at the sound speed in the *co-moving* frame. For a nonrelativistic gas, this effect cannot keep up with the continual decrease in sound speed as the expansion proceeds, so the ultimate expansion velocity is $\sqrt{3}$ times the initial sound speed. The combination of the two effects causes the relativistic gas to convert most of its internal energy to relativistic expansion. The expansion "freezes out" once the "temperature" falls to nonrelativistic values, since the lowered sound speed cuts off communication between different sections of the fluid.

It is revealing to consider the classical problem of a confined jet

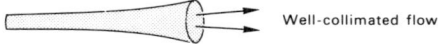

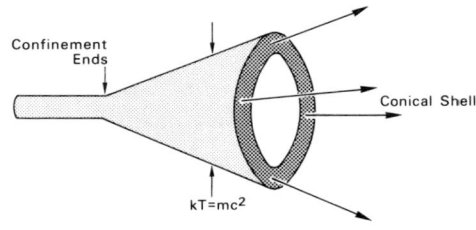
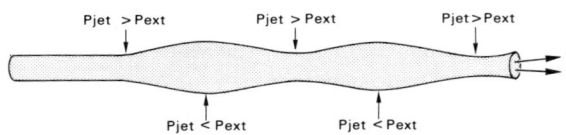

Figure 5 Profiles of relativistic jets experiencing an external pressure drop.

which enters a region of lower external pressure. The outer edge of the jet immediately bends so that the gas at the boundary expands, maintaining pressure equilibrium with its surroundings. A standing rarefaction moves into the jet interior, communicating the need to expand to match the external pressure. If the drop in external pressure is mild, the information is transported via Mach waves. The jet expands, overexpands (because of inertia), and then is recollimated by a compression wave reflected off the boundary (Owczarek 1964). A series of bulges and constrictions is then set up (Figure 5 III). If the drop in external pressure is more severe, standing shocks will intercept the Mach waves, leading to interesting variations and possibly jet disruption (Courant and Friedrichs 1948). This behavior will occur if the gas remains relativistically hot throughout. If the gas has a nonrelativistic "temperature" but relativistic bulk velocity prior to the pressure drop, the flow can be extremely well collimated owing to relativistic aberration of the opening angle and lack of desire to expand much (Figure 5 I).

If the pressure drop is rather large, it is possible for the initially relativistic "temperature" to become nonrelativistic during the expan-

Table 2 Jet experiencing a drop in external pressure

γ_0	$\psi_c - \psi_0$ ($\chi = 5$)	$\psi_c - \psi_0$ ($\chi = 25$)	$\psi_c - \psi_0$ ($\chi \to \infty$)
1.25	21°	36°	65°
1.5	19°	31°	56°
2.0	14°	23°	41°
3.0	9°	15°	27°

ψ_c is the opening half-angle of a jet experiencing a drop in external pressure while still relativistically hot.

sion, such that the expansion freezes out before recollimation can occur. Since the rarefaction still makes it to the jet axis, the entire interior is caught up in the expansion. A nearly hollow conical geometry is thereby set up (see Figure 5 II). The opening half-angle ψ_c of the cone depends on the upstream Lorentz factor γ_0 (again, because of aberration), the opening half-angle before the pressure drop ψ_0, and the ratio χ of upstream external pressure to the value after the pressure drop:

$$\psi_c = \psi_0 + \sqrt{3}\arctan\sqrt{\tfrac{2}{3}\sqrt{\chi}\gamma_0^2 - 1} \; - \; \arctan\sqrt{2\sqrt{\chi}\gamma_0^2 - 3} \\ - \sqrt{3}\arctan\sqrt{\tfrac{2}{3}\gamma_0^2 - 1} \; + \; \arctan\sqrt{2\gamma_0^2 - 3}. \quad (10)$$

[See Königl (1980) for the expressions from which this result is derived.] Table 2 gives values of ψ_c for various choices of γ_0 and χ. As can be seen, the opening angle can be quite large if the pressure drop is precipitous and the upstream Lorentz factor low. (The Lorentz factor after the pressure drop is given by $\gamma = \gamma_0 \chi^{1/4}$, so Lorentz factors $\sim 5\text{--}10$ require very large pressure drops if $\gamma_0 \lesssim 2$.) The result is then a broad, hollow beam. There is a distinct possibility that shocks could arise, and the flow could be considerably more complicated than indicated above. Also, the pressure drop is probably not so sudden in reality, so the transition would not be so abrupt. Still, it is encouraging that broad beams can arise naturally.

Scheuer (1984) has pointed out that broad beams should be observed as obese jets, and I find this to be correct if the beam is circularly symmetric. If, however, the external pressure is not perfectly circularly symmetric after the pressure drop, the jet can squirt out with opening angles which are a function of azimuthal angle. A spectacular "multiple searchlight" pattern is possible, at least in principle. Hence, I think that it is worthwhile to discuss the considerable merits of this scenario rather than give up based on this solitary objection.

The "jet" identified by the observer would be the piece of the beam most closely directed toward the line of sight. The statistical probability of having a piece point within an angle $\sim 1/\gamma$ of the observer is $\sim \gamma\psi_c$

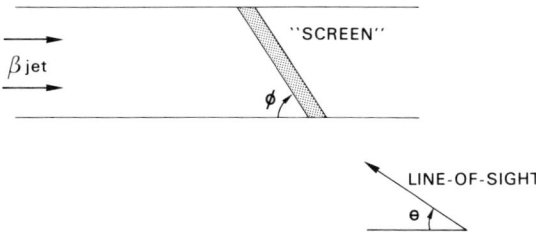

Figure 6 Geometry of a screen model.

times higher than for the narrow beam model. The real beauty of this scenario, however, is that the beam reverts to a well collimated, narrow jet whenever the gas is not relativistically hot at the point of the pressure drop. This might occur most of the time, with the relativistically hot phase being relatively transient. The narrow beam would show up as an arcsecond scale jet which feeds double lobes (if there are any). If the relativistic "temperature" were to turn on gradually, the observed piece of the broad beam would appear to connect smoothly with the narrow jet, as in 3C 120.

The alternately expanding and constricting jet of Figure 5 III would have stationary hotspots near the constrictions. This could explain the appearance of stationary features in jets with superluminal motion, as observed in 4C 39.25 and 3C 395. If so, the superluminal feature should pass through the stationary feature and continue on the other side. Except for values of χ very close to unity, shocks form slightly beyond the first constriction. This leads both to enhanced emission and a disruption of the pattern of oscillating jet boundary.

It is clear that much remains to be learned through theoretical studies of the relativistic gas dynamics in jets.

Superluminal Screens

Roger Blandford (this Workshop, page 310) discusses a possible phase effect which gives rise to a superluminal pattern speed in a jet through the "scissors effect". Superluminal pattern speeds can also be obtained in "screen" models if the screen is oriented properly. A simple geometry of such a screen in a jet is shown in Figure 6. The screen in this example allows superluminal motion if a "signal" (e.g., a shock wave) propagating down the jet strikes the far side of the screen first, and if the angle ϕ is such that the light travel time (toward the observer) from the point struck earlier to a second, closer point, is comparable to the time it takes for the signal to strike that second point. The signal then appears to strike the screen much more rapidly than it actually does.

Mathematically, the pattern speed is

$$\beta_p = \frac{\beta_{\text{jet}}}{1 - \beta_{\text{jet}} \sin\theta \tan\phi}. \tag{11}$$

For example, for $\tan\phi = 2$ ($\phi = 63°$) and a semi-relativistic jet with $\beta_{\text{jet}} = 0.5$, superluminal motion is observed from 87% of all viewing angles, with $\theta = 90°$ giving infinite pattern speed (in this example only).

Of course, it is hard to arrange things so that the source appears as a core-jet, so a workable model is difficult to imagine with my geometry. Still, the basic principle could apply to other geometries, one of which might be more reasonable.

The National Science Foundation partially supported the author's VLBI and theoretical research upon which the above remarks are based, under grants AST 83-15556 and AST 85-16548. NASA partially supported the X-ray observations of NRAO 140 under grants NAG 8-315 (HEAO-2) and NAG 8-518 (EXOSAT). The author is a Guest Observer on the HEAO-2 (Einstein) and EXOSAT programs.

References

Bartel, N., Herring, T. A., Ratner, M. I., Shapiro, I. I., and Corey, B. E. 1986, *Nature*, **319**, 733.
Bezler, M., Kendziorra, E., Staubert, R., Hasinger, G., Pietsch, W., Reppin, C., Trümper, J., and Voges, W. 1984, *Astron. Astrophys.*, **136**, 351.
Biretta, J. A., Moore, R. L., and Cohen, M. H. 1986, *Astrophys. J.*, **308**, 93.
Blandford, R. D., and Königl, A. 1979, *Astrophys. J.*, **232**, 34.
Blandford, R. D., and Rees, M. J. 1978, in *Pittsburgh Conference on BL Lac Objects*, ed. A. M. Wolfe (Pittsburgh: University of Pittsburgh), p. 328.
Burbidge, G. R., Jones, T. W., and O'Dell, S. L. 1974, *Astrophys. J.*, **193**, 43.
Clegg, P. E., Gear, W. K., Ade, P. A. R., Robson, E. I., Smith, M. G., Nolt, I. G., Radostitz, J. V., Glaccum, W., Harper, D. A., and Low, F. J. 1983, *Astrophys. J.*, **273**, 58.
Cotton, W. D., Wittels, J. J., Shapiro, I. I., Marcaide, J., Owen, F. N., Spangler, S. R., Rius, A., Angulo, C., Clark, T. A., and Knight, C. A. 1980, *Astrophys. J. (Letters)*, **238**, L123.
Courant, R., and Friedrichs, K. O. 1948, *Supersonic Flow and Shock Waves* (New York: Interscience).
Gould, R. J. 1979, *Astron. Astrophys.*, **76**, 306.
Halpern, J. P. 1982, Ph. D. thesis, Harvard University.
Hodge, P. E., and Aller, H. D. 1979, *Nature*, **278**, 838.
Jones, T. W., O'Dell, S. L., and Stein, W. A. 1974a, *Astrophys. J.*, **188**, 353.
Jones, T. W., O'Dell, S. L., and Stein, W. A. 1974b, *Astrophys. J.*, **192**, 261.
Kellermann, K. I., and Pauliny-Toth, I. I. K. 1969, *Astrophys. J. (Letters)*, **155**, L71.
Königl, A. 1980, *Phys. Fluids*, **23**, 1083.
Ledden, J. E., and O'Dell, S. L. 1985, *Astrophys. J.*, **298**, 630.
Lynden-Bell, D. 1977, *Nature*, **270**, 396.
Marscher, A. P. 1983, *Astrophys. J.*, **264**, 296.
Marscher, A. P. 1986, in *Continuum Emission in Active Galactic Nuclei*, ed. M. L. Sitko (Tucson: Kitt Peak National Observatory, National Optical Astronomy Observatories), p. 143.
Marscher, A. P., and Broderick, J. J. 1981, *Astrophys. J. (Letters)*, **247**, L49.
Marscher, A. P., and Broderick, J. J. 1985, *Astrophys. J.*, **290**, 735.

Marscher, A. P., and Gear, W. K. 1985, *Astrophys. J.*, **298**, 114.
Milgrom, M., and Bahcall, J. N. 1978, *Nature*, **274**, 349.
Owczarek, J. A. 1964, *Fundamentals of Gas Dynamics* (Scranton: International Textbook Company).
Owen, F. N., Helfand, D. J., and Spangler, S. R. 1981, *Astrophys. J. (Letters)*, **250**, L55.
Petre, R., Mushotzky, R. F., Krolik, J. H., and Holt, S. S. 1984, *Astrophys. J.*, **280**, 499.
Robson, E. I., Gear, W. K., Clegg, P. E., Ade, P. A. R., Smith, M. G., Griffin, M. J., Nolt, I. G., Radostitz, J. V., and Howard, R. J. 1983, *Nature*, **305**, 194.
Sanders, R. H. 1974, *Nature*, **248**, 390.
Scheuer, P. A. G. 1984, in *IAU Symposium 110, VLBI and Compact Radio Sources*, ed. R. Fanti, K. Kellermann, and G. Setti (Dordrecht: Reidel), p. 197.
Terrell, N. J. 1966, *Science*, **154**, 1281.
Urry, C. M. 1986, in *Continuum Emission in Active Galactic Nuclei*, ed. M. L. Sitko (Tucson: Kitt Peak National Observatory, National Optical Astronomy Observatories), p. 91.
Worrall, D. M., Mushotzky, R. F., Boldt, E. A., Holt, S. S., and Serlemitsos, P. J. 1979, *Astrophys. J.*, **232**, 683.

How Fast Can a Blob Go?

E. S. PHINNEY

Theoretical Astrophysics, California Institute of Technology

and

Institute for Theoretical Physics, University of California, Santa Barbara

1. Apparent Speed Limits

The spots of radio emission mapped with VLBI often appear to move across the sky with speeds $v \sim (1\text{--}10)h^{-1}c$. We learned at this workshop from Porcas (page 12) and Pearson, Readhead, and Barthel (page 94) that a large fraction, perhaps a majority, of radio sources with bright cores exhibit this superluminal motion. We also heard of two relatively poorly-studied sources (3C 454.3: Pauliny-Toth, page 55; CTA 102: Bååth, page 206) whose blobs might separate at $v \sim 15h^{-1}c$. Yet there do not appear to be any convincing claims of speeds in excess of $20h^{-1}c$. Here I discuss the significance of a characteristic transverse speed of VLBI knots $v \sim 10h^{-1}c$ and its relevance as a constraint on models of jet acceleration.

Is there really some characteristic maximum speed $\sim 10h^{-1}c$? One response would be that humanity has mapped a grand total of only 100–200 blobs with VLBI. If the blobs moved at random angles to the line of sight, then even if their three-dimensional pattern speeds approached c, the distribution of apparent speeds ($\mathcal{P}(> v/c) \lesssim 2(c/v)^2$) is such that we wouldn't *expect* to have seen any apparent speeds $> 20c$.

This is an unconvincing explanation for two (possibly unconvincing) reasons. First, the statistics of core-to-extended flux ratios (Hough and Readhead, page 114; Browne, page 129) suggest that in the sources with bright cores which are preferentially mapped, the emission from the moving patterns is at least roughly beamed in their directions of motion. If this beaming is over a fraction f of the sky and we map N blobs (with intrinsic pattern speed $\to c$) moving within the beaming cone, the apparent transverse speed of the fastest moving blob will typically be $\bar{v}_{\max} \sim c\sqrt{2N/f}$ and the probability that the actual maximum speed will be $v_{\max} < \bar{v}_{\max}$ is

$$\mathcal{P}(v_{\max}) = \left[1 - \frac{2}{f}\left(\frac{c}{v_{\max}}\right)^2\right]^N . \tag{1}$$

For likely values, $N \sim 100$, $f \lesssim 0.1$, this gives $\overline{v}_{\max} \gtrsim 45c$, and the observed upper limit is quite improbable if $H_0 = 100 \,\mathrm{km\,s^{-1} Mpc^{-1}}$, but not significantly so if $H_0 = 50$!†

A second, more compelling argument is the absence of observational evidence for huge flares (e.g., rapid changes in intensity by more than a factor of 1000) or inferred brightness temperatures much in excess of 10^{15} K. Avoiding trouble with inverse Compton radiation does not require the synchrotron-emitting electrons (as opposed to the *pattern* constrained above) to have bulk Lorentz factors in excess of $5\,h^{-1}$. Even tiny amounts of jet material with a bulk Lorentz factor > 25 would cause easily detectable flares. On the other hand, there appears to be no observational objection to *smooth* jets with Lorentz factors as high as 50, provided their vectors of bulk velocity are spread over enough of the sky (10^{-1}–10^{-3}) to avoid embarrassing space densities of unbeamed counterparts. Although brightness temperatures $> 10^{15}$ K would then be physically possible, they need not occur if the brightness temperature in the comoving frame is typically $\ll 10^{12}$ K (there is no obvious reason why emitting plasma should always be near the inverse Compton limit in its rest frame, though an optically thick synchrotron source cannot have a rest-frame brightness temperature much less than $\sim 10^{10}$ K, whence the upper limit $\gamma \lesssim 50$).

2. The Compton Speed Limit

The particles in a relativistic jet see blue-shifted photons pouring in from the forward direction. The resulting radiation pressure on electrons in the jet will brake any jet whose initial Lorentz factor exceeds some critical value, converting the excess kinetic energy into a directed beam of scattered photons.

---

† The inequalities in the probability distributions in the preceding paragraphs are strictly valid only if it is possible to select blobs whose pattern velocities are randomly oriented. The scarcity of superluminal contractions and 180° misalignments with large-scale structure suggests that the line of sight must always be outside the cone defined by the direction of blob motion and the axis defined by larger-scale structure. Because of this and the general impression that high apparent speeds are "too common", there is interest in models which use beaming to inextricably link the direction of pattern motion to that of the line of sight. In the class of models considered by Scheuer (1984; see also Phinney 1985) the pattern moves at an angle $(1-1/n)$ times closer to the line of sight than the source axis. For randomly oriented source axes, this *increases* the probability of seeing high apparent speeds; however the maximum possible apparent speed is c times the Lorentz factor of the pattern, which is in turn somewhat *less* than the Lorentz factor of the front of particles. Thus, in these simple models, an observed upper limit to v/c constrains γ even more strongly than in simple "cannon-ball" models. It is, however, possible to construct contrived models (which do not show contractions from any viewing angle) in which the pattern speed greatly exceeds that of any physical particles.

The electrons, positrons, or electromagnetically coupled electron-proton pairs making up a (cold) jet in a radiation field of (frequency integrated) specific intensity $I(\theta, \phi)$ will gain or lose energy $p^0 = \gamma mc^2$ according to (Phinney 1982; O'Dell 1981)

$$c\beta\frac{dp^0}{dr} = \frac{dp^0}{dt} = -\sigma\left[\gamma^2\int I(1-\beta\cos\theta)^2\,d\Omega - \int I(1-\beta\cos\theta)\,d\Omega\right] \quad (2)$$

$$c\frac{dp^0}{dr} \simeq \sigma F\left[\frac{1}{4\gamma^2} - k_1\theta^2 - k_2\gamma^2\theta^4\right] - \frac{4}{3}\sigma\gamma^2 U_{\rm ISO} c, \quad (3)$$

where the second approximate equality holds if $\gamma \gg 1$ and the radiation field consists of a flux F confined to a range of angles $\theta \ll 1$ with respect to the direction of particle motion (e.g., at a distance r from a source of size R and isotropic luminosity L, $\theta \sim R/r$ and $F = \frac{L}{4\pi r^2}$), *plus* an isotropic component of energy density $U_{\rm ISO}$. The constants k_1 and k_2 depend on the details of the limb-darkening of the source, but are typically ~ 0.05. The first term on the right of (3) vanishes when $\gamma \sim 1.5/\theta$, i.e., when the particle velocity is such that there is no net momentum flux in the aberrated radiation field of its rest frame.

The radiation field in an active galactic nucleus is probably dominated by emission from some sort of accretion disk with luminosity L_d, inner radius $R_i \sim (2\text{--}10)r_g$ ($r_g \equiv GM_h/c^2$, M_h being the mass of the putative black hole), which emits from the annulus between R and $2R$ a luminosity $L(R) \sim L_d(R_i/R)$. As seen from a point a distance r out along the symmetry axis of the disk, the effective values of $F\theta^2$ and $F\theta^4$ are dominated by that part of the disk with $R \sim r$ ($\theta \sim 1$), and are $\sim L_d R_i/(4\pi r^3)$. In addition, when a broad-line region is present, we expect that a fraction $0.1 f_{-1}$ of the total luminosity $10^{46} L_{46}$ erg s^{-1} will be reradiated roughly isotropically from a radius $R_{\rm BLR} \sim L_{46}^{1/2}$ pc (as lines—there may also be radiation Thomson scattered by the intercloud medium). It then follows from (3) that any jet emerging from the broad-line region will have had its Lorentz factor reduced to $\gamma < \gamma_{\rm BLR}$ (see also Rees 1984) where

$$\gamma_{\rm BLR} \equiv 10^3 f_{-1}^{-1} L_{46}^{-1/2} \left(\frac{m}{m_e}\right). \quad (4)$$

Jets are, however, probably produced far inside the broad-line region. Here drag due to photons from the accretion disk provides a much more severe constraint. For large γ, the term in k_2 in (3) dominates, and using the above estimate for the effective value of $F\theta^4$, we find from (3) that the jet will be rapidly decelerated (or accelerated!) to the equilibrium Lorentz factor if $\gamma > \gamma_d(r)$, where

$$\gamma_d(r) \equiv \frac{1}{k_2}\frac{1}{l_i}\left(\frac{r}{R}\right)^2. \quad (5)$$

Here l_i is the usual compactness parameter,

$$l_i = \frac{\sigma_T L_i}{4\pi mc^3 R_i} = 1836 \left(\frac{L_i}{L_{\text{Edd}}}\right)\left(\frac{r_g}{R_i}\right)\left(\frac{m_e}{m}\right). \tag{6}$$

If a typical quasar has $L_i \sim L_{\text{Edd}}$, $R_i \sim 5r_g$, while a typical radio galaxy has $L_i \sim 10^{-3} L_{\text{Edd}}$, then the constraint (5) becomes $\gamma < 10(r/70r_g)^2$ for an e^+e^- jet in a quasar, and $\gamma < 10(r/2r_g)^2$ for an e–p jet in a quasar or an e^+e^- jet in a radio galaxy.

These already severe constraints would become even more severe in the presence of any of the likely complications to this simple picture—e.g., if a corona above the accretion disk scatters a significant fraction of the disk luminosity, if the outer parts of the disk absorb and reradiate photons from the inner parts, if the jet has a finite opening angle, or if the jet axis is misaligned with the axis of disk symmetry at large radii.

For e^+e^- jets born near R_i in Eddington-limited sources, the first, accelerating term on the right-hand side of (3) can be significant. The jet is then quickly brought to the equilibrium Lorentz factor, which increases as the jet moves outwards until radiation driving is no longer efficient. Solving (3), we find that the terminal Lorentz factor

$$\gamma_f \simeq \left(\frac{l_i}{16k_2}\right)^{1/7} \simeq l_i^{1/7}. \tag{7}$$

For our Eddington-limited disk $\gamma_f \sim 2$–3 for e^+e^-. Note that this is much less than the $\gamma_f \simeq l_i^{1/4} \simeq 5$ which we would have obtained if we had ignored the outer parts of the disk (the relevant radii here being $r \sim 7R_i \sim 35r_g$).

3. Discussion

Quasars very probably have both large compactness parameters l_i and jets with Lorentz factor $\gamma \sim 10$. Compton drag then imposes severe constraints on any model of jet acceleration (equation 5). The speed limit on electron-proton jets accelerated near the black hole in quasars is of the same order as that inferred from observation ($\gamma \sim 10$). There would be no such limit on electron-proton jets in radio galaxies, though electron-positron jets accelerated near the black hole would there also be limited to $\gamma \lesssim 10$.

Electron-positron jets in quasars cannot reach the desired γ until radii $\gtrsim 70r_g$, and must therefore be accelerated slowly and carefully. This seems very difficult to arrange in any local or purely hydrodynamical process unless the kinetic luminosity desired in the jet is a negligible fraction of the total source luminosity.

A slow acceleration can be arranged in magnetohydrodynamic wind models in which large-scale magnetic fields extract rotational energy from a black hole or accretion disk. Flow on rapidly diverging field lines (as might exist near the disk or black hole) is rapidly accelerated, and would be subject to Compton drag, lowering the efficiency with which the wind extracts power and producing a collimated beam of scattered ambient photons (whose polarization properties might constrain the structure of the jet: Begelman and Sikora 1987). But magnetic hoop stresses and the radiation pressure on the particles will tend to collimate the field lines, and on such lines the acceleration of particles (the centrifugal transfer of energy from electromagnetic Poynting flux to particle kinetic energy) is very slow. Even in a quasar an e^+e^- jet could thus escape the effects of Compton drag (detailed models have been constructed by Phinney, in preparation).

Such ideal, axisymmetric jets would not, however, have any internal dissipation and would be invisible until they ran into obstacles. Compton cooling in quasars prevents material not supported by rotation from existing within $\sim 1\,\mathrm{pc}$, so obstacles are not likely to be present inside 1 pc. Since blazars seem to have significantly beamed infrared flux (which can be understood as marginally self-absorbed synchrotron radiation from relativistic electrons at a few gravitational radii), it seems likely that internal dissipation driven by instabilities in the jet is an additional limiting factor.

The author thanks the editors for their patience. His research is supported in part by NSF Presidential Young Investigator award AST 84-51725 and by grants from the Exxon Educational Foundation and the Boeing Company. His stay at the Institute for Theoretical Physics is supported in part by the NSF, under grant PHY 82-17853.

References

Begelman, M. C., and Sikora, M. 1987, *Astrophys. J.*, submitted.
O'Dell, S. L. 1981, *Astrophys. J. (Letters)*, **243**, L147.
Phinney, E. S. 1982, *Monthly Notices Roy. Astron. Soc.*, **198**, 1109.
Phinney, E. S. 1985, in *Astrophysics of Active Galaxies and Quasi-Stellar Objects*, ed. J. S. Miller (Mill Valley: University Science Books), p. 453.
Rees, M. J. 1984, in *IAU Symposium 110, VLBI and Compact Radio Sources*, ed. R. Fanti, K. Kellermann, and G. Setti (Dordrecht: Reidel), p. 207.
Scheuer, P. A. G. 1984, in *IAU Symposium 110, VLBI and Compact Radio Sources*, ed. R. Fanti, K. Kellermann, and G. Setti (Dordrecht: Reidel), p. 197.

The μ–z Diagram

MARSHALL H. COHEN
California Institute of Technology

1. Introduction

Two simple measurable quantities for the variable sources we have been discussing are the redshift z and the internal proper motion μ. After so much discussion about relativistic beams it would be useful, or at least refreshing, to go back to these observables and look at their distribution.

Figure 1 is the μ–z diagram for late 1986. It includes 36 points from 28 sources. (3C 120, 3C 273, 3C 279, and 3C 345 each have more than one point.) The sources include all those with published data, a few from preprints, and three which were presented for the first time at this Workshop (OJ 287: Roberts and Wardle, page 193; 3C 395: Simon et al., page 72, CTA 102: Bååth, page 206). The four arrows represent stable sources with a measured upper limit to μ. The errors in determining μ are not shown; they are large and difficult to assess, partly because μ is derived in differing ways, and in some cases the values are model-dependent. The crosses show candidate sources where the motion needs confirmation, typically because it was determined from only two epochs. The distinction between "candidates" and "true superluminals" is somewhat subjective, and we have been rather conservative in the classification.

The literature contains many references to "superluminal brightening", i.e., cases where the light-travel time is greater than the variability time. None of these is plotted in Figure 1. We only have objects which contain two or more well-separated and identifiable blobs with a measured proper motion.

2. Discussion

An obvious feature of Figure 1 is the strong anticorrelation between μ and z. We consider three explanations for this. Only the second one is in favor at present.

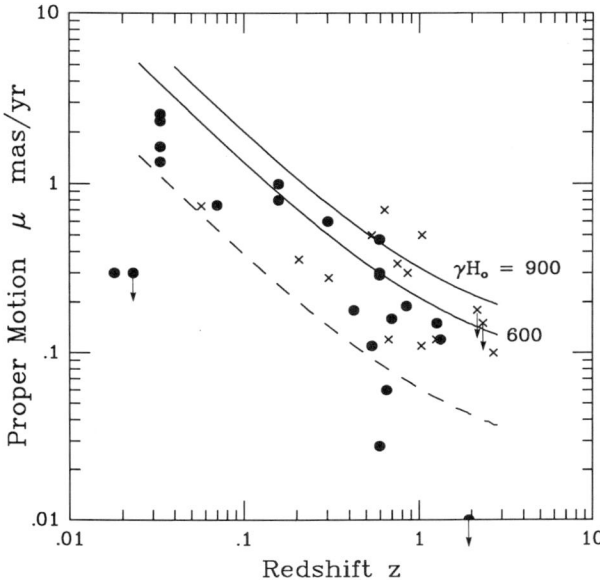

Figure 1 The μ–z diagram for 28 sources with a measured internal proper motion. The solid lines show the limiting proper motion in a Friedmann universe with components moving with Lorentz factor γ. The dashed line shows the proper motion for components with $\gamma \gtrsim 4$ moving at a fixed angle $\theta = 60°$, for $H_0 = 100$ km s^{-1} Mpc^{-1}.

1. Redshift is a Velocity Effect

The pure version of this is to assume that all the objects are at the same distance, r, and the redshift is due to high-speed motion. Assume further that the objects are intrinsically the same; each shoots out luminous blobs with the same Lorentz factor γ, and the nozzles are oriented at random. Then in a set of comoving coordinate systems the proper motions have a fixed upper limit $\mu = (\gamma^2 - 1)^{1/2} r^{-1} c$ and in our coordinate system the distribution of μ with z has an upper envelope which varies approximately as $(1+z)^{-1}$. This varies by a factor of 2.5 between $z = 0$ and $z = 1.5$, whereas the observed data vary by a factor of 10. This model, therefore, cannot explain the data. Unless there is a hidden mechanism which produces an anticorrelation between z and γ, z must be a measure of distance. The lack of observed objects with blueshifts, of course, is another objection to this model.

2. Redshift is Due to Distance, Source Contains a Relativistic Beam

Assume that z measures distance in a Friedmann universe, and that we are dealing with thin, randomly-oriented jets containing luminous blobs moving with constant γ. Then the observed proper motions will have an upper envelope given by

$$\mu \approx 2.11 \times 10^{-4} \gamma H_0 g(q_0, z) \text{ mas yr}^{-1}, \qquad (1)$$

for $\gamma^2 \gg 1$, where $g(q_0, z)$ is given by

$$g(q_0, z) = q_0^2 (1+z) \left(q_0 z + (q_0 - 1)(\sqrt{1 + 2q_0 z} - 1) \right)^{-1}. \qquad (2)$$

Figure 1 shows two curves for this limiting μ, for $\gamma H_0 = 900$ and 600 km s^{-1} Mpc^{-1}, and $q_0 = 0.5$. Unlike case 1, it appears that these curves could reasonably well form upper limits to the data. We expect that in reality there would be a distribution in γ, so a sharp boundary is not expected. The two high points (CTA 102 and 3C 395) have only recently been reported. If the motion of these sources is confirmed, it will require $\gamma H_0 \sim 1500$.

The dashed line in Figure 1 shows $\mu(z)$ for a fixed angle $\theta = 60°$, for $H_0 = 100$ km s^{-1} Mpc^{-1}, $q_0 = 0.5$, and $\gamma = 8$. A set of sources oriented at random would be expected to be half above the dashed line, and half below. This is obviously not the case in Figure 1; so either the parameters are wrong, or the sources are predominantly aligned in our direction. It turns out that the dashed line is insensitive to q_0, to the ratio H_0/θ (it is valid for H_0=50 km s^{-1} Mpc^{-1}, $\theta = 32°$), and to the value of γ provided $\gamma \gtrsim 4$. Now, H_0 is commonly assumed to lie between 50 and 100 km s^{-1} Mpc^{-1}, and for $\gamma < 4$ the dashed line moves down. It therefore is clear that the alignment of the sources in Figure 1 is not random, but preferentially along our line of sight.

Most of the sources in Figure 1 were selected for study because they are particularly strong. Thus, there is a correlation between flux density and angle to the line of sight. Doppler boosting predicts such a correlation, but the appropriate statistical tests of the model cannot be made because the data in Figure 1 do not form a complete sample (Scheuer and Readhead 1979; Orr and Browne 1982). Another obstacle to deriving the state of the universe from Figure 1 is the likelihood that the velocities of the fluid and the pattern are not the same (Lind and Blandford 1985). The conclusions that can be reached now are that the peak value of γ probably lies between 10 and 20, and that beaming plays an important role in these radio sources.

3. Diverging field

The pure version of this model is the magnetic dipole (Sanders 1974; Bahcall and Milgrom 1980) where a wave expands spherically in a magnetic field, and we see the tangent points as they are successively excited. The bound to the observed proper motion is the same as it was for case 2, with $\gamma = 4.4$. However, equation 1 is now a *lower* bound, and the distribution should become sparse as μ increases. This does not fit Figure 1, unless $H_0 \leq 40$ km s^{-1} Mpc^{-1}.

A variant of this model has two observed blobs: a central hot spot, and the tangent wave from below the equatorial plane (Figure 1 of Bahcall and Milgrom 1980). In this case the limit is an *upper* bound with $\gamma = 2.2$. This would require $H_0 \geq 300$ which is unlikely; in addition, the distribution seems wrong because the points should cluster near the limit.

We note that there are strong morphological reasons for preferring a jet model to a diverging field model. The curved core-jet structure does not fit naturally into the diverging field model, and energetic jets are required in any event to explain the outer structure.

I am grateful to many colleagues for sharing some of the data shown in Figure 1 in advance of publication, and to P. D. Barthel and J. A. Zensus for their assistance. This work was supported by NSF grant AST 85-09822.

References

Bahcall, J. N., and Milgrom, M. 1980, *Astrophys. J.*, **236**, 24.
Orr, M. J. L., and Browne, I. W. A. 1982, *Monthly Notices Roy. Astron. Soc.*, **200**, 1067.
Lind, K. R., and Blandford, R. D. 1985, *Astrophys. J.*, **295**, 358.
Sanders, R. H. 1974, *Nature*, **248**, 390.
Scheuer, P. A. G., and Readhead, A. C. S. 1979, *Nature*, **277**, 182.

Grand Unified Models

ROGER D. BLANDFORD
California Institute of Technology

1. Disclaimers

I have been asked by the organizers of this splendid celebration to provide a summary of the many papers discussing superluminal radio sources that we have heard over the past two days. I have also been asked to review ideas about "Grand Unified Models", that is to say attempts to place compact radio sources within a general interpretive framework for galactic nuclear activity. This combined charge allows me to mix quite indiscriminately reporting, comment, and advertisement, a practice which, even if in the best traditions of TV journalism, does not make for good science. The reader must be prepared to separate my idiosyncratic views from more widespread prejudices.

The literature on this subject is extensive, apparently stretching back to the Torah (page 217). No attempt at complete or even fair attribution will be made in this article; preference will be given to papers in this volume. Recent relevant review articles, written from a theoretical perspective, include those of Scheuer (1984), Blandford (1984), Rees (1984), and Phinney (1985).

2. Preamble

The study of superluminal radio sources is at a crossroads. Through an impressive series of technical innovations (that frankly still amaze me), Rees's prediction that the radio-emitting plasma in compact radio sources moves with relativistic speed and exhibits "faster than light" motion (Rees 1966) has been abundantly verified. In fact so successful has been this enterprise that the title of this Workshop has become misleading. We are not just witnessing a few pathological sources which demonstrate a cute piece of freshman physics but have instead discovered a common phenomenon that may well be a characteristic of all compact extragalactic radio components (Impey, page 233). The question we face is where to go from here? The default response is to continue making steadily better maps of more and more sources. However, this may not advance our understanding of the underlying physical conditions. The opportunity is to plan observing strategies to use these sources as the powerful diagnostics they undoubtedly are, to attack the more fundamental issues such as how are jets made, how are active galactic nuclei powered, and what are the physical relationships between and the evolutionary properties of the different types of active galactic nucleus. Only if this challenge is met will we be able to exploit the VLBA to its full potential and justify the development of orbiting VLBI. I have no easy answers. However, in my summary I shall try to emphasize those contributions that seem to me to be pointing the way forward.

3. Superluminal Motion: Illusion or Delusion?

There are now roughly 25 examples of compact extragalactic radio sources with observed expansion speeds reported to be in the range $2h^{-1}c \lesssim v_{\rm app} \lesssim 10h^{-1}c$. Eighteen of these carry Porcas's imprimatur (page 12). Part of the reason why it is so hard to be precise about these numbers is that it is by no means always clear which components in successive epochs should be identified with each other. Another reason is that there is still controversy concerning which mapping algorithms should be trusted and which faint features in a map are "real". At this meeting we have had claims of the presence or the absence of superluminal expansion hotly contested in 3C 120 (Walker, Benson, and Unwin, page 48), 3C 371 (Lind, page 180), NGC 6251 (Jones, page 162), 3C 279 (Unwin, page 34), BL Lac (Mutel and Phillips, page 60), and 3C 454.3 (Pauliny-Toth, page 55). It seems clear that more objective methods must be devised to compare VLBI maps of compact radio sources at different epochs.

As an example, let me suggest one procedure which might be followed when the source is well resolved but is observed too infrequently to follow discrete subcomponents from one observation to the next (Figure 1). Suppose that there is a core that can be easily identified (probably through its flat spectrum) and let us assume that it is at rest (cf. Bartel et al. 1986). If we can resolve and measure the positions of several subcomponents at each epoch of observation t_i then we can assign velocities relative to the core, both positive and negative, for each pair of subcomponents observed at successive times. Next we take all these possible relative velocities and we look for two or more measurements that could be associated with the same uniform expansion or contraction and call these candidate coherent velocities. Finally, we should determine the significance of each candidate velocity by estimating the probability of its occuring by chance from unrelated components. Only those sources yielding single, significant velocities for individual subcomponents should be used for statistical and other arguments. This procedure is surely unnecessary in well-sampled sources like 3C 345 (Biretta

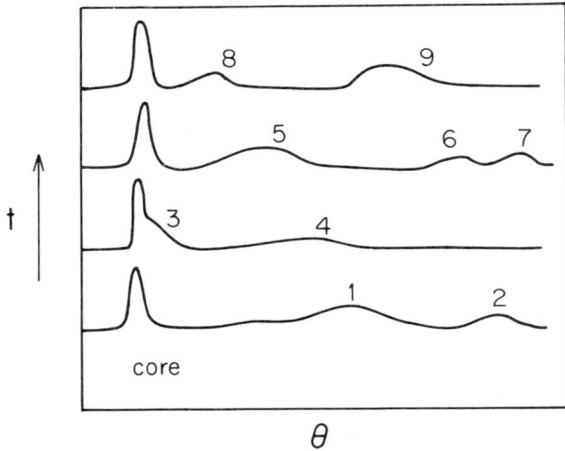

Figure 1 Possible aliasing effect that may occur when VLBI maps are made too infrequently. In this idealized example, strip-scans of surface brightness are aligned so that the cores coincide. The other individual features are identified (1–9). If we associate features 1, 4, 5, 8 with the same component, then there will be a contraction. However, if we relate features 3, 5, 9, there will be an expansion. It is important to compute all possible velocities (e.g., by associating feature 2 with feature 3) and to test candidate coherent motions for their statistical significance. In particular, it is important not to reject contractions *a priori* as some kinematic models predict that they should occur.

and Cohen, page 40) but it may yield some surprises in undersampled sources like 3C 120 (Walker, Benson, and Unwin, page 48).

4. Observed Properties and their Interpretation

If we are prepared to generalize and also to ignore some exceptional cases, then several properties of superluminal sources indicate that relativistic beaming is generally important in compact radio sources.

(a) Superluminal expansion is very common among the brighter sources which have been examined systematically in various surveys (Impey, page 233; Pearson, Readhead, and Barthel, page 94; Porcas, page 12; Witzel, page 83). This has naturally led to the inference that the sources are both bright and superluminally expanding because they contain components that move with relativistic speed at a small angle to the line of sight. It is becoming increasingly likely that in general superluminal sources exhibit "blazar" properties (rapid variability, polarization, etc.) and these are also interpretable on this model (Impey, page 233; Lawrence *et al.*, page 260). Note, though, that not all well-observed sources exhibit faster-than-light motion yet; e.g., 3C 84, which does contain a component expanding with $0.3h^{-1}c$ (Backer, page 76), M87 (Schmitt and Reid 1985), and most compact double sources (Hodges and Mutel, page 168). We might also include in this list the Galactic center, whose average proper motion is $\lesssim 40$ km s^{-1} (Backer and Sramek 1986). By contrast, 4C 39.25, which had been advertised to show superluminal contraction, is now understood to contain a new component shot out of the core towards a stationary outer component (Shaffer and Marscher, page 67).

(b) The expansion speeds are apparently slower ($\lesssim 4h^{-1}c$) on the average in lobe-dominated sources than in core-dominated sources. This conclusion must be regarded as preliminary (Hough and Readhead, page 114; Zensus and Porcas, page 126).† If this trend is substantiated, then it is consistent with the prediction that the lobe-dominated sources are oriented roughly isotropically while the core-dominated sources are pointed towards us.

(c) Extended structure can generally be found surrounding compact sources (Browne, page 129). Obvious asymmetries have the same sense on the small and the large scale.‡ In the case of the compact core-jet sources, any counterjet that may be present has to produce less than a few percent of the flux of the observed jet, a pattern that is familiar from VLA observations of jets in powerful extended sources (e.g., Bridle and Perley 1984). By far the simplest explanation of this one-sidedness is that the outflowing fluid moves with a mildly relativistic speed and Doppler beaming is responsible for the asymmetry on both the small scale and the large scale. (Remember that the flux density from a fluid element moving with speed $\beta \sim 0.85$ at an angle $\sim 30°$ to the line of sight is ~ 100 times that from a similar, but oppositely directed fluid element. Conditions do not have to be extreme to make the counterjet unobservably faint.)

(d) VLBI jets associated with the more obviously superluminal sources are both more curved and more misaligned with the large scale symmetry axes, presumably a result of being directed closer to the observer's line of sight (Readhead et al. 1978). This correlation appears to have stood up fairly well.

(e) Scheuer (page 104) has told us of some recent observations by R. Laing that provide good evidence for relativistic beaming in the one-sided large-scale jets. Laing finds that in seven out of a sample of nine radio galaxies the lobe on the unjetted side is significantly more depolarized than the lobe on the jetted side. This has been interpreted as indicating that the line of sight from the unjetted lobe passes through more gas close to the parent galaxy, and therefore the observed jet is

† Unfortunately, the first such source to be studied, 3C 179 (Porcas, page 12) was found to exhibit a speed $4.8h^{-1}c$. It does, however, have an unusually bright core.
‡ 3C 395 does not conform to this rule (Simon et al., page 72). However, this can be understood as a projection effect.

pointed towards us. Unfortunately, this is not the only interpretation, and devils advocating intrinsic asymmetry of jets can argue that the observed jet is able to propagate into the lobe simply because there is less gas on that side. I believe that this point of view will be hard to sustain because the depolarized lobe must be filled with gas prior to the jet disappearing and it is in the very nature of supersonic flow that conditions in the lobe cannot influence upstream flow.

(f) A quite separate argument for relativistic expansion (Marscher, page 280; Worrall, page 251) comes from setting upper limits on the X-ray fluxes using the observed brightness temperature within the radio cores. If the source were not relativistically expanding, then in many cases the radiation energy density within the core would exceed the magnetic energy density necessary for self-absorption—the usual condition for the inverse Compton catastrophe—and a much larger X-ray flux than is observed should be detected. Relativistic expansion is one way, but not the only way (e.g., Woltjer 1966), to alleviate this constraint.

(g) Wills and Browne (1986) have shown that the width of the Hβ line is inversely correlated with the ratio of the compact component power to the extended component power. This is in line with the beaming model if the emission lines are associated with a disk-like flow perpendicular to the jet axis (as argued here by Oke, page 267).

However, two arguments against relativistic beaming have also been made.

(a) Some sources, e.g., 1928+738 (Simon *et al.*, page 155) and 1721+343 (Barthel, page 148), would have to possess uncommonly large intrinsic sizes if they are expanding sufficiently close to the line of sight to account for the superluminal expansion and the one-sided jets in the simple model (Schilizzi and de Bruyn 1983). I suggest a possible explanation of this phenomenon below (cf. Browne, page 129).

(b) Low-frequency variability can also be interpreted in terms of relativistic beaming, but the source models have a notoriously low energy efficiency. However, it now appears likely that in several sources, interstellar refractive scintillation is responsible for sub-gigahertz variability. The detection of a weak correlation with galactic latitude supports this interpretation (Fanti *et al.* page 200). Small-amplitude centimeter variability ("flickering") of compact sources (Heeschen 1984; Witzel *et al.* 1986) is also consistent with this interpretation. However, in a minority of "blazars", the variation is still believed to be intrinsic and to point to Lorentz factors $\gamma \geq 10$. This may be the case for CTA 102 (Bååth, page 206).

In my opinion the evidence for relativistic motion is far too strong to be ignored and beaming is an inevitable consequence. We must obtain a more detailed understanding of the kinematics of individual sources before we can understand the observational consequences. Some progress has been reported here.

The trajectories followed by individual subcomponents are still a matter of debate. The best data exist for 3C 345 (Biretta and Cohen, page 40). Evidence for both rectilinear motion of the blobs (with the apparent bending perhaps being caused by precession of the source) and genuine curved trajectories has been presented in the past. In more recent observations, component C4 seems to have accelerated from $1.3h^{-1}c$ to $6.5h^{-1}c$ and to have changed its position angle. By contrast, individual components in BL Lac appear to decelerate rapidly to rest when they reach the outer parts of the source (Mutel and Phillips, page 60).

3C 345 is also the best-resolved core-jet source and has a projected opening angle of 26° (Biretta and Cohen, page 40). The jet in 3C 120 can be traced over five decades of radius and the mean volume emissivity appears to scale as $j_\nu \propto r^{-3.3}$ (Walker, Benson, and Unwin, page 48). If we adopt a spectral index $\alpha = -0.5$ and assume equipartition, then the mean magnetic field strength must vary as $B \propto r^{-1}$ as might be expected if a frozen-in transverse field is convected along with the flow and if the pressure scales in the same way as the external pressure $p \propto r^{-2}$ (Blandford and Königl 1979).

The relatively new technique of VLBI polarization measurement can also shed light on the physical conditions in selected bright jets (Roberts and Wardle, page 193). Observations of 3C 345, 3C 120, and OJ 287 seem to be consistent with there being an unpolarized core plus a predominantly parallel field in the jet. The absence of depolarization in the jet allows one to set a (model-dependent) limit on the density of thermal electrons (e.g., Wrobel, page 186).

5. A Simple Source Model

The simple, uniform jet or ballistically moving plasmoid model, in which the observed components move with uniform velocity along the source symmetry axis away from a stationary self-absorbed core, is in trouble on two counts.

(a) The observations that we have just discussed seem to suggest that it is possible for the line of sight to make a substantial angle ($\sim 30°$) to the source symmetry axis and yet allow superluminal motion with speeds up to $v_{\rm app} \sim 10c$ to be seen. This is inconsistent with the naïve ballistic model, in which an observed speed $v_{\rm app}$ can only be measured by an observer within a cone-angle $\lesssim c/v_{\rm app}$.

(b) Theoretically, the problem is that the relativistic electrons will cool rapidly in the expanding flow. The adiabatically expanding source model predicts a decline in the volume emissivity, typically $j_\nu \propto r^{-5}$, more rapid than is observed. This is because most of the energy is transformed into bulk kinetic energy. A model will have a higher radiative efficiency, and a volume emissivity more closely consistent with the observations, if this kinetic energy is converted into magnetic field and relativistic particle energy. By far the most effective means of dissipating kinetic energy is by passing the flow through a shock front, a process that is quite familiar in many other astrophysical contexts (Aller, Hughes, and Aller, page 273; Rudnick, page 217; Marscher, page 280). If shocks are involved, then the emitting gas will move with a velocity that can be quite different from that of the time-averaged jet. This implies that the opening angle of the emission cone may be relatively large, much larger then the opening angle of the jet, just what the observations seem to demand (Lind and Blandford 1985).

So we want to retain rapid superluminal motion on the side projected away from the core. However, we do not need the enormous beaming factors ($\sim$ 1000) characteristic of the ballistic model because most moving components are only observed with dynamic range $\lesssim$ 10. Neither need we yet be too concerned about the transverse elongation likely to be created by a cylindrically symmetric outflow which has worried Peter Scheuer (page 104) as the moving components are generally poorly resolved and in any case the stationary, self-absorbed core may be quite distinct from the apparent origin of the outflow (Marscher, page 280). It seems that the most promising type of explanation is due to Rees (1981) who suggested that the average emissivity decays with increasing angle of observation from the symmetry axis so that we can observe some highly boosted sidelobe. Another possibility, which at the very least emphasizes what a strong amplifier relativistic beaming can be and how we can be "blinded" by relatively small changes in velocity, is to impose a sinusoidal perturbation on the simple ballistic source model (van Groningen, Miley, and Norman 1980; Linfield 1981; Scheuer, page 104). (Personal computers with graphics terminals are quite effective for pedagogic demonstrations of these effects!) Further kinematic suggestions have been reviewed most recently by Scheuer (1984) and Phinney (1985).

Let me give one example of a source model which involves shocks, has the radiative properties required by Rees, and highlights plasma flowing at a fairly small angle to the line of sight on the side of the jet. (We need not observe the same material all the time, just that which is currently beamed towards us; cf. Sanders 1974.) A spherically symmetric blast wave obviously does not have these properties. However, three modifications, which I anticipate to be present in a real source, can combine to produce the desired effect.

(a) The shock speed decreases with angle from the symmetry axis as the density of the plasma in the unshocked outflow increases. For relativistic kinematic reasons, the post-shock gas is then refracted away from the symmetry axis.

(b) The fluid in the jet ahead of the shock expands transversely. This has a similar effect to (a).

(c) The outflow contains a toroidal magnetic field. This reduces the shock compression for a given shock speed and therefore reduces the Doppler factor associated with the post-shock flow.

In Figure 2a I show an example of a source model in which an expanding magnetized jet is illuminated by a non-spherical blast wave. (The radial shock speed is $(\cos\theta)^{0.1}c$.) In Figure 2b I have plotted $\delta^{2.5}$ (where δ is the usual Doppler factor)—a rough guess of the (model-dependent) surface brightness—as observed at an inclination angle of 30° to the symmetry axis. It turns out that the brightest part of the source moves with $v_{\text{app}} \sim 8c$, far larger than the values 1.7–3.7c expected for the ballistic model. Contours are shown down to 25% of the peak component brightness. Emission from closer to the origin will probably be self-absorbed and constitute the core.

Of course I do not propose this as a serious model as it is neither self-consistent dynamically nor an accurate description of the radiative transfer. It is just another example which shows how subtle relativistic kinematic effects are likely to be in a real source.

6. Unified Models

The existence of relativistic motion leads inevitably to the conclusion that our view of these sources should depend upon our orientation with respect to them (Impey, page 233). The observational consequences of this beaming have been widely discussed and are still far from understood. In particular, we have not been able to agree about the identity of the superluminal sources beamed away from us. A model that addresses this issue is described as "unified". "Grand unified" models are more ambitious and attempt to account for the differences between all types of nuclear activity in physical terms. The many observational problems in developing unified models were discussed by Rudnick (page 217). At the risk of oversimplification, there are two extreme positions. The unbeamed compact radio sources may be the radio-quiet quasars (Scheuer and Readhead 1979). Alternatively, they may be the steep-spectrum, lobe-dominated, radio-loud quasars and galaxies (Orr and Browne 1982).

It seems to me that the arguments against the former model are now so strong that it should be buried (albeit with full honors as it did

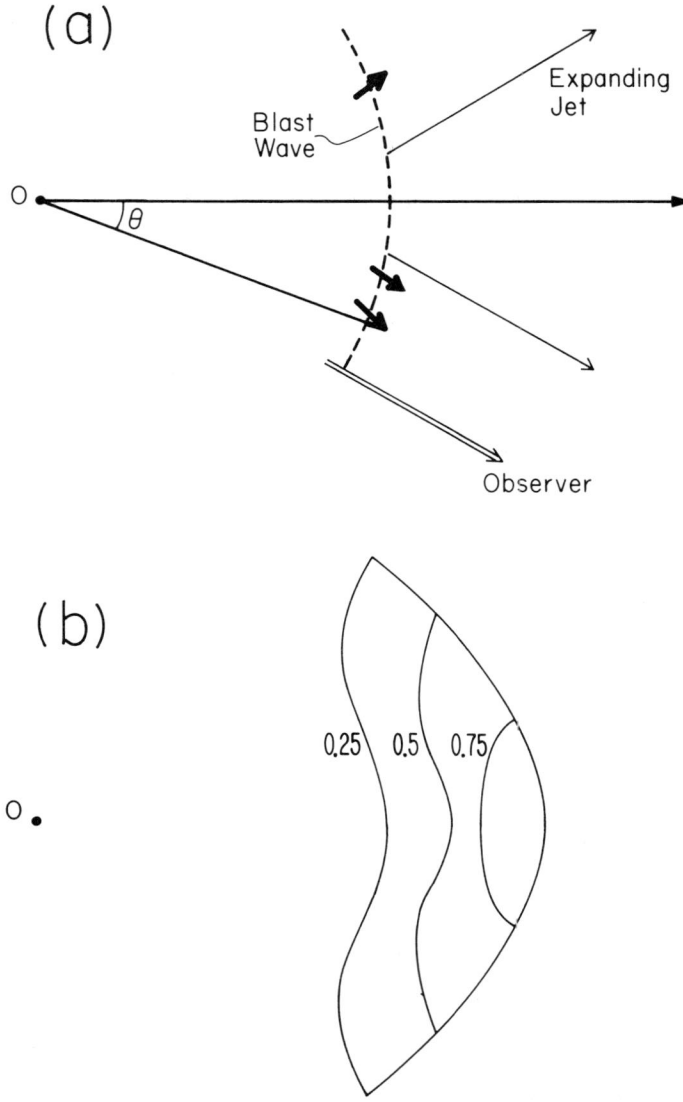

Figure 2 (a) Simple model of an expanding, magnetized jet flow (*light arrows*) illuminated by a cylindrically symmetric blast wave (*dashed curve*) moving with speed $(\cos\theta)^{0.1}c$. The emission comes from the post-shock flows (*bold arrows*). (The location of the blast wave is shown at a fixed coordinate time, not observer time.) O marks the origin of the blast wave which may but need not coincide with the stationary, self-absorbed core. (b) Contours of constant surface brightness from the blast wave, as observed from an inclination angle of 30°. The lowest contour shown has a surface brightness equal to 25% of the peak. Fainter emission that surrounds O originates at small radii and will probably be self-absorbed.

stimulate a lot of thought about the important questions). As pointed out by Strittmatter *et al.* (1980) and others, we do not see the predicted low power radio sources, corresponding to superluminal sources pointed away from us, the number of which should increase with decreasing flux S according to $N(> S) \propto S^{-1/2}$ if the spectrum is flat. Instead the distribution of radio powers is roughly bimodal—representing the radio-loud and the radio-quiet quasars. Even more damning is the discovery of significant extended (and presumably isotropic) radio emission around the radio-loud quasars, (by definition) absent from the radio-quiet quasars (Browne, page 129).

The second hypothesis must be right at some level. The problem with it has always been that there seem to be far too few lobe-dominated quasars to go with the core-dominated quasars. (These must be compared at the same flux of an isotropically emitted component like the broad emission lines, e.g., Phinney 1985.) However, the arithmetic may not be as unfavorable as at one stage seemed to be the case, for two reasons. First, the number of unbeamed sources per beamed source may well be much less than $(v_{\rm app}/c)^2 \sim 50$ for the reasons discussed above. Secondly, recent studies of the evolution of the apparent luminosity functions of flat-spectrum and steep-spectrum quasars (e.g., Peacock and Dunlop 1986) show that the fraction of flat-spectrum sources declines quite dramatically beyond a redshift $z \sim 2$. If this hypothesis does eventually make quantitative sense, then the radio emission from the radio-quiet quasars would presumably be just due to weak, stifled jets as seems to be the case in the Seyfert galaxies.

7. Parabolic Spectra: an Interpretation

We have heard here from Rudnick (page 217) and Impey (page 233) how the radio-through-ultraviolet spectra of blazars, expressed as $\log(\nu S_\nu)$, may be fitted by quadratic, as opposed to the usual linear, functions of $\log \nu$. The shape of these spectra is supposed to be preserved during variation. While I am a bit skeptical about the significance of some of the fits, it is clearly true that many sources do produce most of their power in the far infrared, with significant, though minor, contributions from the radio, millimeter, and optical regions of the spectrum. This is entirely consistent with the most elementary version of a jet model (e.g., Blandford and Königl 1979).

The key feature of this model, which probably survives the complicating factors introduced in the above discussion of superluminal expansion, is that, at a given radius, a jet is probably only able to emit efficiently over a restricted range of frequency. At low frequencies self-absorption causes the emitted spectrum to turn over quite abruptly,

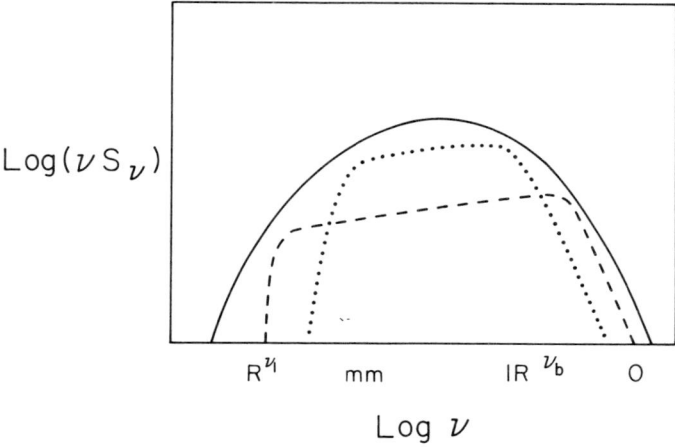

Figure 3 Possible explanation of "parabolic" spectra as a superposition of power-law synchrotron spectra emitted over a range of jet radii. The spectrum drawn with a dashed line originates from an intermediate jet radius. Emission at frequencies less than ν_l is suppressed through synchrotron self-absorption. The electrons emitting at frequencies above ν_b have a reduced density on account of radiative cooling. The spectrum drawn with a dotted line originates from a smaller jet radius. Variability at frequencies ν_l and ν_b might be correlated.

and at high frequencies particle acceleration of the synchrotron-emitting electrons is incapable of keeping up with expansion losses and this causes a similar apparent cut-off. Now the range of frequencies emitted is likely to increase with jet radius, and superposing all these finite power-law spectra can give a roughly parabolic total spectrum (Figure 3). To give a quantitative example, suppose that locally the electrons have a power-law spectrum with equal energy density per octave of energy, similar to what is observed in well-resolved jets. The spectral index of optically thin synchrotron emission is then $\alpha \sim -0.5$ or $\nu S_\nu \propto \nu^{0.5}$. When the radiation becomes optically thick, we expect the local spectrum to be $\nu S_\nu \propto \nu^{3.5}$. A rough fit to the observed intensity variation in resolved jets gives a magnetic field variation $B \propto r^{-1}$. This implies that most of the emission at frequency ν comes from radii $r \propto \nu^{-1}$. Integrating along the jet gives a flux density variation $\nu S_\nu \propto \nu^2$.

Next, let us suppose that the particle acceleration time balances the expansion time, $\sim r/c$, independent of particle energy so that at moderate energy a stationary spectrum is achieved. At high energies, the synchroton and inverse Compton losses will be relatively more important and there will be a maximum energy to which particles can be accelerated. For illustration, we assume that $\gamma_{max} \propto r^{\frac{2}{3}}$, a slightly flatter variation than the relation $\gamma_{max} \propto r$ obtained by equating the flow time to the synchrotron radiation time. This implies that radiation at high frequency ν is dominated by emission from jet radii greater than $r_{min} \propto \nu^3$. Again, integrating over radius gives a high-frequency spectrum $\nu S_\nu \propto \nu^{-1}$. Although this spectrum is roughly triangular in profile, it is not hard to imagine that relativistic beaming is less effective at large radii. With a minimal modelling effort, a strictly parabolic spectrum could undoubtedly be reproduced. What is perhaps more relevant is that this interpretation associates emission at two separated frequencies with the same location. Near-infrared and millimeter variability might be correlated together with (separately) optical and centimeter variations. Of course there will also be some contamination of the spectrum from other synchrotron-emitting components and perhaps also inverse Compton emission which could complicate matters, but it is feasible to search for correlated variation. Blazar X-ray emission is most naturally attributed to inverse Compton scattering on this model (Königl 1981).

8. Black Holes and Grand Unified Models

Although there has been relatively little direct discussion of it at this Workshop, most of the participants tacitly accept the black hole model as the best working hypothesis for explaining general activity in the nuclei of galaxies and radio emission in particular. The (comparatively weak) arguments in its favor have been rehearsed on numerous occasions and in several reviews (e.g., Blandford 1987; Begelman 1985; Rees 1984) and will not be reproduced here. Prospects for direct verification of this model really seem no brighter than they did at the time of the discovery of quasars. The problem is essentially the "no-hair" theorem which forbids any observable signatures of black holes other than mass and angular momentum. (Astrophysical black holes cannot carry gravitationally significant amounts of charge.) VLBI comes closest to resolving black holes but even here our angular resolution corresponds to ~ 100 Schwarzschild radii at best.

Nevertheless, the indirect arguments have strengthened. Foremost amongst these is the efficiency argument (Sołtan 1982; Phinney 1983) which takes the estimated bolometric flux density of all the observed quasars (dominated by the optical emission from $\sim 21^m$ quasars) and converts this into an energy radiated per bright galaxy ($\sim 3\times 10^5 M_\odot c^2$). If this energy is gravitational in origin then, assuming an efficiency factor $\epsilon \sim 0.1$ we find that the central objects must have average masses $\sim 3 \times 10^6 (10\epsilon)^{-1} M_\odot$. Already this is comparable with upper limits on and reported measurements of central masses of galaxies like our own, M31, and M87 (e.g., Kormendy 1987). Any substantial reduction in efficiency cannot be tolerated within the confines of conventional physics.

The mechanics of spinars, star clusters, etc. are even less well specified than those of black holes but it is hard for me at least to imagine that they could ever convert more than 1% of their mass into radiant energy. If so, they are ruled out as prime movers for the optical quasars (although it is just conceivable that they power the less numerous radio quasars that we are discussing here and which account for roughly 5–10% of the total quasar energy radiated). After the Hubble Space Telescope is operational and the central velocity distributions of nearby galaxies are probed with unprecedented resolution, we should be in a position to tighten up this argument substantially.

If the black hole interpretation is adopted, then it is inevitable that we speculate upon what dictates the differences between the different types of active galactic nuclei.† I have already discussed, under the rubric of unified models, the nature of the unbeamed compact sources and argued that these have to be the lobe-dominated quasars. However, from a radio perspective, the lobe-dominated quasars are related to the radio galaxies, objects that are mostly nearby. (There are of course some important differences between these two classes of objects, but these are attributable to the radio galaxies being located in more isolated surroundings.) Now, in the case of the largest radio galaxies, the energies in the radio lobes can exceed $10^6 M_\odot c^2$. Adopting an efficiency of 10%, this translates into a minimum mass of $10^7 M_\odot$ for the central black hole. The associated Eddington limit is 10^{45} erg s^{-1}, far larger than the bolometric luminosity observed from the centers of these sources. Giant radio galaxies are therefore presumably *not* accreting at the Eddington limit. In fact in the case of a source like Cen A, which has a nuclear bolometric luminosity of only $\sim 10^{43}$ erg s^{-1}, we require a larger mean power supply to the lobes. All of this points to the black hole itself as the source of the energy. This is an attractive possibility because a rapidly spinning black hole has up to $\sim 10^{61}(M/10^8 M_\odot)$ erg of extractable energy, and this is most likely to be liberated in the form of relativistic particles and magnetic field, the raw materials of radio sources.

By contrast, the optical quasars, and their low-luminosity counterparts the type 1 Seyfert galaxies, seem to radiate most of their power in the ultraviolet (albeit with a substantial infrared-to-optical power-law component). This is consistent with the escaping radiation being ther-

---

† As has proved to be the case in particle physics, genuine progress on these "grand unified" models is likely to be slower than advances towards the more limited goal of "unification". If we wish to be encouraged in this enterprise, we can inspect the early Hertzsprung–Russell diagrams for stars constructed by Hertzsprung (1911) and Russell (1914; for a reproduction see Snow 1983) and contrast them with a contemporary version. Alternatively, we can depress ourselves by studying the diagram of absolute radio magnitude versus linear diameter by Shklovskii (1962) and consider how much subsequent observational discoveries have altered our view of radio galaxies.

malized under optically thick conditions around a black hole accreting at the Eddington limit. Now the most powerful quasars in the universe require central black hole masses in excess of $\sim 3\times 10^9 M_\odot$ if they radiate at the Eddington limit. The peak comoving density of these objects is roughly $10^{-8} h^3$ Mpc^{-3}. That is to say, the nearest fossil quasar of this class should be $\sim 500 h^{-1}$ Mpc away and be essentially unrecognizable as such (although it could well be an energetic radio galaxy). If, by contrast, quasars radiate at 1% of the Eddington limit, the nearest black hole of this mass should be 50 Mpc away and this should be in the range of the Hubble Space Telescope. The best argument for optical quasars and type 1 Seyfert galaxies radiating at the Eddington limit is that, in many cases, gas is likely to be supplied to the nucleus at rates much in excess of the Eddington rate, and we know that the accretion cannot be too inefficient for the reasons given above. Therefore, the accretion has to be limited and radiation pressure is the best way to accomplish this.

If the radio galaxies are identified with highly sub-critical accretion and the optical quasars with critical accretion, then the radio quasars, which display features of both classes, are most simply associated with holes that are accreting moderately sub-critically and are therefore capable of radiating significant powers in the relativistic particle/electromagnetic field channel as well as the ultraviolet channel. (Browne summarized arguments against the alternative hypothesis that the radio quasars were pole-on radio galaxies; page 111.) The blazars comprise optically violently variable quasars, in which the anisotropic jet emission dominates the more isotropic radiation from the central accretion disc, and the BL Lac objects which are lower-luminosity radio galaxies similarly beamed towards us (Impey, page 233). Objects in which the large-scale low-frequency emission associated with the outer jet is beamed toward us, but the beamed optical emission from the inner jet is beamed away from us, may be the compact steep-spectrum sources (Fanti and Fanti, page 174).

9. Evolutionary Considerations

In an even more speculative vein, we can attempt to incorporate the apparent evolutionary properties of active galactic nuclei into this grand interpretation. Recent counts of quasars (e.g., Boyle et al. 1987) seem to show quasar detections increasing at a rate of eight per magnitude down to $\sim 19^m$, where roughly three quasars are found per square degree. At fainter magnitudes, the counts level off to a total of roughly 100 per square degree at 22^m, a total of four million over the whole sky. Quasars appear to be roughly uniformly distributed in redshift for $0.5 < z < 2.5$, implying strong cosmological evolution. Surveys designed to detect quasars at greater redshifts have discovered that they are much less plentiful than anticipated. It seems that the high-redshift quasars that have been found are preferentially of large luminosity, although it is not possible to quantify this at present. By contrast, we know that the Seyfert galaxies can hardly evolve at all. The strongest constraint comes from the X-ray background to which an unevolving population of Seyferts must contribute at roughly the 10% level.

Many astronomers have interpreted these observations as evidence for luminosity evolution. That is to say most quasars are born at some redshift $z \gtrsim 3$ and subsequently fade with constant comoving density. However, this cannot fit the high-redshift survey data. In addition, it is quite in conflict with the interpretation of active galactic nuclei presented above. Black holes should grow with cosmic time, and if quasar luminosities are to scale with the Eddington luminosity, then they should brighten not fade as they age. The only way to reconcile this theoretical expectation with the observations is to invoke strong density evolution. In other words, black holes are increasingly dormant as time goes on. Now if we combine the notion that the energy released in building up black hole masses should be converted into radiant energy with high ($\sim 10\%$) efficiency with the view that the powerful quasars are approximately Eddington-limited, then we are led to conclude that it is the rate of supply of fuel to the black hole that controls the density evolution.

In the simplest quantitative model (Blandford 1986 and in preparation), it is assumed that gas is supplied either at the Eddington rate or not at all with a probability that is a decreasing function of cosmic time (Figure 4). Black holes grow continually from modest initial masses $\sim 100 M_\odot$ with an e-folding time $\sim 10^8$ yr until the rate of gas supply is insufficient to allow them to grow further. The black holes that are formed first, presumably associated with the first galaxies, are able to grow to masses $\sim 10^9$–$10^{10} M_\odot$ in $\sim 10^9$ yr and correspond to the brightest high-redshift quasars while they are accreting and the most energetic radio sources thereafter. Black holes that start to grow somewhat later are only able to achieve masses $\sim 10^8$–$10^9 M_\odot$ and come to fruition at $z \sim 2$ and make the majority of quasars. The last galaxies that form, perhaps preferentially the spirals in lower density surroundings, are only able to grow central black holes to masses $\sim 10^6$–$10^7 M_\odot$. These are now actively accreting for about 1–10% of the time and are the Seyfert galaxies. When they are accreting subcritically, these galaxies will be emission line galaxies, liners, etc. Most galaxies (probably including our own) should have been Seyferts in the past. Compact radio sources, associated with intrinsically lower-power sources and smaller black-hole masses, evolve less than the extended radio sources.

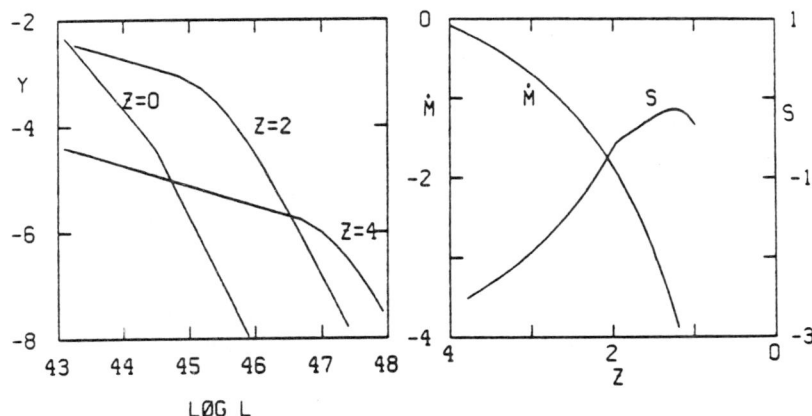

Figure 4 Simple simulation of the evolution of active galactic nuclei (Blandford 1986). *Left:* the differential luminosity function for quasars and Seyfert galaxies is shown schematically expressed as Y, the fraction of bright galaxies (comoving density $\sim 10^{-2} h^3$ Mpc^{-3}) with bolometric luminosity L erg s^{-1} in the interval $d \ln L$. *Right:* this luminosity function can be reproduced if small central black holes form in the nuclei of galaxies at the rate S. The mean fueling rate $\dot{M}$ declines dramatically with cosmic epoch giving the appearance of strong luminosity-dependent density evolution at late times.

10. Conclusion

Let me conclude, as I began, by saying that the study of superluminal sources is at a difficult but interesting stage. It has survived attacks by those whom Marshall Cohen referred to (not unaffectionately) as the flat-earthers, scrutiny from other astronomers skeptical about VLBI results, and internal differences concerning image processing procedures. Superluminal expansion now seems to be such a common feature of compact radio sources that it cannot be ignored in general discussions of active galactic nuclei. It is not only an indication that relativistically deep potential wells are present, but also a warning that the luminosity functions of compact radio sources and blazars (at optical and X-ray wavelengths) are probably telling us more about beaming than about source densities and evolution.

The opportunities for future research are considerable. On the observational side, the advent of the VLBA and parallel developments in Europe should bring about a dramatic improvement in the sampling of the most variable sources. There is also the possibility of major improvements in the dynamic range of individual maps (Wilkinson, page 211) which will presumably confirm that the sources are far too complex to be described by the simple ballistic model. Looking further into the future, if QUASAT is ever launched, then we should be able to start resolving the cores of the core-jet sources and to see if their positions are frequency-dependent as predicted. It is also possible that we will actually see jets being collimated although the best current indications are that this is accomplished much closer to the central black hole.

From the theoretical standpoint, the major unsolved problem is to understand how and where jets are collimated. Phinney (page 301) has explained how there are much stronger constraints on the nature of jets and their confinement than is generally appreciated. It can be hoped that, perhaps by a process of elimination, one general mechanism will be shown to operate.

Let me conclude by suggesting that we all agree to meet in five years to celebrate Marshall's sixty-fifth birthday and to discuss the first results from the almost complete VLBA.

I thank Marshall Cohen and Sterl Phinney for helpful discussions, and Tim Pearson and Tony Zensus for substantial editorial assistance. This work was supported through grants from the National Science Foundation (AST 84-15355) and the National Aeronautics and Space Administration (NAS-4-PMY143).

References

Backer, D. C., and Sramek, R. A. 1986, in *Proceedings of the Symposium on the Galactic Center*, ed. D. C. Backer (New York: American Institute of Physics), p. 161.
Bartel, N., Herring, T. A., Ratner, M. I., Shapiro, I. I., and Corey, B. E. 1986, *Nature*, **319**, 733.
Begelman, M. C. 1985, in *Astrophysics of Active Galaxies and Quasi-Stellar Objects*, ed. J. S. Miller (Mill Valley: University Science Books), p. 411.
Blandford, R. D. 1984, in *IAU Symposium 110, VLBI and Compact Radio Sources*, ed. R. Fanti, K. Kellermann, and G. Setti (Dordrecht: Reidel), p. 215.
Blandford, R. D. 1986, in *IAU Symposium 119, Quasars*, ed. G. Swarup and V. K. Kapahi (Dordrecht: Reidel), p. 359.
Blandford, R. D. 1987, in *Three Hundred Years of Gravitation*, ed. S. Hawking and W. Israel (Cambridge: Cambridge University Press), in press.
Blandford, R. D., and Königl, A. 1979, *Astrophys. J.*, **232**, 34.
Boyle, B. J., Fong, R., Shanks, T., and Peterson, B. A. 1987, *Monthly Notices Roy. Astron. Soc.*, in press.
Bridle, A. H., and Perley, R. A. 1984, *Ann. Rev. Astron. Astrophys.*, **22**, 319.
Heeschen, D. S. 1984, *Astron. J.*, **89**, 1111.
Hertzsprung, E. 1911, *Publ. Astrophys. Obs. Potsdam*, **22**, Nr. 63.
Königl, A. 1981, *Astrophys. J.*, **243**, 700.
Kormendy, J. 1987, in *Supermassive Black Holes*, ed. M. Kafatos (Cambridge: Cambridge University Press), in press.
Lind, K. R., and Blandford, R. D. 1985, *Astrophys. J.*, **295**, 358.
Linfield, R. 1981, *Astrophys. J.*, **250**, 464.
Orr, M. J. L., and Browne, I. W. A. 1982, *Monthly Notices Roy. Astron. Soc.*, **200**, 1067.
Peacock, J. A., and Dunlop, J. S. 1986, in *IAU Symposium 119, Quasars*, ed. G. Swarup and V. K. Kapahi (Dordrecht: Reidel), p. 455.
Phinney, E. S. 1983, Ph. D. thesis, University of Cambridge.
Phinney, E. S. 1985, in *Astrophysics of Active Galaxies and Quasi-Stellar Objects*, ed. J. S. Miller (Mill Valley: University Science Books), p. 453.
Readhead, A. C. S., Cohen, M. H., Pearson, T. J., and Wilkinson, P. N. 1978, *Nature*, **276**, 768.
Rees, M. J. 1966, *Nature*, **211**, 468.

Rees, M. J. 1981, in *IAU Symposium 94, Origin of Cosmic Rays*, ed. G. Setti, G. Spada, and A. W. Wolfendale (Dordrecht: Reidel), p. 139.
Rees, M. J. 1984, *Ann. Rev. Astron. Astrophys.*, **22**, 471.
Russell, H. R. 1914 (published 1918), *Publications of the American Astronomical Society*, **3**, 22; and 1914, *Popular Astronomy*, **22**, 275.
Sanders, R. H. 1974, *Nature*, **248**, 390.
Scheuer, P. A. G. 1984, in *IAU Symposium 110, VLBI and Compact Radio Sources*, ed. R. Fanti, K. Kellermann, and G. Setti (Dordrecht: Reidel), p. 197.
Scheuer, P. A. G., and Readhead, A. C. S. 1979, *Nature*, **277**, 182.
Schilizzi, R. T., and de Bruyn, A. G. 1983, *Nature*, **303**, 26.
Schmitt, J. H. M. M., and Reid, J. M. 1985, *Astrophys. J.*, **289**, 120.
Shklovskii, I. S. 1962, *Astron. Zh.*, **39**, 591. English translation: *Soviet Astron.*, **6**, 465.
Snow, T. P. 1983, *The Dynamic Universe* (St. Paul: West Publishing), p. 297.
Sołtan, A. 1982, *Monthly Notices Roy. Astron. Soc.*, **200**, 115.
Strittmatter, P. A., Hill, P., Pauliny-Toth, I. I. K., Steppe, H., and Witzel, A. 1980, *Astron. Astrophys.*, **88**, L12.
van Groningen, E., Miley, G. K., and Norman, C. A. 1980, *Astron. Astrophys.*, **90**, L7.
Wills, B. J., and Browne, I. W. A. 1986, *Astrophys. J.*, **302**, 56.
Witzel, A., Heeschen, D. S., Schalinski, C., and Krichbaum, Th. 1986, *Mitt. Astron. Ges.*, **65**, 239.
Woltjer, L. 1966, *Astrophys. J.*, **146**, 597.

Bibliography

The following bibliography lists all the books and articles referred to in the text. The italic numbers in square brackets are the starting page numbers of the papers in which the item is referenced.

Alef, W., Götz, M. M. A., Preuss, E., and Kellermann, K. I. 1987, "Evidence for superluminal motion in the double radio galaxy 3C 390.3", *Astron. Astrophys.*, submitted. *[12]*

Alef, W., and Porcas, R. W. 1986, "VLBI fringe-fitting with antenna-based residuals", *Astron. Astrophys.*, **168**, 365. *[12, 55]*

Aller, H. D., Aller, M. F., and Hodge, P. E. 1981, "The polarization of outbursts in extragalactic variable sources at centimeter wavelengths", *Astron. J.*, **86**, 325. *[217]*

Aller, H. D., Aller, M. F., and Hughes, P. A. 1985, "Polarized radio outbursts in BL Lacertae. I. Polarized emission from a compact jet", *Astrophys. J.*, **298**, 296. *[34, 60, 273]*

Aller, H. D., Aller, M. F., Latimer, G. E., and Hodge, P. E. 1985, "Spectra and linear polarizations of extragalactic variable sources at centimeter wavelengths", *Astrophys. J. Suppl.*, **59**, 513. *[67]*

Altschuler, D. R., Broderick, J. J., Condon, J. J., Dennison, B., Mitchell, K. J., O'Dell, S. L., and Payne, H. E. 1984, "Multifrequency light curves of low-frequency variable radio sources", *Astron. J.*, **89**, 1784. *[200]*

Angel, J. R. P., and Stockman, H. S. 1980, "Optical and infrared polarization of active extragalactic objects", *Ann. Rev. Astron. Astrophys.*, **18**, 321. *[233, 251]*

Antonucci, R. R. J. 1984, "Optical spectropolarimetry of radio galaxies", *Astrophys. J.*, **278**, 499. *[233]*

Antonucci, R. R. J. 1986, "Deep radio maps of BL Lacertae and 3C 446", *Astrophys. J.*, **304**, 634. *[60]*

Antonucci, R. R. J., Hickson, P., Olszewski, E. W., and Miller, J. S. 1986, "Deep radio maps of high-frequency selected BL Lac objects", *Astron. J.*, **92**, 1. *[89]*

Antonucci, R. R. J., and Ulvestad, J. S. 1985, "Extended radio emission and the nature of blazars", *Astrophys. J.*, **294**, 158. *[129]*

Avni, Y., Sołtan, A., Tananbaum, H., and Zamorani, G. 1980, "A method for determining luminosity functions incorporating both flux measurements and flux upper limits, with applications to the average X-ray to optical luminosity ratio for quasars", *Astrophys. J.*, **238**, 800. *[251]*

Bååth, L. B. 1984, "VLBI monitoring of BL Lac-objects", in *IAU Symposium 110, VLBI and Compact Radio Sources*, ed. R. Fanti, K. Kellermann, and G. Setti (Dordrecht: Reidel), p. 127. *[12, 83]*

Bååth, L. B., Cotton, W. D., Counselman, C. C., Shapiro, I. I., Wittels, J. J., Hinterregger, H. F., Knight, C. A., Rogers, A. E. E., Whitney, A. R., Clark, T. A., Hutton, L. K., and Niell, A. E. 1980, "Time-dependent radio fine structure of the compact sources NRAO 150 and 4C 39.25", *Astron. Astrophys.*, **86**, 364. *[55]*

Backer, D. C. 1984, "Extragalactic VLBI at 89 GHz", in *IAU Symposium 110, VLBI and Compact Radio Sources*, ed. R. Fanti, K. Kellermann, and G. Setti (Dordrecht: Reidel), p. 31. *[32, 76]*

Backer, D. C. 1984, "Very long baseline interferometry at 90 GHz", in *URSI International Symposium on Millimeter and Submillimeter Wave Radio Astronomy*, Granada, Spain, September 11-14, 1984 (Union Radio-Scientifique Internationale), p. 93. *[76]*

Backer, D. C., and Sramek, R. A. 1986, "Proper motion of the compact, nonthermal radio source in the Galactic center", in *Proceedings of the Symposium on the Galactic Center*, ed. D. C. Backer (New York: American Institute of Physics), p. 161. *[310]*

Bahcall, J. N., and Milgrom, M. 1980, "The magnetic dipole model for superluminal radio sources", *Astrophys. J.*, **236**, 24. *[1, 306]*

Baldwin, J. A., Carswell, R. F., Wampler, E. J., Smith, H. E., Burbidge, E. M., and Boksenberg, A. 1980, "The nebulosity associated with 3C 120", *Astrophys. J.*, **236**, 388. *[48]*

Balick, B., and Heckman, T. M. 1982, "Extranuclear clues to the origin and evolution of activity in galaxies", *Ann. Rev. Astron. Astrophys.*, **20**, 431. *[233]*

Balick, B., Heckman, T. M., and Crane, P. C. 1982, "The large-scale radio structure of 3C 120", *Astrophys. J.*, **254**, 483. *[48]*

Balonek, T. J. 1986, "Radio-optical broadband spectra and variability of quasars", in *Continuum Emission in Active Galactic Nuclei*, ed. M. L. Sitko (Tucson: Kitt Peak National Observatory, National Optical Astronomy Observatories), p. 161. *[217]*

Bare, C., Clark, B. G., Kellermann, K. I., Cohen, M. H., and Jauncey, D. L. 1967, "Interferometer experiment with independent local oscillators", *Science*, **157**, 189. *[1]*

Bartel, N., Herring, T. A., Ratner, M. I., Shapiro, I. I., and Corey, B. E. 1986, "VLBI limits on the proper motion of the 'core' of the superluminal quasar 3C 345", *Nature*, **319**, 733. *[12, 40, 94, 280, 310]*

Bartel, N., Shapiro, I. I., Corey, B. E., Marcaide, J. M., Rogers, A. E. E., Whitney, A. R., Cappallo, R. J., Kühr, H., Graham, D. A., and Bååth, L. B. 1984, "The compact radio source 2021+614: simultaneous 2.3 and 8.3 GHz Mark III VLBI observations", *Astrophys. J.*, **279**, 116. *[94]*

Bartel, N., Shapiro, I. I., Huchra, J. P., and Kühr, H. 1984, "The compact radio source 2021+614: a peculiar narrow-line radio galaxy", *Astrophys. J.*, **279**, 112. *[12, 168]*

Barthel, P. D. 1984, "Radio structure in quasars", Ph. D. thesis, Rijksuniversiteit Leiden. *[129, 148]*

Barthel, P. D., Miley, G. K., Schilizzi, R. T., and Preuss, E. 1984, "Compact radio cores in extended quasars", *Astron. Astrophys.*, **140**, 399. *[148]*

Barthel, P. D., Miley, G. K., Schilizzi, R. T., and Preuss, E. 1985, "Two-epoch observations of the core radio structure in extended quasars", *Astron. Astrophys.*, **151**, 131. *[12, 148]*

Barthel, P. D., Pearson, T. J., Readhead, A. C. S., and Canzian, B. J. 1986, "0850+581: another superluminal radio source", *Astrophys. J. (Letters)*, **310**, L7. *[12, 94, 114, 129, 148]*

Barthel, P. D., Schilizzi, R. T., Miley, G. K., Jägers, W. J., and Strom, R. G. 1985, "The large and small scale radio structure of 3C 236", *Astron. Astrophys.*, **148**, 243. *[148]*

Bechtold, J., Green, R. F., Weymann, R. J., Schmidt, M., Estabrook, F. B., Sherman, R. D., Wahlquist, H. D., and Heckman, T. M. 1984, "*IUE* observations of high-redshift quasars", *Astrophys. J.*, **281**, 76. *[233]*

Begelman, M. C. 1985, "Accretion disks in active galactic nuclei", in *Astrophysics of Active Galaxies and Quasi-Stellar Objects*, ed. J. S. Miller (Mill Valley: University Science Books), p. 411. *[310]*

Begelman, M. C., Blandford, R. D., and Rees, M. J. 1980, "Massive black hole binaries in active galactic nuclei", *Nature*, **287**, 307. *[40]*

Begelman, M. C., Blandford, R. D., and Rees, M. J. 1984, "Theory of extragalactic radio sources", *Rev. Mod. Phys.*, **56**, 255. *[186]*

Begelman, M. C., and Sikora, M. 1987, "Inverse Compton scattering of ambient radiation by a cold relativistic jet: a source of beamed, polarized continuum in blazars?", *Astrophys. J.*, submitted. *[301]*

Bezler, M., Kendziorra, E., Staubert, R., Hasinger, G., Pietsch, W., Reppin, C., Trümper, J., and Voges, W. 1984, "The high energy X-ray spectrum of 3C 273", *Astron. Astrophys.*, **136**, 351. *[280]*

Biermann, P., Duerbeck, H., Eckart, A., Fricke, K., Johnston, K. J., Kühr, H., Liebert, J., Pauliny-Toth, I. I. K., Schleicher, H., Stockman, H., Strittmatter, P. A., and Witzel, A. 1981, "Observations of six flat spectrum sources from the 5 GHz survey", *Astrophys. J. (Letters)*, **247**, L53. *[233]*

Bignell, R. C. 1982, "Polarimetry", Lecture No. 6 in *Synthesis Mapping*, Proceedings of the NRAO–VLA Workshop held in Socorro, New Mexico, June 21–25, 1982, ed. A. R. Thompson and L. R. D'Addario (Green Bank: National Radio Astronomy Observatory). *[186]*

Biretta, J. A., Cohen, M. H., Hardebeck, H. E., Kaufmann, P., Abraham, Z., Perfetto, A. A., Scalise Jr., E., Schaal, R. E., and Silva, P. M. 1985, "Observations of 3C 273 with high north-south resolution", *Astrophys. J. (Letters)*, **292**, L5. *[26, 40, 129]*

Biretta, J., Cohen, M., and Moore, R. 1986, "Evolution of the compact radio source 3C 345", in *IAU Symposium 119, Quasars*, ed. G. Swarup and V. K. Kapahi (Dordrecht: Reidel), p. 157. *[129]*

Biretta, J. A., Moore, R. L., and Cohen, M. H. 1986, "The evolution of the compact radio source in 3C 345. I. VLBI observations", *Astrophys. J.*, **308**, 93. *[12, 40, 60, 72, 129, 155, 186, 233, 280]*

Biretta, J. A., Schneider, D. P., and Gunn, J. E. 1985, "Optical identifications of six compact double radio sources", *Astron. J.*, **90**, 2508. *[168]*

Blandford, R. D. 1984, "Theoretical models of active galactic nuclei", in *IAU Symposium 110, VLBI and Compact Radio Sources*, ed. R. Fanti, K. Kellermann, and G. Setti (Dordrecht: Reidel), p. 215. *[310]*

Blandford, R. D. 1986, "Black hole models of quasars", in *IAU Symposium 119, Quasars*, ed. G. Swarup and V. K. Kapahi (Dordrecht: Reidel), p. 359. *[310]*

Blandford, R. D. 1987, "Astrophysical black holes", in *Three Hundred Years of Gravitation*, ed. S. Hawking and W. Israel (Cambridge: Cambridge University Press), in press. *[310]*

Blandford, R. D., and Königl, A. 1979, "Relativistic jets as compact radio sources", *Astrophys. J.*, **232**, 34. *[1, 104, 114, 126, 155, 168, 186, 193, 273, 280, 310]*

Blandford, R. D., McKee, C. F., and Rees, M. J. 1977, "Super-luminal expansion in extragalactic radio sources", *Nature*, **267**, 211. *[1]*

Blandford, R. D., and Rees, M. J. 1978, "Some comments on radiation mechanisms in lacertids", in *Pittsburgh Conference on BL Lac Objects*, ed. A. M. Wolfe (Pittsburgh: University of Pittsburgh), p. 328. *[1, 233, 280]*

Bònoli, F., Braccesi, A., Federici, L., Zitelli, V., and Formiggini, L. 1979, "A study of the optical variability of the quasistellar objects in the $13^h + 36°$ field—I: The blue magnitudes", *Astron. Astrophys. Suppl.*, **35**, 391. *[233]*

Boroson, T. A., and Oke, J. B. 1984, "Spectroscopy of the nebulosity around eight high-luminosity QSOs", *Astrophys. J.*, **281**, 535. *[233]*

Boroson, T. A., Persson, S. E., and Oke, J. B. 1985, "More spectroscopy of the fuzz around QSOs: additional evidence for two types of QSO", *Astrophys. J.*, **293**, 120. *[233]*

Borra, E. F., and Corriveau, G. 1984, "A search for faint highly polarized objects", *Astrophys. J.*, **276**, 449. *[233]*

Boyle, B. J., Fong, R., Shanks, T., and Peterson, B. A. 1987, "The evolution of optically selected QSOs", *Monthly Notices Roy. Astron. Soc.*, in press. *[310]*

Bregman, J. N. 1986, "Compact continuum emission from optically violently variable quasars and BL Lacertae objects", in *Continuum Emission in Active Galactic Nuclei*, ed. M. L. Sitko (Tucson: Kitt Peak National Observatory, National Optical Astronomy Observatories), p. 102. *[217, 233]*

Bregman, J. N., Glassgold, A. E., Huggins, P. J., and Kinney, A. L. 1986, "Variability of Lyman-alpha and the ultraviolet continuum of 3C 446", *Astrophys. J.*, **301**, 698. *[233]*

Bregman, J. N., Glassgold, A. E., Huggins, P. J., Neugebauer, G., Soifer, B. T., Matthews, K., Elias, J., Webb, J., Pollock, J. T., Pica, A. J., Leacock, R. J., Smith, A. G., Aller, H. D., Aller, M. F., Hodge, P. E., Dent, W. A., Balonek, T. J., Barvainis, R. E., Roellig, T. P. L., Wiśniewski, W. Z., Rieke, G. H., Lebofsky, M. J., Wills, B. J., Wills, D., Ku, W. H.-M., Bregman, J. D., Witteborn, F. C., Lester, D. F., Impey, C. D., and Hackwell, J. A. 1986, "Multifrequency observations of the superluminal quasar 3C 345", *Astrophys. J.*, **301**, 708. *[40, 217, 251, 260]*

Bridle, A. H., and Perley, R. A. 1984, "Extragalactic radio jets", *Ann. Rev. Astron. Astrophys.*, **22**, 319. *[148, 168, 310]*

Broderick, J. J., Vitkevich, V. V., Jauncey, D. L., Efanov, V. A., Kellermann, K. I., Clark, B. G., Kogan, L. R., Kostenko, V. I., Cohen, M. H., Matveenko, L. I., Moiseev, I. G., Payne, J., and Hansson, B. 1970, "Nablyudeniya kompaktnyukh radioistochnikov na radiointerferometre s bazoj grin behnk–krym", *Astron. Zh.*, **47**, 784. English translation: 1971, "Observations of compact radio sources with a radio interferometer having a Green Bank–Crimea baseline", *Soviet Astron.*, **14**, 627. *[1]*

Broten, N. W., Legg, T. H., Locke, J. L., McLeish, C. W., Richards, R. S., Chisholm, R. M., Gush, H. P., Yen, J. L., and Galt, J. A. 1967, "Observations of quasars using interferometer baselines up to 3,074 km", *Nature*, **215**, 38. *[1]*

Brown, R. L., Johnston, K. J., Briggs, F. H., Wolfe, A. M., Neff, S. G., and Walker, R. C. 1981, "The variable radio structure of 3C 446", *Astrophys. Letters*, **21**, 105. *[12]*

Browne, I. W. A., Clark, R. R., Moore, P. K., Muxlow, T. W. B., Wilkinson, P. N., Cohen, M. H., and Porcas, R. W. 1982, "MERLIN observations of superluminal radio sources", *Nature*, **299**, 788. *[40, 67, 129, 155]*

Browne, I. W. A., and Murphy, D. W. 1987, "Beaming and the X-ray, optical and radio properties of quasars", *Monthly Notices Roy. Astron. Soc.*, in press. *[251]*

Browne, I. W. A., and Orr, M. J. L. 1981, "Observations of core-dominated sources with MTRLI", in *Optical Jets in Galaxies*, Proceedings of the second ESO/ESA workshop on the use of the Space Telescope and co-ordinated ground based research, ESA SP-162, (Noordwijk: European Space Agency), p. 87. *[129, 155, 180]*

Browne, I. W. A., Orr, M. J. L., Davis, R. J., Foley, A., Muxlow, T. W. B., and Thomasson, P. 1982, "Jodrell Bank MTRLI observations of nine core-dominated sources at 408 MHz", *Monthly Notices Roy. Astron. Soc.*, **198**, 673. *[55, 186]*

Browne, I. W. A., and Perley, R. A. 1986, "Extended radio emission round core-dominated quasars—constraints on relativistic beaming models", *Monthly Notices Roy. Astron. Soc.*, **222**, 149. *[129]*

Browne, I. W. A., and Wright, A. E. 1985, "Relativistic beaming and the optical magnitudes of QSOs", *Monthly Notices Roy. Astron. Soc.*, **213**, 97. *[233]*

Burbidge, G. R., Jones, T. W., and O'Dell, S. L. 1974, "Physics of compact nonthermal sources. III. Energetic considerations", *Astrophys. J.*, **193**, 43. *[60, 280]*

Burns, J. O., Basart, J. P., De Young, D. S., and Ghiglia, D. C. 1984, "Radio jets in classical double radio sources with strong cores", *Astrophys. J.*, **283**, 515. *[155]*

Capps, R. W., Sitko, M. L., and Stein, W. A. 1982, "The spectral flux distributions of sources in an optically selected sample of QSOs: 10^{13}–10^{15} Hz", *Astrophys. J.*, **255**, 413. *[233]*

Carter, W. E., Robertson, D. S., and MacKay, J. R. 1985, "Geodetic radio interferometric surveying: applications and results", *J. Geophys. Res.*, **90**, 4577. *[83]*

Carvalho, J. C. 1985, "An evolution model for compact double radio sources", *Monthly Notices Roy. Astron. Soc.*, **215**, 463. *[168]*

Cawthorne, T. V., Scheuer, P. A. G., Morison, I., and Muxlow, T. W. B. 1986, "A sample of powerful radio sources for VLBI studies", *Monthly Notices Roy. Astron. Soc.*, **219**, 883; erratum *Monthly Notices Roy. Astron. Soc.*, **222**, 895. *[1, 94, 104]*

Clark, B. G. 1973, "The NRAO tape-recorder interferometer system", *Proc. Inst. Elec. Electron. Engrs.*, **61**, 1242. *[55]*

Clark, B. G. 1981, *How to Use the VLA for VLB*, internal document, National Radio Astronomy Observatory. *[186]*

Clegg, P. E., Gear, W. K., Ade, P. A. R., Robson, E. I., Smith, M. G., Nolt, I. G., Radostitz, J. V., Glaccum, W., Harper, D. A., and Low, F. J. 1983, "Millimeter and submillimeter observations of 3C 273", *Astrophys. J.*, **273**, 58. *[280]*

Cohen, M. H. 1969, "High-resolution observations of radio sources", *Ann. Rev. Astron. Astrophys.*, **7**, 619. *[1]*

Cohen, M. H. 1985, "Inverse Compton X-rays and VLBI radio structures", in *Extragalactic Energetic Sources*, ed. V. K. Kapahi (Bangalore: Indian Academy of Sciences), p. 1. *[60]*

Cohen, M. H. 1986, "Relativistic motion in quasars", in *Highlights of Modern Astrophysics*, ed. S. L. Shapiro and S. A. Teukolsky (New York: Wiley), p. 299. *[233]*

Cohen, M. H., Cannon, W., Purcell, G. H., Shaffer, D. B., Broderick, J. J., Kellermann, K. I., and Jauncey, D. L. 1971, "The small-scale structure of radio galaxies and quasi-stellar sources at 3.8 centimeters", *Astrophys. J.*, **170**, 207. *[1, 12, 26]*

Cohen, M. H., and Gundermann, E. J. 1967, "Angular diameter of 3C 279 from interplanetary scintillations", *Astrophys. J. (Letters)*, **148**, L49. *[1]*

Cohen, M. H., Kellermann, K. I., Shaffer, D. B., Linfield, R. P., Moffet, A. T., Romney, J. D., Seielstad, G. A., Pauliny-Toth, I. I. K., Preuss, E., Witzel, A., Schilizzi, R. T., and Geldzahler, B. J. 1977, "Radio sources with superluminal velocities", *Nature*, **268**, 405. *[12, 26]*

Cohen, M. H., Moffet, A. T., Romney, J. D., Schilizzi, R. T., Seielstad, G. A., Kellermann, K. I., Purcell, G. H., Shaffer, D. B., Pauliny-Toth, I. I. K., Preuss, E., Witzel, A., and Rinehart, R. 1976, "Rapid increase in the size of 3C 345", *Astrophys. J. (Letters)*, **206**, L1. *[40]*

Cohen, M. H., Moffet, A. T., Romney, J. D., Schilizzi, R. T., Shaffer, D. B., Kellermann, K. I., Purcell, G. H., Grove, G., Swenson Jr., G. W., Yen, J. L., Pauliny-Toth, I. I. K., Preuss, E., Witzel, A., and Graham, D. 1975, "Observations with a VLB array. I. Introduction and procedures", *Astrophys. J.*, **201**, 249. *[55]*

Cohen, M. H., Pearson, T. J., Readhead, A. C. S., Seielstad, G. A., Simon, R. S., and Walker, R. C. 1979, "Superluminal variations in 3C 120, 3C 273, and 3C 345", *Astrophys. J.*, **231**, 293. *[26]*

Cohen, M. H., and Unwin, S. C. 1984, "Superluminal effects and bulk relativistic motion", in *IAU Symposium 110, VLBI and Compact Radio Sources*, ed. R. Fanti, K. Kellermann, and G. Setti (Dordrecht: Reidel), p. 95. *[55, 60, 114, 155, 162, 251]*

Cohen, M. H., Unwin, S. C., Lind, K. R., Moffet, A. T., Simon, R. S., Wilkinson, P. N., Spencer, R. E., Booth, R. S., Nicolson, G. D., Niell, A. E., and Young, L. E. 1983, "VLBI maps of 3C 273 and 3C 345 at 2.3 GHz", *Astrophys. J.*, **272**, 383. *[26]*

Cohen, M. H., Unwin, S. C., Pearson, T. J., Seielstad, G. A., Simon, R. S., Linfield, R. P., and Walker, R. C. 1983, "Rectilinear motions in 3C 345", *Astrophys. J. (Letters)*, **269**, L1. *[40]*

Cohen, M. H., Zensus, J. A., Biretta, J. A., Comoretto, G., Kaufmann, P., and Abraham, Z. 1987, "Evolution of 3C 273 at 10.7 GHz", *Astrophys. J. (Letters)*, **315**, 489. *[12, 26]*

Condon, J. J., Ledden, J. E., O'Dell, S. L., and Dennison, B. 1979, "318-MHz variability of complete samples of extragalactic radio sources", *Astron. J.*, **84**, 1. *[200]*

Condon, J. J., O'Dell, S. L., Puschell, J. J., and Stein, W. A. 1981, "Radio emission from bright, optically selected quasars", *Astrophys. J.*, **246**, 624. *[233]*

Conway, R. G., Burn, B. J., and Vallée, J. P. 1977, "Measurements of structure and polarization of 72 sources from the 4C catalogue", *Astron. Astrophys. Suppl.*, **27**, 155. *[155]*

Conway, R. G., and Kronberg, P. P. 1969, "Interferometric measurement of polarization distributions in radio sources", *Monthly Notices Roy. Astron. Soc.*, **142**, 11. *[193]*

Cornwell, T. J., and Wilkinson, P. N. 1981, "A new method for making maps with unstable radio interferometers", *Monthly Notices Roy. Astron. Soc.*, **196**, 1067. *[34, 67]*

Cornwell, T. J., and Wilkinson, P. N. 1984, "Selfcalibration", in *Indirect Imaging*, ed. J. A. Roberts (Cambridge: Cambridge University Press), p. 207. *[211]*

Cotton, W. D. 1979, "A method of mapping compact structure in radio sources using VLBI observations", *Astron. J.*, **84**, 1122. *[211]*

Cotton, W. D., Counselman III, C. C., Geller, R. B., Shapiro, I. I., Wittels, J. J., Hinteregger, H. F., Knight, C. A., Rogers, A. E. E., Whitney, A. R., and Clark, T. A. 1979, "3C 279: the case for 'superluminal' expansion", *Astrophys. J. (Letters)*, **229**, L115. *[12, 34, 129]*

Cotton, W. D., Geldzahler, B. J., Marcaide, J. M., Shapiro, I. I., Sanroma, M., and Rius, A. 1984, "VLBI observations of the polarized radio emission from the quasar 3C 454.3", *Astrophys. J.*, **286**, 503. *[186]*

Cotton, W. D., Owen, F. N., Geldzahler, B. J., Johnston, K., Bååth, L., and Romney, J. 1984, "High-resolution observations of the steep spectrum source 2147+145", *Astrophys. J. (Letters)*, **277**, L41. *[206]*

Cotton, W. D., Wittels, J. J., Shapiro, I. I., Marcaide, J., Owen, F. N., Spangler, S. R., Rius, A., Angulo, C., Clark, T. A., and Knight, C. A. 1980, "The very flat radio spectrum of 0735+178: a cosmic conspiracy?", *Astrophys. J. (Letters)*, **238**, L123. *[217, 280]*

Couderc, P. 1939, "Les auréoles lumineuses des novae", *Ann. Astrophys.*, **2**, 271. *[1]*

Courant, R., and Friedrichs, K. O. 1948, *Supersonic Flow and Shock Waves* (New York: Interscience). *[280]*

Davis, R. J. 1986, "Radio jet and lobe of 3C 273", in *IAU Symposium 119, Quasars*, ed. G. Swarup and V. K. Kapahi (Dordrecht: Reidel), p. 211. *[233]*

Davis, R. J., Muxlow, T. W. B., and Conway, R. G. 1985, "Radio emission from the jet and lobe of 3C 273", *Nature*, **318**, 343. *[26, 129]*

de Bruyn, A. G., and Schilizzi, R. T. 1984, "Extended structure around superluminal and other core-dominated radio sources", in *IAU Symposium 110, VLBI and Compact Radio Sources*, ed. R. Fanti, K. Kellermann, and G. Setti (Dordrecht: Reidel), p. 165. *[48, 233]*

de Bruyn, A. G., and Schilizzi, R. T. 1986, "WSRT observations of core-dominated radio sources", in *IAU Symposium 119, Quasars*, ed. G. Swarup and V. K. Kapahi (Dordrecht: Reidel), p. 203. *[129]*

Dennison, B., Broderick, J. J., O'Dell, S. L., Mitchell, K. J., Altschuler, D. R., Payne, H. E., and Condon, J. J. 1984, "The spectral evolution of low-frequency variable radio sources", *Astrophys. J. (Letters)*, **281**, L55. *[206]*

Dennison, B., and Condon, J. J. 1981, "A search for interstellar scintillations in a large sample of low-frequency variable sources", *Astrophys. J.*, **246**, 91. *[200]*

Dent, W. A. 1965, "Quasi-stellar sources: variations in the radio emission of 3C 273", *Science*, **148**, 1458. *[1]*

Dent, W. A. 1966, "Variation in the radio emission from the Seyfert galaxy NGC 1275", *Astrophys. J.*, **144**, 843. *[76]*

Dent, W. A., O'Dea, C. P., Balonek, T. J., Hobbs, R. W., and Howard, R. J. 1983, "A rapid millimetre wave outburst in the nucleus of NGC 1275", *Nature*, **306**, 41. *[76]*

de Pater, I., and Perley, R. A. 1983, "The radio structure of 3C 279", *Astrophys. J.*, **273**, 64. *[34, 129]*

de Waard, G. J. 1986, "Thermal-nonthermal relationships in active galactic nuclei", Ph. D. thesis, Rijksuniversiteit Leiden. *[1, 148]*

Dulk, G. A., McLean, D. J., Manchester, R. N., Ostry, D. I., and Rogers, P. G. 1984, "Image restoration techniques—how good are they?", in *Indirect Imaging*, ed. J. A. Roberts (Cambridge: Cambridge University Press), p. 355. *[211]*

Dyson, J. E. (ed.) 1985, *Active Galactic Nuclei* (Manchester: Manchester University Press). *[217]*

Eckart, A. 1983, "Untersuchung der Eigenschaften extragalaktischer Radioquellen", Ph. D. thesis, Westfälische Wilhelms-Universität Münster, Westfalen. *[83]*

Eckart, A., Witzel, A., Biermann, P., Johnston, K. J., Simon, R., Schalinski, C., and Kühr, H. 1986, "Investigation of a complete sample of flat spectrum radio sources from the S5 survey. I. Analysis", *Astron. Astrophys.*, **168**, 17. *[83]*

Eckart, A., Witzel, A., Biermann, P., Johnston, K. J., Simon, R., Schalinski, C., and Kühr, H. 1987, "Investigation of a complete sample of flat spectrum radio sources from the S5 survey. II. Results", *Astron. Astrophys. Suppl.*, **67**, 121. *[83]*

Eckart, A., Witzel, A., Biermann, P., Pearson, T. J., Readhead, A. C. S., and Johnston, K. J. 1985, "Milliarcsecond structure of 1928+738: apparent superluminal motion along an extended jet?", *Astrophys. J. (Letters)*, **296**, L23. *[12, 83, 94, 114, 129, 148, 155, 251]*

Ennis, D. J., Neugebauer, G., and Werner, M. 1982, "1 millimeter continuum observations of quasars", *Astrophys. J.*, **262**, 460. *[217]*

Fanaroff, B. L., and Riley, J. M. 1974, "The morphology of extragalactic radio sources of high and low luminosity", *Monthly Notices Roy. Astron. Soc.*, **167**, 31P. *[155]*

Fanti, C., Fanti, R., Parma, P., Schilizzi, R. T., and van Breugel, W. J. M. 1985, "Compact steep spectrum 3CR radio sources. VLBI observations at 18 cm", *Astron. Astrophys.*, **143**, 292. *[168, 174]*

Fanti, C., Fanti, R., Schilizzi, R. T., Spencer, R. E., and van Breugel, W. J. M. 1986, "The peculiar morphology of the compact steep spectrum radio source 3C 119", *Astron. Astrophys.*, **170**, 10. *[174]*

Fanti, R., Ficarra, A., Mantovani, F., Padrielli, L., and Weiler, K., 1979, "Variability of extragalactic radiosources at 408 MHz. Results of a 3 year monitoring program", *Astron. Astrophys. Suppl.*, **36**, 359. *[55]*

Feretti, L., Giovannini, G., Gregorini, L., Parma, P., and Zamorani, G. 1984, "Statistical properties of the radio cores in elliptical galaxies", *Astron. Astrophys.*, **139**, 55. *[174]*

Fiedler, R. L., Dennison, B., Johnston, K. J., and Hewish, A. 1987, "Occultations of extragalactic radio sources", *Nature*, submitted. *[217]*

Fiedler, R. L., Waltman, E. B., Spencer, J. H., Johnston, K. J., Angerhoffer, P. E., Florkowski, D. R., Josties, F. J., Klepczynski, F. J., McCarthy, D. D., and Matsakis, D. N. 1987, "Daily observations of compact extragalactic radio sources at 2695 and 8085 MHz, 1979 to 1985", *Astrophys. J. Suppl.*, submitted. *[217]*

Filippenko, A. V. 1986, "Studies of narrow emission lines in AGNs", in *IAU Symposium 119, Quasars*, ed. G. Swarup and V. K. Kapahi (Dordrecht: Reidel), p. 289. *[60]*

Flatters, C., and Conway, R. G. 1985, "The radio jet of the quasar 3C 273", *Nature*, **314**, 425. *[186]*

Foley, A. R. 1982, "MERLIN observations of core dominated quasars", Ph. D. thesis, The Victoria University of Manchester. *[114, 129]*

Gear, W. K., Brown, L. M. J., Robson, E. I., Ade, P. A. R., Griffin, M. J., Smith, M. G., Nolt, I. G., Radostitz, J. V., Veeder, G., and Lebofsky, L. 1986, "Multifrequency observations of blazars. II. The variability of the 1 μm to 2 mm continuum", *Astrophys. J.*, **304**, 295. *[217]*

Gear, W. K., Gee, G., Robson, E. I., and Nolt, I. G. 1987, "Thermal and non-thermal emission from NGC 1275 (3C 84)", *Monthly Notices Roy. Astron. Soc.*, in press. *[76]*

Ghisellini, G., Maraschi, L., Tanzi, E. G., and Treves, A. 1986, "Spectral properties of blazars. I. Objects observed in the far-ultraviolet", *Astrophys. J.*, **310**, 317. *[233, 251]*

Giovannini, G. 1985, "Properties of the radio cores in elliptical galaxies", in *Active Galactic Nuclei*, ed. J. E. Dyson (Manchester: Manchester University Press), p. 93. *[168]*

Giuricin, G., Mardirossian, F., Mezzetti, M., and Ramella, M. (ed.) 1986, *Structure and Evolution of Active Galactic Nuclei* (Dordrecht: Reidel). *[217]*

Götz, M. M. A., Alef, W., Preuss, E., and Kellermann, K. I. 1987, "Strong structural variability in the lobe-dominated radio galaxy 3C 111", *Astron. Astrophys.*, in press. *[12, 55]*

Gould, R. J. 1979, "Compton and synchrotron processes in spherically-symmetric non-thermal sources", *Astron. Astrophys.*, **76**, 306. *[40, 280]*

Greenstein, J. L., and Schmidt, M. 1964, "The quasi-stellar radio sources 3C 48 and 3C 273", *Astrophys. J.*, **140**, 1. *[26]*

Gubbay, J., Legg, A. J., Robertson, D. S., Moffet, A. T., Ekers, R. D., and Seidel, B. 1969, "Variations of small quasar components at 2,300 MHz", *Nature*, **224**, 1094. *[1]*

Gubbay, J., Legg, A. J., Robertson, D. S., Moffet, A. T., and Seidel, B. 1969, "Trans-Pacific interferometer measurements at 2,300 MHz", *Nature*, **222**, 730. *[1]*

Halpern, J. P. 1982, "X-ray spectra of active galactic nuclei", Ph. D. thesis, Harvard University. *[34, 280]*

Halpern, J. P. 1985, "X-ray spectrum and variability of 3C 120", *Astrophys. J.*, **290**, 130. *[251]*

Hardee, P. E. 1986, "A helically twisted relativistic jet: nonradial motions and accelerations in 3C 345", *Can. J. Phys.*, **64**, 484. *[40]*

Harvey, P. M., Wilking, B. A., and Joy, M. 1982, "Far-infrared photometry of compact extragalactic objects: detection of 3C 345", *Astrophys. J. (Letters)*, **254**, L29. *[217]*

Harwit, M. 1981, *Cosmic Discovery* (New York: Basic Books), p. 140. *[12]*

Hazard, C., Mackey, M. B., and Shimmins, A. J. 1963, "Investigation of the radio source 3C 273 by the method of lunar occultations", *Nature*, **197**, 1037. *[26]*

Heeschen, D. S. 1984, "Flickering of extragalactic radio sources", *Astron. J.*, **89**, 1111. *[310]*

Hertzsprung, E. 1911, "Über die Verwendung photographischer effektiver Wellenlängen zur Bestimmung von Farbenäquivalenten", *Publ. Astrophys. Obs. Potsdam*, **22**, Nr. 63. *[310]*

Hintzen, P., Ulvestad, J., and Owen, F. 1983, "Are wide-angle radio-tail QSOs members of clusters of galaxies? I. VLA maps at 20 cm of 117 radio quasars", *Astron. J.*, **88**, 709. *[129]*

Hodge, P. E., and Aller, H. D. 1979, "Circular polarisation and the magnetic dipole model", *Nature*, **278**, 838. *[280]*

Hodges, M. W., Mutel, R. L., and Phillips, R. B. 1984, "A VLBI survey of peaked spectrum radio sources", *Astron. J.*, **89**, 1327. *[168]*

Hough, D. H. 1986, "Parsec-scale structure in the nuclei of double-lobed radio quasars", Ph. D. thesis, California Institute of Technology. *[114, 129, 148]*

Hoyle, F., and Burbidge, G. R. 1966, "On the nature of the quasi-stellar objects", *Astrophys. J.*, **144**, 534. *[1]*

Hoyle, F., Burbidge, G. R., and Sargent, W. L. W. 1966, "On the nature of the quasi-stellar sources", *Nature*, **209**, 751. *[1]*

Hughes, P. A., Aller, H. D., and Aller, M. F. 1985, "Polarized radio outbursts in BL Lacertae. II. The flux and polarization of a piston-driven shock", *Astrophys. J.*, **298**, 301. *[34, 60, 217, 273]*

Hughes, P. A., Aller, H. D., and Aller, M. F. 1986, "Shocks as the origin of radio-source variability", *Can. J. Phys.*, **64**, 466. *[273]*

Hunstead, R. W. 1972, "Four variable radio sources at 408 MHz", *Astrophys. Letters*, **12**, 193. *[55, 200]*

Hutchings, J. B., Crampton, D., and Campbell, B. 1984, "Optical imaging of 78 quasars and host galaxies", *Astrophys. J.*, **280**, 41. *[104]*

Impey, C. D., and Brand, P. W. J. L. 1982, "The calibration of a radio-independent search for BL Lac objects", *Monthly Notices Roy. Astron. Soc.*, **201**, 849. *[233]*

Impey, C. D., Brand, P. W. J. L., Wolstencroft, R. D., and Williams, P. M. 1982, "Infrared polarimetry and photometry of BL Lac objects", *Monthly Notices Roy. Astron. Soc.*, **200**, 19. *[233]*

Jägers, W. J., van Breugel, W. J. M., Miley, G. K., Schilizzi, R. T., and Conway, R. G. 1982, "Radio observations of the giant quasar 4C 34.47", *Astron. Astrophys.*, **105**, 278. *[155]*

Jenkins, C. J., Pooley, G. G., and Riley, J. M. 1977, "Observations of 104 extragalactic radio sources with the Cambridge 5-km telescope at 5 GHz", *Mem. Roy. Astron. Soc.*, **84**, 61. *[174]*

Johnston, K. J., Simon, R. S., Eckart, A., Biermann, P., Schalinski, C., Witzel, A., and Strom, R. G. 1987, "1928+738: a superluminal source with large-scale structure", *Astrophys. J. (Letters)*, **313**, L85. *[83, 155]*

Johnston, K. J., Spencer, J. H., Witzel, A., and Fomalont, E. B. 1983, "3C 395—a quasar with asymmetrical radio structure", *Astrophys. J. (Letters)*, **265**, L43. *[72, 168]*

Jones, D. L. 1986, "Evidence for intrinsically asymmetrical radio structure in the nucleus of NGC 6251", *Astrophys. J. (Letters)*, **309**, L5. *[12, 162]*

Jones, D. L., Unwin, S. C., Readhead, A. C. S., Sargent, W. L. W., Seielstad, G. A., Simon, R. S., Walker, R. C., Benson, J. M., Perley, R. A., Bridle, A. H., Pauliny-Toth, I. I. K., Romney, J., Witzel, A., Wilkinson, P. N., Bååth, L. B., Booth, R. S., Fort, D. N., Galt, J. A., Mutel, R. L., and Linfield, R. P. 1986, "High dynamic range VLBI observations of NGC 6251", *Astrophys. J.*, **305**, 684. *[162, 180]*

Jones, T. W. 1982, "Numerical modelling of compact radio jet dynamics", *Bull. Am. Astron. Soc.*, **14**, 963. *[273]*

Jones, T. W. 1986, "Magnetic-field configurations in compact extragalactic jets", *Can. J. Phys.*, **64**, 463. *[217]*

Jones, T. W. (ed.) 1986, "The Minnesota Lectures on Active Galactic Nuclei", *Publ. Astron. Soc. Pacific*, **98**, 129. *[217]*

Jones, T. W., and Burbidge, G. R. 1973, "On the problem of explaining decimeter flux variations in 3C 454.3 and other extragalactic sources", *Astrophys. J.*, **186**, 791. *[200]*

Jones, T. W., and O'Dell, S. L. 1977, "Transfer of polarized radiation in self-absorbed synchrotron sources. I. Results for a homogeneous source", *Astrophys. J.*, **214**, 522. *[217]*

Jones, T. W., and O'Dell, S. L. 1977, "Transfer of polarized radiation in self-absorbed synchrotron sources. II. Treatment of inhomogeneous media and calculation of emergent polarization", *Astrophys. J.*, **215**, 236. *[273]*

Jones, T. W., O'Dell, S. L., and Stein, W. A. 1974, "Physics of compact nonthermal sources. I. Theory of radiation processes", *Astrophys. J.*, **188**, 353. *[280]*

Jones, T. W., O'Dell, S. L., and Stein, W. A. 1974, "Physics of compact nonthermal sources. II. Determination of physical parameters", *Astrophys. J.*, **192**, 261. *[280]*

Jones, T. W., Rudnick, L., Aller, H. D., Aller, M. F., Hodge, P. E., and Fiedler, R. L. 1985, "Magnetic field structures in active compact radio sources", *Astrophys. J.*, **290**, 627. *[186, 217, 273]*

Jones, T. W., Rudnick, L., and Landau, R. 1986, "Evidence for a universal, simple model for active extragalactic sources", in *Continuum Emission in Active Galactic Nuclei*, ed. M. L. Sitko (Tucson: Kitt Peak National Observatory, National Optical Astronomy Observatories), p. 122. *[217]*

Jones, T. W., Rudnick, L., Owen, F. N., Puschell, J. J., Ennis, D. J., and Werner, M. W. 1981, "The broad-band spectra and variability of compact nonthermal sources", *Astrophys. J.*, **243**, 97. *[217]*

Königl, A. 1980, "Relativistic gasdynamics in two dimensions", *Phys. Fluids*, **23**, 1083. *[280]*

Königl, A. 1981, "Relativistic jets as X-ray and gamma-ray sources", *Astrophys. J.*, **243**, 700. *[34, 40, 310]*

Königl, A., and Choudhuri, A. R. 1985, "Force-free equilibria of magnetized jets", *Astrophys. J.*, **289**, 173. *[193]*

Kühr, H., Nauber, U., Pauliny-Toth, I. I. K., and Witzel, A. 1979, "A catalogue of radio sources", Preprint No. 55 (Bonn: Max-Planck-Institut für Radioastronomie). *[251]*

Kühr, H., Pauliny-Toth, I. I. K., Witzel, A., and Schmidt, J. 1981, "The 5-GHz strong source surveys. V. Survey of the area between declinations 70° and 90°", *Astron. J.*, **86**, 854. *[83]*

Kühr, H., Witzel, A., Pauliny-Toth, I. I. K., and Nauber, U. 1981, "A catalogue of extragalactic radio sources having flux densities greater than 1 Jy at 5 GHz", *Astron. Astrophys. Suppl.*, **45**, 367. *[83, 94, 233]*

Kellermann, K. 1970, "The Russian-American VLB experiment (parts I–III)", *The Observer* (Green Bank: National Radio Astronomy Observatory), **10**, No. 1, p. 9; No. 2, p. 7; No. 4, p. 5. *[1]*

Kellermann, K. I. 1972, "Radio emission from compact objects", in *IAU Symposium 44, External Galaxies and Quasi-Stellar Objects*, ed. D. S. Evans (Dordrecht: Reidel), p. 190. *[1]*

Kellermann, K. I., Clark, B. G., Bare, C. C., Rydbeck, O., Ellder, J., Hansson, B., Kollberg, E., Hoglund, B., Cohen, M. H., and Jauncey, D. L. 1968, "High-resolution interferometry of small radio sources using intercontinental base lines", *Astrophys. J. (Letters)*, **153**, L209. *[1]*

Kellermann, K. I., Clark, B. G., Shaffer, D. B., Cohen, M. H., Jauncey, D. L., Broderick, J. J., and Niell, A. E. 1974, "Further observations of apparent changes in the structure of 3C 273 and 3C 279", *Astrophys. J. (Letters)*, **189**, L19. *[26]*

Kellermann, K. I., and Pauliny-Toth, I. I. K. 1967, "Variations in the flux density of some quasi-stellar sources", *Nature*, **213**, 977. *[55]*

Kellermann, K. I., and Pauliny-Toth, I. I. K. 1969, "The spectra of opaque radio sources", *Astrophys. J. (Letters)*, **155**, L71. *[280]*

Kellermann, K. I., and Pauliny-Toth, I. I. K. 1981, "Compact radio sources", *Ann. Rev. Astron. Astrophys.*, **19**, 373. *[174]*

Kellermann, K. I., Shaffer, D. B., Purcell, G. H., Pauliny-Toth, I. I. K., Preuss, E., Witzel, A., Graham, D., Schilizzi, R. T., Cohen, M. H., Moffet, A. T., Romney, J. D., and Niell, A. E. 1977, "Very high-resolution observations of the radio sources NRAO 150, OJ 287, 3C 273, M87, 1633+38, BL Lacertae, and 3C 454.3", *Astrophys. J.*, **211**, 658. *[26]*

Kellermann, K., Sramek, R., Shaffer, D., Green, R., and Schmidt, M. 1986, "Radio emission from optically selected quasars", in *IAU Symposium 119, Quasars*, ed. G. Swarup and V. K. Kapahi (Dordrecht: Reidel), p. 95. *[1]*

Kembhavi, A., Feigelson, E. D., and Singh, K. P. 1986, "X-ray and radio core emission in radio quasars", *Monthly Notices Roy. Astron. Soc.*, **220**, 51. *[251]*

Kidger, M. R., and Beckman, J. E. 1986, "The I.A.C./Q.M.C. catalogue of quasar multiband spectra", in *Structure and Evolution of Active Galactic Nuclei*, ed. G. Giuricin, F. Mardirossian, M. Mezzetti, and M. Ramella (Dordrecht: Reidel), p. 591. *[217]*

Kinney, A. L., Huggins, P. J., Bregman, J. N., and Glassgold, A. E. 1985, "The ultraviolet spectra of intermediate-redshift quasars", *Astrophys. J.*, **291**, 128. *[233, 251]*

Knight, C. A., Robertson, D. S., Rogers, A. E. E., Shapiro, I. I., Whitney, A. R., Clark, T. A., Goldstein, R. M., Marandino, G. E., and Vandenberg, N. R. 1971, "Quasars: millisecond-of-arc structure revealed by very-long-baseline interferometry", *Science*, **172**, 52. *[1]*

Komesaroff, M. M., Roberts, J. A., Milne, D. K., Rayner, P. T., Cooke, D. J. 1984, "Circular and linear polarization variations of compact radio sources", *Monthly Notices Roy. Astron. Soc.*, **208**, 409. *[217]*

Kormendy, J. 1987, "Evidence for a nuclear black hole in M31", in *Supermassive Black Holes*, ed. M. Kafatos (Cambridge: Cambridge University Press), in press. *[310]*

Ku, W. H.-M., Helfand, D. J., and Lucy, L. B. 1980, "X-ray properties of quasars", *Nature*, **288**, 323. *[40, 251]*

Laing, R. A. 1981, "Multifrequency observations of 40 powerful extragalactic sources with the 5-km telescope", *Monthly Notices Roy. Astron. Soc.*, **195**, 261. *[186]*

Laing, R. A. (attributed) 1984, "Symmetries of jets and hot spots", in *Physics of Energy Transport in Extragalactic Radio Sources*, NRAO Workshop No. 9, ed. A. H. Bridle and J. A. Eilek (Green Bank: National Radio Astronomy Observatory), p. 128. *[148]*

Laing, R. A., Riley, J. M., and Longair, M. S. 1983, "Bright radio sources at 178 MHz: flux densities, optical identifications and the cosmological evolution of powerful radio galaxies", *Monthly Notices Roy. Astron. Soc.*, **204**, 151. *[104, 114, 129]*

Landau, R., Golisch, B., Jones, T. J., Jones, T. W., Pedelty, J., Rudnick, L., Sitko, M. L., Kenney, J., Roellig, T., Salonen, E., Urpo, S., Schmidt, G., Neugebauer, G., Matthews, K., Elias, J. H., Impey, C., Clegg, P., and Harris, S. 1986, "Active extragalactic sources: nearly simultaneous observations from 20 centimeters to 1400 Å", *Astrophys. J.*, **308**, 78. *[217, 233]*

Landau, R., Jones, T. W., Epstein, E. E., Neugebauer, G., Soifer, B. T., Werner, M. W., Puschell, J. J., and Balonek, T. J. 1983, "Extragalactic 1 millimeter sources: simultaneous observations at centimeter, millimeter, and visual wavelengths", *Astrophys. J.*, **268**, 68. *[217]*

Lawrence, C. R., Pearson, T. J., Readhead, A. C. S., and Unwin, S. C. 1986, "New redshifts of strong radio sources", *Astron. J.*, **91**, 494. *[12, 83, 155]*

Lawrence, C. R., Readhead, A. C. S., Linfield, R. P., Payne, D. G., Preston, R. A., Schilizzi, R. T., Porcas, R. W., Booth, R. S., and Burke, B. F. 1985, "Strong source VLBI observations at 22 GHz", *Astrophys. J.*, **296**, 458. *[26]*

Ledden, J. E., and O'Dell, S. L. 1983, "A search for X-ray emission from optically quiet, compact radio sources", *Astrophys. J.*, **270**, 434. *[251]*

Ledden, J. E., and O'Dell, S. L. 1985, "The radio–optical–X-ray spectral flux distributions of blazars", *Astrophys. J.*, **298**, 630. *[217, 251, 280]*

Legg, T. H. 1984, "Evidence of Doppler beaming in variable radio sources", in *IAU Symposium 110, VLBI and Compact Radio Sources*, ed. R. Fanti, K. Kellermann, and G. Setti (Dordrecht: Reidel), p. 183. *[217]*

Lind, K. R., and Blandford, R. D. 1985, "Semidynamical models of radio jets: relativistic beaming and source counts", *Astrophys. J.*, **295**, 358. *[40, 55, 104, 129, 306, 310]*

Linfield, R. 1981, "A precessing jet model of compact radio sources", *Astrophys. J.*, **250**, 464. *[310]*

Linfield, R. P. 1986, "A modified hybrid mapping technique for VLBI data", *Astron. J.*, **92**, 213. *[180]*

Lloyd, C. 1984, "Optical monitoring of radio sources", *Monthly Notices Roy. Astron. Soc.*, **209**, 697. *[267]*

Lovell, B. 1973, *Out of the Zenith* (Oxford: Oxford University Press), chapter 6. *[1]*

Lynden-Bell, D. 1977, "Hubble's constant determined from super-luminal radio sources", *Nature*, **270**, 396. *[1, 280]*

Madau, P., Ghisellini, G., and Persic, M. 1987, "Beaming in blazars", *Monthly Notices Roy. Astron. Soc.*, **224**, 257. *[251]*

Maddox, J. 1984, "Dispute over scale of universe", *Nature*, **307**, 313. *[12]*

Madejski, G. M. 1985, "X-ray studies of BL Lacertae objects", Ph. D. thesis, Harvard University. *[233, 251]*

Madejski, G. M., and Schwartz, D. A. 1983, "X-ray studies of BL Lacertae objects with the *Einstein Observatory*: confrontation with the synchrotron self-Compton predictions", *Astrophys. J.*, **275**, 467. *[251]*

Malkan, M. A. 1984, "The underlying galaxies of quasars. II. Imaging of a radio-loud sample", *Astrophys. J.*, **287**, 555. *[233]*

Maraschi, L., Ghisellini, G., Tanzi, E. G., and Treves, A. 1986, "Spectral properties of blazars. II. An X-ray observed sample", *Astrophys. J.*, **310**, 325. *[251]*

Marcaide, J. M., Bartel, N., Gorenstein, M. V., Shapiro, I. I., Corey, B. E., Rogers, A. E. E., Webber, J. C., Clark, T. A., Romney, J. D., and Preston, R. A. 1985, "Quasar 4C 39.25 is not contracting", *Nature*, **314**, 424. *[55]*

Marcaide, J. M., and Shapiro, I. I. 1984, "VLBI study of 1038+528 A and B: discovery of wavelength dependence of peak brightness location", *Astrophys. J.*, **276**, 56. *[12]*

Marcaide, J. M., Shapiro, I. I., Corey, B. E., Cotton, W. D., Gorenstein, M. V., Rogers, A. E. E., Romney, J. D., Schild, R. E., Bååth, L., Bartel, N., Cohen, N. L., Clark, T. A., Preston, R. A., Ratner, M. I., and Whitney, A. R. 1985, "The quasars 1038+528 A and B", *Astron. Astrophys.*, **142**, 71. *[12]*

Marscher, A. P. 1977, "Structure of radio sources with remarkably flat spectra: PKS 0735+178", *Astron. J.*, **82**, 781. *[40]*

Marscher, A. P. 1979, "Absorption models for low-frequency variability in compact radio sources", *Astrophys. J.*, **228**, 27. *[200]*

Marscher, A. P. 1983, "Accurate formula for the self-Compton X-ray flux density from a uniform, spherical, compact radio source", *Astrophys. J.*, **264**, 296. *[280]*

Marscher, A. P. 1986, "On the connection between high-frequency radio and X-ray emission in quasars", in *Continuum Emission in Active Galactic Nuclei*, ed. M. L. Sitko (Tucson: Kitt Peak National Observatory, National Optical Astronomy Observatories), p. 143. *[251, 280]*

Marscher, A. P., and Broderick, J. J. 1981, "Distance-independent evidence for relativistic motion in the quasar NRAO 140", *Astrophys. J. (Letters)*, **247**, L49. *[280]*

Marscher, A. P., and Broderick, J. J. 1981, "X-ray and VLBI radio observations of the quasars NRAO 140 and NRAO 530", *Astrophys. J.*, **249**, 406. *[251]*

Marscher, A. P., and Broderick, J. J. 1982, "Apparent superluminal motion in the quasar NRAO 140", *Astrophys. J. (Letters)*, **255**, L11. *[129, 251]*

Marscher, A. P., and Broderick, J. J. 1985, "Multifrequency radio VLBI observations of the superluminal, low-frequency variable quasar NRAO 140", *Astrophys. J.*, **290**, 735. *[12, 114, 280]*

Marscher, A. P., and Gear, W. K. 1985, "Models for high-frequency radio outbursts in extragalactic sources, with application to the early 1983 millimeter-to-infrared flare of 3C 273", *Astrophys. J.*, **298**, 114. *[217, 280]*

Marscher, A. P., Marshall, F. E., Mushotzky, R. F., Dent, W. A., Balonek, T. J., and Hartman, M. F. 1979, "Search for X-ray emission from bursting radio sources", *Astrophys. J.*, **233**, 498. *[251]*

Marscher, A. P., and Scott, J. S. 1980, "Superluminal motion in compact radio sources", *Publ. Astron. Soc. Pacific*, **92**, 127. *[1]*

Meisenheimer, K., and Heavens, A. F. 1986, "Particle acceleration in the hotspot of the jet of quasar 3C 273", *Nature*, **323**, 419. *[211]*

Miley, G. 1980, "The structure of extended extragalactic radio sources", *Ann. Rev. Astron. Astrophys.*, **18**, 165. *[174]*

Miley, G. 1983, "Optical emission from jets", in *Astrophysical Jets*, ed. A. Ferrari and A. G. Pacholczyk (Dordrecht: Reidel), p. 99. *[148]*

Milgrom, M., and Bahcall, J. N. 1978, "Apparent superluminal expansion velocities in the dipole magnetic field model", *Nature*, **274**, 349. *[280]*

Miller, J. S. 1975, "The composite nature of the N galaxy 3C 371", *Astrophys. J. (Letters)*, **200**, L55. *[180, 186]*

Miller, J. S. (ed.) 1985, *Astrophysics of Active Galaxies and Quasi-Stellar Objects*, (Mill Valley, CA: University Science Books). *[217]*

Miller, J. S., French, H. B., and Hawley, S. A. 1978, "Optical spectra of BL Lacertae objects", in *Pittsburgh Conference on BL Lac Objects*, ed. A. M. Wolfe (Pittsburgh: University of Pittsburgh), p. 176. *[60]*

Moffet, A. T., Gubbay, J., Robertson, D. S., and Legg, A. J. 1972, "High-resolution observations of variable radio sources", in *IAU Symposium 44, External Galaxies and Quasi-Stellar Objects*, ed. D. S. Evans (Dordrecht: Reidel), p. 228. *[1, 34]*

Moore, P. K., Browne, I. W. A., Daintree, E. J., Noble, R. G., and Walsh, D. 1981, "A statistical study of flat-spectrum radio sources at 966 MHz", *Monthly Notices Roy. Astron. Soc.*, **197**, 325. *[129]*

Moore, R. L., McGraw, J. T., Angel, J. R. P., Duerr, R., Lebofsky, M. J., Rieke, G. H., Wiśniewski, W. Z., Axon, D. J., Bailey, J., Hough, J. M., Thompson, I., Breger, M., Schulz, H., Clayton, G. C., Martin, P. G., Miller, J. S., Schmidt, G. D., Africano, J., and Miller, H. R. 1982, "The noise of BL Lacertae", *Astrophys. J.*, **260**, 415. *[217]*

Moore, R. L., Schmidt, G. D., and West, S. C. 1987, "The hiss of BL Lacertae", *Astrophys. J.*, **314**, 176. *[217]*

Moore, R. L., and Stockman, H. S. 1981, "The class of highly polarized quasars: observations and description", *Astrophys. J.*, **243**, 60. *[34, 40]*

Moore, R. L., and Stockman, H. S. 1984, "A comparison of the properties of highly polarized QSOs versus low-polarization QSOs", *Astrophys. J.*, **279**, 465. *[217, 233]*

Mutel, R. L., Aller, H. D., and Phillips, R. B. 1981 "Milliarcsecond structure of BL Lac during outburst", *Nature*, **294**, 236. *[60]*

Mutel, R. L., Hodges, M. W., and Phillips, R. B. 1985, "The structure of three compact double radio sources at 5 GHz", *Astrophys. J.*, **290**, 86. *[168]*

Mutel, R. L., and Phillips, R. B. 1984, "Superluminal motion in BL Lacertae: 10.65 GHz VLBI observations from 1981.7 to 1982.7", in *IAU Symposium 110, VLBI and Compact Radio Sources*, ed. R. Fanti, K. Kellermann, and G. Setti (Dordrecht: Reidel), p. 117. *[12, 60, 114, 129]*

Narlikar, J. V. 1986, "Noncosmological redshifts", in *IAU Symposium 119, Quasars*, ed. G. Swarup and V. K. Kapahi (Dordrecht: Reidel), p. 463. *[12]*

Neugebauer, G., Miley, G. K., Soifer, B. T., and Clegg, P. E. 1986, "Quasars measured by the *Infrared Astronomical Satellite*", *Astrophys. J.*, **308**, 815. *[217, 233]*

Neugebauer, G., Oke, J. B., Becklin, E. E., and Matthews, K. 1979, "Absolute spectral energy distribution of quasi-stellar objects from 0.3 to 10 microns", *Astrophys. J.*, **230**, 79. *[40, 233]*

Niell, A. E., Kellermann, K. I., Clark, B. G., and Shaffer, D. B. 1975, "Milli-arcsecond structure of 3C 84, 3C 273, and 3C 279 at 2 centimeter wavelength", *Astrophys. J. (Letters)*, **197**, L109. *[26]*

Norman, C., and Miley, G. 1984, "Jets and emission-line regions", *Astron. Astrophys.*, **141**, 85. *[168]*

Norman, C., and Silk, J. 1983, "The dynamics and fueling of active nuclei", *Astrophys. J.*, **266**, 502. *[168]*

O'Dea, C. P., Balonek, T. J., Dent, W. A., and Kapitzky, J. E. 1982, "Opacity effects at low frequencies", in *Low Frequency Variability of Extragalactic Radio Sources*, ed. W. D. Cotton and S. R. Spangler (Green Bank: National Radio Astronomy Observatory), p. 115. *[217]*

O'Dea, C. P., Dent, W. A., and Balonek, T. J. 1984, "The 20 year spectral evolution of the radio nucleus of NGC 1275", *Astrophys. J.*, **278**, 89. *[76]*

O'Dea, C. P., Dent, W. A., Kinzel, W. M., and Balonek, T. J. 1986, "Multifrequency radio observations of the variable quasars 0133+476, 0235+164, 1749+096, and 2131−021", *Astron. J.*, **92**, 1262. *[217]*

O'Dell, S. L. 1981, "Radiation force on a relativistic plasma and the Eddington limit", *Astrophys. J. (Letters)*, **243**, L147. *[301]*

Oke, J. B. 1967, "Optical variations in the radio galaxy 3C 371", *Astrophys. J. (Letters)*, **150**, L5. *[180]*

Oke, J. B. 1978, "The redshift and other properties of I Zw 1727+5015", *Astrophys. J. (Letters)*, **219**, L97. *[180]*

Oke, J. B., and Gunn, J. E. 1983, "Secondary standard stars for absolute spectrophotometry", *Astrophys. J.*, **266**, 713. *[260]*

Oke, J. B., Neugebauer, G., and Becklin, E. E. 1970, "Absolute spectral energy distribution of quasi-stellar objects from 0.3 to 2.2 microns", *Astrophys. J.*, **159**, 341. *[217]*

Olsen, E. T. 1970, "Optical identification of radio sources selected from the 4C catalogue", *Astron. J.*, **75**, 764. *[155]*

Orr, M. J. L., and Browne, I. W. A. 1982, "Relativistic beaming and quasar statistics", *Monthly Notices Roy. Astron. Soc.*, **200**, 1067. *[1, 55, 126, 129, 233, 251, 306, 310]*

Owczarek, J. A. 1964, *Fundamentals of Gas Dynamics* (Scranton: International Textbook Company). *[280]*

Owen, F. N., Helfand, D. J., and Spangler, S. R. 1981, "The correlation of X-ray emission with strong millimeter activity in extragalactic sources", *Astrophys. J. (Letters)*, **250**, L55. *[251, 280]*

Owen, F. N., Porcas, R. W., and Neff, S. G. 1978, "Interferometer observations of quasars from the Jodrell Bank 966-MHz survey", *Astron. J.*, **83**, 1009. *[126]*

Owen, F. N., and Puschell, J. J. 1984, "VLA observations of Jodrell Bank radio quasars", *Astron. J.*, **89**, 932. *[126]*

Ozernoj, L. M., and Sazonov, V. N. 1969, "Spektr i polyarizatsiya istochnika sinkhrotronnogo izlucheniya pri relyativistskom razlete ego komponent", *Astrophys. Space Sci.*, **3**, 365. English translation: "The spectrum and polarization of a source of synchrotron emission with components flying apart at relativistic velocities", *Astrophys. Space Sci.*, **3**, 395. *[1]*

Padrielli, L. 1982, "Results from the Bologna 408 MHz monitoring program", in *Low Frequency Variability of Extragalactic Radio Sources*, ed. W. D. Cotton and S. R. Spangler (Green Bank: National Radio Astronomy Observatory), p. 1. *[206]*

Padrielli, L., Aller, M. F., Aller, H. D., Fanti, C., Fanti, R., Ficarra, A., Gregorini, L., Mantovani, F., and Nicolson, G. 1987, "Multifrequency observations of low frequency variable sources: a statistical analysis", *Astron. Astrophys. Suppl.*, **67**, 63. *[200]*

Padrielli, L., Romney, J. D., Bartel, N., Fanti, R., Ficarra, A., Mantovani, F., Matveyenko, L., Nicolson, G. D., and Weiler, K. W. 1986, "Two epoch VLBI observations of a sample of low frequency variable sources", *Astron. Astrophys.*, **165**, 53. *[12, 200]*

Palmer, H. P., Rowson, B., Anderson, B., Donaldson, W., Miley, G. K., Gent, H., Adgie, R. L., Slee, O. B., and Crowther, J. H. 1967, "Radio diameter measurements with interferometer baselines of one million and two million wavelengths", *Nature*, **213**, 789. *[1]*

Pauliny-Toth, I. I. K. 1987, "3C 454.3: a new kind of superluminal source?", in *IAU Symposium 121, Observational Evidences of Activity in Galaxies*, ed. E. Khachikian, G. Melnick, and B. Fricke (Dordrecht: Reidel), p. 295. *[12, 55]*

Pauliny-Toth, I. I. K., and Kellermann, K. I. 1968, "Repeated outbursts in the radio galaxy 3C 120", *Astrophys. J. (Letters)*, **152**, L169. *[217]*

Pauliny-Toth, I. I. K., Porcas, R. W., Zensus, A., and Kellermann, K. I. 1984, "The structural variations of 3C 454.3 and 2134+004", in *IAU Symposium 110, VLBI and Compact Radio Sources*, ed. R. Fanti, K. Kellermann, and G. Setti (Dordrecht: Reidel), p. 149. *[12, 129]*

Pauliny-Toth, I. I. K., Preuss, E., Witzel, A., Graham, D., Kellermann, K. I., and Rönnäng, B. 1981, "6-cm VLBI observations of compact radio sources", *Astron. J.*, **86**, 371. *[26, 34]*

Pauliny-Toth, I. I. K., Witzel, A., Preuss, E., Kühr, H., Kellermann, K. I., Fomalont, E. B., and Davis, M. M. 1978, "The 5 GHz strong source surveys. IV. Survey of the area between declination 35 and 70 degrees and summary of source counts, spectra and optical identifications", *Astron. J.*, **83**, 451. *[94]*

Peacock, J. A., and Dunlop, J. S. 1986, "The statistics of radio galaxies & quasars at high redshift", in *IAU Symposium 119, Quasars*, ed. G. Swarup and V. K. Kapahi (Dordrecht: Reidel), p. 455. *[310]*

Peacock, J. A., and Wall, J. V. 1982, "Bright extragalactic radio sources at 2.7 GHz— II. Observations with the Cambridge 5-km telescope", *Monthly Notices Roy. Astron. Soc.*, **198**, 843. *[94]*

Pearson, T. J., Barthel, P. D., Lawrence, C. R., and Readhead, A. C. S. 1986, "1642+690: a superluminal quasar", *Astrophys. J. (Letters)*, **300**, L25. *[12, 94, 114, 129, 148, 155]*

Pearson, T. J., Barthel, P. D., Readhead, A. C. S., and Lawrence, C. R. 1986, "New superluminal quasars", in *IAU Symposium 119, Quasars*, ed. G. Swarup and V. K. Kapahi (Dordrecht: Reidel), p. 163. *[94]*

Pearson, T. J., Perley, R. A., and Readhead, A. C. S. 1985, "Compact radio sources in the 3C catalog", *Astron. J.*, **90**, 738. *[72, 94, 129, 155, 174, 186]*

Pearson, T. J., and Readhead, A. C. S. 1981, "The milli-arcsecond structure of a complete sample of radio sources. I. VLBI maps of seven sources", *Astrophys. J.*, **248**, 61. *[94, 186]*

Pearson, T. J., and Readhead, A. C. S. 1984, "Image formation by self-calibration in radio astronomy", *Ann. Rev. Astron. Astrophys.*, **22**, 97. *[1, 55, 60, 211]*

Pearson, T. J., and Readhead, A. C. S. 1984, "VLBI survey of a complete sample of active nuclei and quasars", in *IAU Symposium 110, VLBI and Compact Radio Sources*, ed. R. Fanti, K. Kellermann, and G. Setti (Dordrecht: Reidel), p. 15. *[83, 94, 168, 180, 233, 260]*

Pearson, T. J., Unwin, S. C., Cohen, M. H., Linfield, R. P., Readhead, A. C. S., Seielstad, G. A., Simon, R. S., and Walker, R. C. 1981, "Superluminal expansion of quasar 3C 273", *Nature*, **290**, 365. *[26, 211]*

Perley, R. A. 1982, "The positions, structures, and polarizations of 404 compact radio sources", *Astron. J.*, **87**, 859. *[72, 83, 180]*

Perley, R. A. 1984, "Evidence for relativistic motion based on arcsecond structure", in *IAU Symposium 110, VLBI and Compact Radio Sources*, ed. R. Fanti, K. Kellermann, and G. Setti (Dordrecht: Reidel), p. 153. *[186]*

Perley, R. A. 1986, "High-fidelity imaging", in *Synthesis Imaging*, NRAO Workshop No. 13, ed. R. A. Perley, F. R. Schwab, and A. H. Bridle (Green Bank: National Radio Astronomy Observatory), p. 161. *[211]*

Perley, R. A., Bridle, A. H., and Willis, A. G. 1984, "High-resolution VLA observations of the radio jet in NGC 6251", *Astrophys. J. Suppl.*, **54**, 291. *[104, 211]*

Perley, R. A., Fomalont, E. B., and Johnston, K. J. 1980, "Compact radio sources with faint components", *Astron. J.*, **85**, 649. *[26, 72, 83, 114, 180, 186]*

Perley, R. A., Fomalont, E. B., and Johnston, K. J. 1982, "The extended radio structure of compact extragalactic sources", *Astrophys. J. (Letters)*, **255**, L93. *[155]*

Petre, R., Mushotzky, R. F., Krolik, J. H., and Holt, S. S. 1984, "Soft X-ray spectral observations of quasars and high X-ray luminosity Seyfert galaxies", *Astrophys. J.*, **280**, 499. *[280]*

Phillips, R. B., and Mutel, R. L. 1980, "High-resolution observations of the compact radio sources CTD 93 and 3C 395 at 1671 megahertz", *Astrophys. J.*, **236**, 89. *[72]*

Phillips, R. B., and Mutel, R. L. 1982, "On symmetric structure in compact radio sources", *Astron. Astrophys.*, **106**, 21. *[94, 168]*

Phillips, R. B., and Mutel, R. L. 1982, "Rapid expansion of BL Lacertae", *Astrophys. J. (Letters)*, **257**, L19. *[60, 129, 217]*
Phillips, R. B., and Shaffer, D. B. 1983, "VLBI maps of 3C 147, 3C 286, 3C 380, NRAO 150, CTD 93, and 3C 395 at 2.3 GHz", *Astrophys. J.*, **271**, 32. *[72, 168]*
Phinney, E. S. 1982, "Acceleration of a relativistic plasma by radiation pressure", *Monthly Notices Roy. Astron. Soc.*, **198**, 1109. *[301]*
Phinney III, E. S. 1983, "A theory of radio sources", Ph. D. thesis, University of Cambridge. *[310]*
Phinney, E. S. 1985, "Central radio sources", in *Astrophysics of Active Galaxies and Quasi-Stellar Objects*, ed. J. S. Miller (Mill Valley: University Science Books), p. 453. *[217, 233, 301, 310]*
Pilbratt, G. 1986, "High resolution observations of superluminal radio sources", Ph. D. thesis (Technical report No. 166, School of Electrical and Computer Engineering), Chalmers University of Technology, Göteborg. *[34]*
Pollock, J. T., Pica, A. J., Smith, A. G., Leacock, R. J., Edwards, P. L., and Scott, R. L. 1979, "Long-term optical variations of 20 violently variable extragalactic radio sources", *Astron. J.*, **84**, 1658. *[40]*
Porcas, R. W. 1981, "Superluminal quasar 3C 179 with double radio lobes", *Nature*, **294**, 47. *[12, 114, 126]*
Porcas, R. 1983, "Astronomers still puzzled", *Nature*, **302**, 753. *[12]*
Porcas, R. W. 1984, "Superluminal motion in weak quasar cores", in *IAU Symposium 110, VLBI and Compact Radio Sources*, ed. R. Fanti, K. Kellermann, and G. Setti (Dordrecht: Reidel), p. 157. *[114, 129, 155]*
Porcas, R. W. 1986, "Compact radio structure of active galactic nuclei", *Mitt. Astron. Ges.*, **65**, 95. *[12]*
Porcas, R. W. 1986, "Compact radio structure of quasars", in *IAU Symposium 119, Quasars*, ed. G. Swarup and V. K. Kapahi (Dordrecht: Reidel), p. 131. *[26, 129]*
Preuss, E., Alef, W., Whyborn, N., Wilkinson, P. N., and Kellermann, K. I. 1984, "The milliarcsecond core of 3C 147 at 6 cm", in *IAU Symposium 110, VLBI and Compact Radio Sources*, ed. R. Fanti, K. Kellermann, and G. Setti (Dordrecht: Reidel), p. 29. *[12]*
Readhead, A. C. S., Cohen, M. H., Pearson, T. J., and Wilkinson, P. N. 1978, "Bent beams and the overall size of extragalactic radio sources", *Nature*, **276**, 768. *[26, 40, 114, 310]*
Readhead, A. C. S., Hough, D. H., Ewing, M. S., Walker, R. C., and Romney, J. D. 1983, "Asymmetric structure in the nuclei of NGC 1275 and 3C 345", *Astrophys. J.*, **265**, 107. *[76, 114, 129]*
Readhead, A. C. S., Masson, C. R., Moffet, A. T., Pearson, T. J., Seielstad, G. A., Woody, D. P., Backer, D. C., Plambeck, R. L., Welch, W. J., Wright, M. C. H., Rogers, A. E. E., Webber, J. C., Shapiro, I. I., Moran, J. M., Goldsmith, P. F., Predmore, C. R., Bååth, L., and Rönnäng, B. 1983, "Very long baseline interferometry at a wavelength of 3.4 mm", *Nature*, **303**, 504. *[32, 76]*
Readhead, A. C. S., and Pearson, T. J. 1982, "The milliarcsecond structure of radio galaxies and quasars", in *IAU Symposium 97, Extragalactic Radio Sources*, ed. D. S. Heeschen and C. M. Wade (Dordrecht: Reidel), p. 279. *[162]*
Readhead, A. C. S., Pearson, T. J., Cohen, M. H., Ewing, M. S., and Moffet, A. T. 1979, "Hybrid maps of the milli-arcsecond structures of 3C 120, 3C 273, and 3C 345", *Astrophys. J.*, **231**, 299. *[26, 60]*
Readhead, A. C. S., Pearson, T. J., and Unwin, S. C. 1984, "Multi-epoch observations of survey sources", in *IAU Symposium 110, VLBI and Compact Radio Sources*, ed. R. Fanti, K. Kellermann, and G. Setti (Dordrecht: Reidel), p. 131. *[12, 94, 168, 180, 186]*
Readhead, A. C. S., and Wilkinson, P. N. 1978, "The mapping of compact radio sources from VLBI data", *Astrophys. J.*, **223**, 25. *[26, 193, 211]*
Rees, M. J. 1966, "Appearance of relativistically expanding radio sources", *Nature*, **211**, 468. *[1, 148, 310]*

Rees, M. J. 1967, "Studies in radio source structure I. A relativistically expanding model for variable quasi-stellar radio sources", *Monthly Notices Roy. Astron. Soc.*, **135**, 345. *[1]*

Rees, M. J. 1981, "Nuclei of galaxies: the origin of plasma beams", in *IAU Symposium 94, Origin of Cosmic Rays*, ed. G. Setti, G. Spada, and A. W. Wolfendale (Dordrecht: Reidel), p. 139. *[310]*

Rees, M. J. 1984, "Black hole models for active galactic nuclei", *Ann. Rev. Astron. Astrophys.*, **22**, 471. *[310]*

Rees, M. J. 1984, "Physics of relativistic jets on sub-milliarcsecond scales", in *IAU Symposium 110, VLBI and Compact Radio Sources*, ed. R. Fanti, K. Kellermann, and G. Setti (Dordrecht: Reidel), p. 207. *[301]*

Richstone, D. O., and Schmidt, M. 1980, "The spectral properties of a large sample of quasars", *Astrophys. J.*, **235**, 361. *[233]*

Rickett, B. J. 1986, "Refractive interstellar scintillation of radio sources", *Astrophys. J.*, **307**, 564. *[55, 200]*

Rickett, B. J., Coles, W. A., and Bourgois, G. 1984, "Slow scintillation in the interstellar medium", *Astron. Astrophys.*, **134**, 390. *[55, 200, 206]*

Rieke, G. H., and Lebofsky, M. J. 1979, "Infrared emission of extragalactic radio sources", *Ann. Rev. Astron. Astrophys.*, **17**, 477. *[217]*

Roberts, D. H., Potash, R. I., Wardle, J. F. C., Rogers, A. E. E., and Burke, B. F. 1984, "Milliarcsecond polarization measurements", in *IAU Symposium 110, VLBI and Compact Radio Sources*, ed. R. Fanti, K. Kellermann, and G. Setti (Dordrecht: Reidel), p. 35. *[193]*

Robson, E. I., Gear, W. K., Clegg, P. E., Ade, P. A. R., Smith, M. G., Griffin, M. J., Nolt, I. G., Radostitz, J. V., and Howard, R. J. 1983, "A flare in the millimetre to IR spectrum of 3C 273", *Nature*, **305**, 194. *[280]*

Röser, H.-J., and Meisenheimer, K. 1986, "CCD photo-polarimetry of the jet of 3C 273", *Astron. Astrophys.*, **154**, 15. *[26]*

Rogers, A. E. E., Cappallo, R. J., Hinteregger, H. F., Levine, J. I., Nesman, E. F., Webber, J. C., Whitney, A. R., Clark, T. A., Ma, C., Ryan, J., Corey, B. E., Counselman, C. C., Herring, T. A., Shapiro, I. I., Knight, C. A., Shaffer, D. B., Vandenberg, N. R., Lacasse, R., Mauzy, R., Rayhrer, B., Schupler, B. R., and Pigg, J. C. 1983, "Very-long-baseline radio interferometry: the Mark III system for geodesy, astrometry, and aperture synthesis", *Science*, **219**, 51. *[114]*

Rogers, A. E. E., Hinteregger, H. F., Whitney, A. R., Counselman, C. C., Shapiro, I. I., Wittels, J. J., Klemperer, W. K., Warnock, W. W., Clark, T. A., Hutton, L. K., Marandino, G. E., Rönnäng, B. O., and Rydbeck, O. E. H. 1974, "The structure of radio sources 3C 273B and 3C 84 deduced from the 'closure' phases and visibility amplitudes observed with three-element interferometers", *Astrophys. J.*, **193**, 293. *[26]*

Rogers, A. E. E., Moffet, A. T., Backer, D. C., and Moran, J. M. 1984, "Coherence limits in VLBI observations at 3-millimeter wavelength", *Radio Sci.*, **19**, 1552. *[76]*

Rogers, A. E. E., and Morrison, P. 1972, "Long-baseline interferometry", *Science*, **175**, 218. *[1]*

Rogora, A., Padrielli, L., and de Ruiter, H. R. 1986, "VLA observations of B2 quasars. I. Extended sources", *Astron. Astrophys. Suppl.*, **64**, 557. *[129]*

Rogora, A., Padrielli, L., and de Ruiter, H. R. 1987, "VLA observations of B2 quasars. II. Compact sources", *Astron. Astrophys. Suppl.*, **67**, 267. *[129]*

Romney, J. D., Alef, W., Pauliny-Toth, I. I. K., Preuss, E., and Kellermann, K. I. 1984, "The nucleus of NGC 1275", in *IAU Symposium 110, VLBI and Compact Radio Sources*, ed. R. Fanti, K. Kellermann, and G. Setti (Dordrecht: Reidel), p. 137. *[12, 76]*

Romney, J., Padrielli, L., Bartel, N., Weiler, K. W., Ficarra, A., Mantovani, F., Bååth, L. B., Kogan, L., Matveenko, L., Moiseev, I. G., and Nicolson, G.

1984, "The milliarcsecond scale structure of low frequency variable sources", *Astron. Astrophys.*, **135**, 289. *[200, 206]*
Rossi, B. B. 1971, "Report of the Rumford committee", *Records Am. Acad. Arts Sci.*, 1970–1971, p. 14. *[1]*
Rothschild, R. E., Baity, W. A., Marscher, A. P., and Wheaton, W. A. 1981, "Nonthermal hard X-ray emission from the nucleus of NGC 1275", *Astrophys. J. (Letters)*, **243**, L9. *[76]*
Rudnick, L. 1982, "Nuclear ejection—one side at a time", in *IAU Symposium 97, Extragalactic Radio Sources*, ed. D. S. Heeschen and C. M. Wade (Dordrecht: Reidel), p. 47. *[162]*
Rudnick, L., and Edgar, B. K. 1984, "Alternating-side ejection in extragalactic radio sources", *Astrophys. J.*, **279**, 74. *[162]*
Rudnick, L., and Jones, T. W. 1982, "Compact radio sources: the dependence of variability and polarization on spectral shape", *Astrophys. J.*, **255**, 39. *[217]*
Rudnick, L., and Jones, T. W. 1983, "Rotation measures for compact variable radio sources", *Astron. J.*, **88**, 518. *[186, 217]*
Rudnick, L., Jones, T. W., Aller, H. D., Aller, M. F., Hodge, P. E., Owen, F. N., Fiedler, R. L., Puschell, J. J., and Bignell, R. C. 1985, "Broad-band polarization observations of active compact radio sources", *Astrophys. J. Suppl.*, **57**, 693. *[217]*
Rudnick, L., Jones, T. W., and Fiedler, R. 1986, "Weak nuclei of powerful radio sources—spectra and polarizations", *Astron. J.*, **91**, 1011. *[217]*
Rudnick, L., Owen, F. N., Jones, T. W., Puschell, J. J., and Stein, W. A. 1978, "Coordinated centimeter, millimeter, infrared, and visual polarimetry of compact nonthermal sources", *Astrophys. J. (Letters)*, **225**, L5. *[217]*
Rudnick, L., Sitko, M. L., and Stein, W. A. 1984, "The nature of radio-quiet QSO's—VLA observations of 0026+129, 0205+024, and 1351+640", *Astron. J.*, **89**, 753. *[217]*
Rusk, R., and Rusk, A. C. M. 1986, "Alignment of milliarcsecond with arcsecond scale structure in core-dominated radio sources", *Can. J. Phys.*, **64**, 440. *[94, 129]*
Rusk, R., and Seaquist, E. R. 1985, "Alignment of radio and optical polarization with VLBI structure", *Astron. J.*, **90**, 30. *[217, 233]*
Rusk, R. E., and Seaquist, E. R. 1986, "The arcsecond brightness and polarization structure of 0850+581", *Bull. Am. Astron. Soc.*, **18**, 994. *[94]*
Russell, H. R. 1914 (published 1918), "Relations between the spectra and other characteristics of the stars", *Publications of the American Astronomical Society*, **3**, 22; and 1914, "Relations between the spectra and other characteristics of the stars", *Popular Astronomy*, **22**, 275. *[310]*
Saikia, D. J., Swarup, G., and Kodali, P. D. 1985, "Polarization properties of radio cores in galaxies and quasars", *Monthly Notices Roy. Astron. Soc.*, **216**, 385. *[217]*
Sandage, A. 1966, "Redshifts of nine radio galaxies including the abnormal system 3C 305", *Astrophys. J.*, **145**, 1. *[180, 186]*
Sanders, R. H. 1974, "Super-relativistic phase velocities of radio source components", *Nature*, **248**, 390. *[1, 280, 306, 310]*
Schalinski, C. J., Alef, W., Campbell, J., Schuh, H., and Witzel, A. 1986, "Erste Ergebnisse zur radioastronomischen Auswertung geodätischer VLBI-Experimente", in *Die Arbeiten des Sonderforschungsbereiches 78 Satellitengeodäsie der Technischen Universität München 1984 und 1985*, ed. M. Schneider (München: Bayerische Akademie der Wissenschaften), p. 292. *[83]*
Schalinski, C. J., Biermann, P., Eckart, A., Johnston, K. J., Krichbaum, T. Ph., and Witzel, A. 1987, "Bulk relativistic motion in a complete sample of radio selected AGN", in *IAU Symposium 121, Observational Evidences of Activity in Galaxies*, ed. E. Khachikian, G. Melnick, and K. Fricke (Dordrecht: Reidel), p. 287. *[12]*

Scheuer, P. A. G. 1974, "Models of extragalactic radio sources with a continuous energy supply from a central object", *Monthly Notices Roy. Astron. Soc.*, **166**, 513. *[168]*

Scheuer, P. A. G. 1976, "AO 0235+16.4—another source exceeding the speed limit", *Monthly Notices Roy. Astron. Soc.*, **177**, 1P. *[12]*

Scheuer, P. A. G. 1984, "Explanations of superluminal motion", in *IAU Symposium 110, VLBI and Compact Radio Sources*, ed. R. Fanti, K. Kellermann, and G. Setti (Dordrecht: Reidel), p. 197. *[1, 280, 301, 310]*

Scheuer, P. A. G., and Readhead, A. C. S. 1979, "Superluminally expanding radio sources and the radio-quiet QSOs", *Nature*, **277**, 182. *[1, 104, 114, 126, 251, 306, 310]*

Schilizzi, R. T., Cohen, M. H., Romney, J. D., Shaffer, D. B., Kellermann, K. I., Swenson Jr., G. W., Yen, J. L., and Rinehart, R. 1975, "Observations with a VLB array. III. The sources 3C 120, 3C 273B, 2134+004, and 3C 84", *Astrophys. J.*, **201**, 263. *[26]*

Schilizzi, R. T., and de Bruyn, A. G. 1983, "Large-scale radio structures of superluminal sources", *Nature*, **303**, 26. *[1, 40, 48, 114, 129, 155, 310]*

Schmidt, M. 1963, "3C 273: a star-like object with large red-shift", *Nature*, **197**, 1040. *[26]*

Schmidt, M. 1974, "Optical spectra and redshifts of 4C quasi-stellar radio sources", *Astrophys. J.*, **193**, 505. *[155]*

Schmitt, J. H. M. M., and Reid, J. M. 1985, "VLBI observations of the nucleus of M87 at two epochs", *Astrophys. J.*, **289**, 120. *[12, 310]*

Schwab, F. R., and Cotton, W. D. 1983, "Global fringe search techniques for VLBI", *Astron. J.*, **88**, 688. *[162]*

Schwartz, D. A., and Ku, W. H.-M. 1983, "Studies of BL Lacertae objects with the *Einstein* X-ray observatory: the absolute volume density", *Astrophys. J.*, **266**, 459. *[60]*

Segal, I. E. 1986, "Complete quasar samples and comparative cosmology", in *IAU Symposium 119, Quasars*, ed. G. Swarup and V. K. Kapahi (Dordrecht: Reidel), p. 493. *[12]*

Seielstad, G. A., Cohen, M. H., Linfield, R. P., Moffet, A. T., Romney, J. D., Schilizzi, R. T., and Shaffer, D. B. 1979, "Further monitoring of the structure of superluminal radio sources", *Astrophys. J.*, **229**, 53. *[26, 48]*

Seielstad, G. A., Pearson, T. J., and Readhead, A. C. S. 1983, "10.8-GHz flux density variations among a complete sample of sources from the NRAO-Bonn S4 survey", *Publ. Astron. Soc. Pacific*, **95**, 842. *[148]*

Shaffer, D. B., Kellermann, K. I., Purcell, G. H., Pauliny-Toth, I. I. K., Preuss, E., Witzel, A., Graham, D., Schilizzi, R. T., Cohen, M. H., Moffet, A. T., Romney, J. D., and Niell, A. E. 1977, "The compact radio sources in 4C 39.25 and 3C 345", *Astrophys. J.*, **218**, 353. *[12, 55]*

Shaffer, D. B., and Marscher, A. P. 1985, "Multi-epoch VLBI observations of the quasar 4C 39.25: superluminal motion sandwiched by stationary structure", *Bull. Am. Astron. Soc.*, **17**, 609. *[12]*

Shaffer, D. B., Marscher, A. P., Marcaide, J., and Romney, J. D. 1987, "Multi-epoch VLBI observations of 4C 39.25: superluminal motion amid stationary structure", *Astrophys. J. (Letters)*, **314**, L1. *[55, 67, 217]*

Shapiro, I. I. 1967, "New method for the detection of light deflection by solar gravity", *Science*, **157**, 806. *[1]*

Shapirovskaya, N. Ya. 1978, "O peremennosti vnegalakticheskikh istochnikov v detsimetrovom diapazone", *Astron. Zh.*, **55**, 953. English translation: 1979, "Variability of extragalactic decimeter radio sources", *Soviet Astron.*, **22**, 544. *[200]*

Shklovskii, I. S. 1962, "O prirode radiogalaktik", *Astron. Zh.*, **39**, 591. English translation: 1963, "On the nature of radio galaxies", *Soviet Astron.*, **6**, 465. *[310]*

Sholomitskii, G. B. 1965, "Flyuktuatsii potoka CTA 102 na volne 32.5 cm", *Astron. Zh.*, **42**, 673. English translation: 1965, "Fluctuations in the 32.5-cm flux of CTA 102", *Soviet Astron.*, **9**, 516. *[206]*

Shone, D. L., Porcas, R. W., and Zensus, J. A. 1985, "Combined MERLIN/VLA observations of the superluminal quasar 3C 179", *Nature*, **314**, 603. *[12, 129]*
Sieber, W. 1982, "Causal relationship between pulsar long-term intensity variations and the interstellar medium", *Astron. Astrophys.*, **113**, 311. *[200]*
Simard-Normandin, M., Kronberg, P. P., and Button, S. 1981, "The Faraday rotation measures of extragalactic radio sources", *Astrophys. J. Suppl.*, **45**, 97. *[186]*
Simon, R. S., Spencer, J. H., and Johnston, K. J. 1987, "Deep 21 cm radio imaging of 0.6 degree fields around 12 core dominated radio sources", *Astron. J.*, submitted. *[186]*
Simon, R. S., Hall, J., Johnston, K. J., Spencer, J. H., Waak, J. A., and Mutel, R. L. 1987, "Superluminal motion toward a stationary knot in the radio core of the quasar 3C 395", *Astrophys. J. (Letters)*, submitted. *[72, 155]*
Simonetti, J. H., Cordes, J. M., and Heeschen, D. S. 1985, "Flicker of extragalactic radio sources at two frequencies", *Astrophys. J.*, **296**, 46. *[217]*
Sitko, M. L. (ed.) 1986, *Continuum Emission in Active Galactic Nuclei* (Tucson: Kitt Peak National Observatory, National Optical Astronomy Observatories). *[217]*
Slish, V. I. 1963, "Angular size of radio stars", *Nature*, **199**, 682. *[206]*
Smith, D. R. 1986, "Diffusive shock acceleration and quasar variability", in *Continuum Emission in Active Galactic Nuclei*, ed. M. L. Sitko (Tucson: Kitt Peak National Observatory, National Optical Astronomy Observatories), p. 111. *[217]*
Smith, M. D., and Norman, C. A. 1981, "Extragalactic jets—I. Trajectories", *Monthly Notices Roy. Astron. Soc.*, **194**, 771. *[40]*
Smith, P. S., Balonek, T. J., Heckert, P. A., and Elston, R. 1986, "The optical and near-infrared polarization properties of the OVV quasar 3C 345", *Astrophys. J.*, **305**, 484. *[260]*
Snow, T. P. 1983, *The Dynamic Universe* (St. Paul: West Publishing), p. 297. *[310]*
Soboleva, N. S., Berlin, A. B., Nizhel'skij, N. A., and Spangenberg, E. E. 1982, "Poisk protyazhennykh struktur vblizi radioistochnikov 3C 120 i 3C 273", *Pis'ma Astron. Zh.*, **8**, 205. English translation: "A search for extended structures near the radio sources 3C 120 and 3C 273", *Soviet Astron. Lett.*, **8**, 108. *[48]*
Soltan, A. 1982, "Masses of quasars", *Monthly Notices Roy. Astron. Soc.*, **200**, 115. *[310]*
Spangler, S. R., and Cotton, W. D. 1981, "Broadband radio observations of low-frequency variable sources", *Astron. J.*, **86**, 730. *[200]*
Spencer, R. E., and Junor, W. 1986, "A compact radio source in the nucleus of M87", *Nature*, **321**, 753. *[211]*
Staubert, R., Brunner, H., and Worrall, D. M. 1986, "*EXOSAT* observations of 3C 371", *Astrophys. J.*, **310**, 694. *[180, 186]*
Stocke, J. T., Liebert, J., Schmidt, G., Gioia, I. M., Maccacaro, T., Schild, R. E., Maccagni, D., and Arp, H. C. 1985, "Optical and radio properties of X-ray selected BL Lacertae objects", *Astrophys. J.*, **298**, 619. *[217, 251]*
Stockman, H. S., Moore, R. L., and Angel, J. R. P. 1984, "The optical polarization properties of 'normal' quasars", *Astrophys. J.*, **279**, 485. *[233]*
Strittmatter, P. A. (attributed) 1985, "Optical and infrared studies of active galactic nuclei", in *Extragalactic Energetic Sources*, ed. V. K. Kapahi (Bangalore: Indian Academy of Sciences), p. 13. *[60]*
Strittmatter, P. A., Hill, P., Pauliny-Toth, I. I. K., Steppe, H., and Witzel, A. 1980, "Radio observations of optically selected quasars", *Astron. Astrophys.*, **88**, L12. *[310]*
Strom, R. G., and Willis, A. G. 1980, "Multifrequency observations of very large radio galaxies. II. 3C 236", *Astron. Astrophys.*, **85**, 36. *[155]*
Stubbs, P. 1971, "Red shift without reason", *New Scientist*, **50**, 254. *[1, 12]*
Terrell, N. J. 1966, "Quasi-stellar objects: possible local origin", *Science*, **154**, 1281. *[280]*
Ulrich, M.-H. 1984, "The infrared, optical and ultraviolet properties of active nuclei", in *IAU Symposium 110, VLBI and Compact Radio Sources*, ed. R. Fanti, K. Kellermann, and G. Setti (Dordrecht: Reidel), p. 73. *[233]*

Ulvestad, J. S., and Antonucci, R. R. J. 1986, "Blazars with arcminute-scale radio halos", *Astron. J.*, **92**, 6. *[217]*

Ulvestad, J. S., and Johnston, K. J. 1984, "A search for arcminute-scale radio emission in BL Lacertae objects", *Astron. J.*, **89**, 189. *[129, 186]*

Ulvestad, J., Johnston, K. J., Perley, R., and Fomalont, E. 1981, "A VLA survey of strong radio sources", *Astron. J.*, **86**, 1010. *[83]*

Unwin, S. C. 1986, "Superluminal motion in the quasar 3C 279", in *IAU Symposium 119, Quasars*, ed. G. Swarup and V. K. Kapahi (Dordrecht: Reidel), p. 161. *[12, 129]*

Unwin, S. C., and Biretta, J. A. 1984, "VLBI observations of the superluminal sources 3C 273 and 3C 279", in *IAU Symposium 110, VLBI and Compact Radio Sources*, ed. R. Fanti, K. Kellermann, and G. Setti (Dordrecht: Reidel), p. 105. *[129]*

Unwin, S. C., Cohen, M. H., Biretta, J. A., Pearson, T. J., Seielstad, G. A., Walker, R. C., Simon, R. S., and Linfield, R. P. 1985, "VLBI monitoring of the superluminal quasar 3C 273, 1977–1982", *Astrophys. J.*, **289**, 109. *[12, 26, 34, 40, 251]*

Unwin, S. C., Cohen, M. H., Pearson, T. J., Seielstad, G. A., Simon, R. S., Linfield, R. P., and Walker, R. C. 1983, "Superluminal motion in the quasar 3C 345", *Astrophys. J.*, **271**, 536. *[34, 40, 251]*

Unwin, S. C., Mutel, R. L., Phillips, R. B., and Linfield, R. P. 1982, "Multifrequency VLBI observations of the nucleus of NGC 1275", *Astrophys. J.*, **256**, 83. *[76]*

Urry, C. M. 1984, "X-ray and ultraviolet observations of BL Lacertae objects", Ph. D. thesis, The Johns Hopkins University. *[233]*

Urry, C. M. 1986, "X-ray spectra of BL Lacertae objects", in *Continuum Emission in Active Galactic Nuclei*, ed. M. L. Sitko (Tucson: Kitt Peak National Observatory, National Optical Astronomy Observatories), p. 91. *[217, 251, 280]*

Urry, C. M., and Shafer, R. A. 1984, "Luminosity enhancement in relativistic jets and altered luminosity functions for beamed objects", *Astrophys. J.*, **280**, 569. *[233]*

van Breugel, W. 1984, "Steep spectrum radio cores and far-sightedness", in *IAU Symposium 110, VLBI and Compact Radio Sources*, ed. R. Fanti, K. Kellermann, and G. Setti (Dordrecht: Reidel), p. 59. *[168]*

van Breugel, W., Miley, G., and Heckman, T. 1984, "Studies of kiloparsec-scale, steep-spectrum radio cores. I. VLA maps", *Astron. J.*, **89**, 5. *[72, 174]*

van der Laan, H. 1966, "A model for variable extragalactic radio sources", *Nature*, **211**, 1131. *[1, 200, 217]*

van der Laan, H., Zieba, S., and Noordam, J. E. 1984, "Faint extended radio sources surrounding active galaxy nuclei", in *IAU Symposium 110, VLBI and Compact Radio Sources*, ed. R. Fanti, K. Kellermann, and G. Setti (Dordrecht: Reidel), p. 9. *[180, 186]*

van Groningen, E., Miley, G. K., and Norman, C. A. 1980, "One-sided jets in extragalactic radiosources", *Astron. Astrophys.*, **90**, L7. *[310]*

Véron, M. P., Véron, P., and Witzel, A. 1974, "The spectra of 373 radio sources", *Astron. Astrophys. Suppl.*, **13**, 1. *[72]*

Véron-Cetty, M.-P., and Véron, P. 1985, *A Catalogue of Quasars and Active Nuclei (2nd Edition)*, European Southern Observatory Scientific Report No. 4. (Garching: European Southern Observatory). *[251]*

Waak, J. A., Spencer, J. H., Johnston, K. J., and Simon, R. S. 1985, "Superluminal resupply of a stationary hot spot in 3C 395?", *Astron. J.*, **90**, 1989. *[12, 67, 72, 129, 155]*

Walker, R. C. 1984, "3C 120: a continuous link between moving features and a large scale radio jet", in *Physics of Energy Transport in Extragalactic Radio Sources*, NRAO Workshop No. 9, ed. A. H. Bridle and J. A. Eilek (Green Bank: National Radio Astronomy Observatory), p. 20. *[48]*

Walker, R. C. 1986, "The radio jet in 3C 120 at very long baseline interferometry and Very Large Array scales", *Can. J. Phys.*, **64**, 452. *[12, 129, 211]*

Walker, R. C., Benson, J. M., Seielstad, G. A., and Unwin, S. C. 1984, "Observations of superluminal motions in 3C 120", in *IAU Symposium 110, VLBI and Compact Radio Sources*, ed. R. Fanti, K. Kellermann, and G. Setti (Dordrecht: Reidel), p. 121. *[48, 193, 251]*

Walker, R. C., Benson, J. M., and Unwin, S. C. 1987, "The radio morphology of 3C 120 on scales from 0.5 pc to 400 kpc", *Astrophys. J.*, **316**, 546. *[48, 129, 155, 193]*

Walker, R. C., Seielstad, G. A., Simon, R. S., Unwin, S. C., Cohen, M. H., Pearson, T. J., and Linfield, R. P. 1982, "Rapid structural variations in 3C 120", *Astrophys. J.*, **257**, 56. *[48]*

Wardle, J. F. C. 1977, "Upper limits on the Faraday rotation in variable radio sources", *Nature*, **269**, 563. *[217]*

Wardle, J. F. C., and Potash, R. I. 1984, "Observations of large scale jets in quasars and the sidedness problem", in *Physics of Energy Transport in Extragalactic Radio Sources*, NRAO Workshop No. 9, ed. A. H. Bridle and J. A. Eilek (Green Bank: National Radio Astronomy Observatory), p. 30. *[148, 155]*

Wardle, J. F. C., and Roberts, D. H. 1986, "Very long baseline interferometry polarization studies of quasars and active galactic nuclei", *Can. J. Phys.*, **64**, 434. *[186]*

Wardle, J. F. C., Roberts, D. H., Potash, R. I., and Rogers, A. E. E. 1986, "The linear polarization of 3C 345 at milliarcsecond resolution", *Astrophys. J. (Letters)*, **304**, L1. *[40, 186, 193]*

Whitney, A. R., Shapiro, I. I., Rogers, A. E. E., Robertson, D. S., Knight, C. A., Clark, T. A., Goldstein, R. M., Marandino, G. E., and Vandenberg, N. R. 1971, "Quasars revisited: rapid time variations observed via very-long-baseline interferometry", *Science*, **173**, 225. *[1, 12, 26]*

Wiita, P. J. 1985, "Active galactic nuclei I. Observations and fundamental interpretations", *Physics Reports*, **123**, 117. *[217]*

Wilkes, B., and Elvis, M. 1986, "The diverse soft X-ray slopes of QSOs", in *Continuum Emission in Active Galactic Nuclei*, ed. M. L. Sitko (Tucson: Kitt Peak National Observatory, National Optical Astronomy Observatories), p. 56. *[233]*

Wilkes, B. J., and Elvis, M. 1987, "QED-1: soft X-ray spectra of quasars", *Astrophys. J.*, submitted. *[251]*

Wilkinson, A., Hine, R. G., and Sargent, W. L. W. 1981, "Optical and radio properties of 4C galaxies", *Monthly Notices Roy. Astron. Soc.*, **196**, 669. *[104]*

Wilkinson, P. N. 1983, "The attainment of higher quality maps from VLBI", in *Techniques d'Interférométrie à Très Grande Base*, Proceedings of an International Conference on Very Long Baseline Interferometry Techniques organized by the Centre National d'Études Spatiales (Toulouse: Cepadues-Éditions), p. 375. *[211]*

Wilkinson, P. N., Booth, R. S., Cornwell T. J., and Clark, R. R. 1984, "Peculiar radio structure in the quasar 3C 380", *Nature*, **308**, 619. *[174]*

Wilkinson, P. N., Cornwell, T. J., Kus, A. J., Readhead, A. C. S., and Pearson, T. J. 1984, "Bending in the first few hundred parsecs", in *Physics of Energy Transport in Extragalactic Radio Sources*, NRAO Workshop No. 9, ed. A. H. Bridle and J. A. Eilek (Green Bank: National Radio Astronomy Observatory), p. 76. *[94]*

Wilkinson, P. N., Kus, A. J., Pearson, T. J., Readhead, A. C. S., and Cornwell, T. J. 1986, "The nuclear jets in 3C 309.1 & 3C 380", in *IAU Symposium 119, Quasars*, ed. G. Swarup and V. K. Kapahi (Dordrecht: Reidel), p. 165. *[94, 174]*

Wilkinson, P. N., Readhead, A. C. S., Anderson, B., and Purcell, G. H. 1979, "VLBI observations of compact radio sources at 609 megahertz", *Astrophys. J.*, **232**, 365. *[200]*

Wilkinson, P. N., Spencer, R. E., Readhead, A. C. S., Pearson, T. J., and Simon, R. S. 1984, "Distortions in compact steep-spectrum radio sources", in *IAU Symposium 110, VLBI and Compact Radio Sources*, ed. R. Fanti, K. Kellermann, and G. Setti (Dordrecht: Reidel), p. 25. *[94]*

Willis, A. G., Strom, R. G., Perley, R. A., and Bridle, A. H. 1982, "Recent WSRT and VLA observations of the jet radio galaxy NGC 6251", in *IAU Symposium 97, Extragalactic Radio Sources*, ed. D. S. Heeschen and C. M. Wade (Dordrecht: Reidel), p. 141. *[162]*

Wills, B. 1987, "A broad 3-micron emission feature in QSOs", *Astrophys. J.*, submitted. *[104]*

Wills, B. J., and Browne, I. W. A. 1986, "Relativistic beaming and quasar emission lines", *Astrophys. J.*, **302**, 56. *[233, 260, 310]*

Wills, B. J., Netzer, H., and Wills, D. 1985, "Broad emission features in QSOs and active galactic nuclei. II. New observations and theory of Fe II and H I emission", *Astrophys. J.*, **288**, 94. *[260]*

Wills, D., and Wills, B. J. 1976, "Spectroscopy of 206 QSO candidates and radio galaxies", *Astrophys. J. Suppl.*, **31**, 143. *[12]*

Wittels, J. J., Cotton, W. D., Counselman III, C. C., Shapiro, I. I., Hinteregger, H. F., Knight, C. A., Rogers, A. E. E., Whitney, A. R., Clark, T. A., Hutton, L. K., Rönnäng, B. O., Rydbeck, O. E. H., and Neill, A. E. 1976, "Apparent 'superrelativistic' expansion of the extragalactic radio source 3C 345", *Astrophys. J. (Letters)*, **206**, L75. *[40]*

Witzel, A., Heeschen, D. S., Schalinski, C., and Krichbaum, Th. 1986, "Kurzzeit-Variabilität extragalaktischer Radioquellen", *Mitt. Astron. Ges.*, **65**, 239. *[310]*

Woltjer, L. 1966, "Inverse Compton radiation in quasi-stellar objects", *Astrophys. J.*, **146**, 597. *[310]*

Worrall, D. M. 1986, "BL Lac objects and relativistic beaming", in *Continuum Emission in Active Galactic Nuclei*, ed. M. L. Sitko (Tucson: Kitt Peak National Observatory, National Optical Astronomy Observatories), p. 97. *[217, 233]*

Worrall, D. M., Giommi, P., Tananbaum, H., and Zamorani, G. 1987, "X-ray studies of quasars with the Einstein Observatory. IV. X-ray dependence on radio emission", *Astrophys. J.*, **313**, 596. *[251]*

Worrall, D. M., Mushotzky, R. F., Boldt, E. A., Holt, S. S., and Serlemitsos, P. J. 1979, "The X-ray spectrum of 3C 273", *Astrophys. J.*, **232**, 683. *[251, 280]*

Worrall, D. M., Puschell, J. J., Bruhweiler, F. C., Miller, H. R., Rudy, R. J., Ku, W. H.-M., Aller, M. F., Aller, H. D., Hodge, P. E., Matthews, K., Neugebauer, G., Soifer, B. T., Webb, J. R., Pica, A. J., Pollock, J. T., Smith, A. G., and Leacock, R. J. 1984, "Two multifrequency observations of 3C 371", *Astrophys. J.*, **278**, 521. *[180, 186]*

Yee, H. K. C. 1980, "Optical continuum and emission-line luminosity of active galactic nuclei and quasars", *Astrophys. J.*, **241**, 894. *[104]*

Yee, H. K. C., and Oke, J. B. 1978, "Photoelectric spectrophotometry of radio galaxies", *Astrophys. J.*, **226**, 753. *[104]*

Yee, H. K. C., and Oke, J. B. 1981, "Optical spectral variability of the N galaxies 3C 382 and 3C 390.3", *Astrophys. J.*, **248**, 472. *[267]*

Zamorani, G. 1986, "Results from X-ray satellites", in *IAU Symposium 119, Quasars*, ed. G. Swarup and V. K. Kapahi (Dordrecht: Reidel), p. 223. *[251]*

Zamorani, G., Henry, J. P., Maccacaro, T., Tananbaum, H., Sołtan, A., Avni, Y., Liebert, J., Stocke, J., Strittmatter, P. A., Weymann, R. J., Smith, M. G., and Condon, J. J. 1981, "X-ray studies of quasars with the *Einstein* observatory. II.", *Astrophys. J.*, **245**, 357. *[34, 251]*

Zensus, J. A. 1984, "Radioastronomische Untersuchung der Zentralregionen extragalaktischer Strahlungsquellen", Ph. D. thesis, Westfälische Wilhelms-Universität Münster. *[126]*

Zensus, J. A., Hough, D. H., and Porcas, R. W. 1987, "Superluminal motion in the double-lobed quasar 3C 263", *Nature*, **325**, 36. *[12, 114, 126, 129, 148]*

Zensus, J. A., and Porcas, R. W. 1984, "VLBI observations of weak cores in extended quasars", in *IAU Symposium 110, VLBI and Compact Radio Sources*, ed. R. Fanti, K. Kellermann, and G. Setti (Dordrecht: Reidel), p. 163. *[126, 233]*

Zensus, J. A., and Porcas, R. W. 1986, "Search for superluminal motion in the weak cores of extended quasars", in *IAU Symposium 119, Quasars*, ed. G. Swarup and V. K. Kapahi (Dordrecht: Reidel), p. 167.   *[12, 114, 126]*

Index of Authors

Aller, H. D., Hughes, P. A., and Aller, M. F.: "Evidence for Shocks in Relativistic Jets"   273
Aller, M. F.   *see* Aller, H. D.
Bååth, L. B.: "Superluminal Motion in CTA 102"   206
Backer, D. C.: "Subluminal Expansion in NGC 1275"   76
Barthel, P. D.: "Feeling Uncomfortable"   148
Barthel, P. D.   *see also* Pearson, T. J.
Benson, J. M.   *see* Walker, R. C.
Biermann, P.   *see* Simon, R. S.
Biretta, J. A., and Cohen, M. H.: "Investigations of 3C 345"   40
Blandford, R. D.: "Grand Unified Models"   310
Browne, I. W. A.: "Extended Structure of Superluminal Radio Sources"   129
Cohen, M. H.: "The μ–z Diagram"   306
Cohen, M. H.   *see also* Biretta, J. A.
Eckart, A.   *see* Simon, R. S.
Fanti, C., and Fanti, R.: "VLBI Observations of Compact Steep-Spectrum Radio Sources"   174
Fanti, R., Gregorini, L., Padrielli, L., and Spangler, S.: "The Low Frequency Variability of Extragalactic Radio Sources: a Relativistic Effect or Galactic Scintillation?"   200
Fanti, R.   *see also* Fanti, C.
Gregorini, L.   *see* Fanti, R.
Hall, J.   *see* Simon, R. S.
Hodges, M. W., and Mutel, R. L.: "Are Compact Doubles Misaligned Superluminals?"   168
Hough, D. H., and Readhead, A. C. S.: "Relativistic Beaming and the Nuclei of Double-Lobed Quasars"   114
Hughes, P. A.   *see* Aller, H. D.
Impey, C.: "Infrared, Optical, UV, and X-ray Properties of Superluminal Radio Sources"   233
Johnston, K. J.   *see* Simon, R. S.
Jones, D. L.: "Intrinsic Asymmetry in NGC 6251"   162
Lawrence, C. R., Readhead, A. C. S., Pearson, T. J., and Unwin, S. C.: "Optical Spectra of Superluminal Sources"   260
Lind, K. R.: "VLBI Observations of the Suspected Superluminal 3C 371"   180
Marscher, A. P.: "Synchro-Compton Emission from Superluminal Sources"   280
Marscher, A. P.   *see also* Shaffer, D. B.
Moffet, A. T., and Readhead, A. C. S.: "Observations of 3C 273 at 3 mm Wavelength"   32
Mutel, R. L., and Phillips, R. B.: "Superluminal Motion in BL Lac: Evidence for Deceleration in Two Events"   60
Mutel, R. L.   *see also* Hodges, M. W.
Oke, J. B.: "Emission-Line Profile Changes in 3C 390.3"   267
Padrielli, L.   *see* Fanti, R.
Pauliny-Toth, I. I. K.: "Structural Variations in the Quasar 3C 454.3"   55
Pearson, T. J., Readhead, A. C. S., and Barthel, P. D.: "The Quest for Superluminal Sources"   94
Pearson, T. J., and Zensus, J. A.: "Superluminal Radio Sources: Introduction"   1

Pearson, T. J. *see also* Lawrence, C. R.
Phillips, R. B. *see* Mutel, R. L.
Phinney, E. S.: "How Fast Can a Blob Go?" 301
Porcas, R. W.: "Summary of Known Superluminal Sources" 12
Porcas, R. W. *see also* Zensus, J. A.
Readhead, A. C. S. *see* Hough, D. H., Lawrence, C. R., Moffet, A. T., *and* Pearson, T. J.
Roberts, D. H., and Wardle, J. F. C.: "Milliarcsecond Polarization of Superluminal Sources" 193
Rudnick, L.: "A Different Perspective on Superluminal Sources" 217
Schalinski, C. *see* Simon, R. S.
Scheuer, P. A. G.: "Tests of Beaming Models" 104
Shaffer, D. B., and Marscher, A. P.: "4C 39.25: Superluminal Motion Between Stationary Components" 67
Simon, R. S., Johnston, K. J., Eckart, A., Biermann, P., Schalinski, C., Witzel A., and Strom, R. G.: "The Arcminute Structure of 1928+738" 155
Simon, R. S., Johnston, K. J., Hall, J., Spencer, J. H., and Waak, J. A.: "Superluminal Motion Towards a Stationary Component in Quasar 3C 395" 72
Spangler, S. *see* Fanti, R.
Spencer, J. H. *see* Simon, R. S.
Strom, R. G. *see* Simon, R. S.
Unwin, S. C.: "Superluminal Motion in the Quasar 3C 279" 34
Unwin, S. C. *see also* Lawrence, C. R., *and* Walker, R. C.
Waak, J. A. *see* Simon, R. S.
Walker, R. C., Benson, J. M., and Unwin, S. C.: "3C 120" 48
Wardle, J. F. C. *see* Roberts, D. H.
Wilkinson, P. N.: "Imaging Superluminal Sources: Prospects for the Next Decade" 211
Witzel, A.: "Superluminal Motion and Other Indications of Bulk Relativistic Motion in a Complete Sample of Radio Sources from the S5 Survey" 83
Witzel, A. *see also* Simon, R. S.
Worrall, D. M.: "Superluminal Radio Sources: What Does X-ray Emission Tell Us?" 251
Wrobel, J. M.: "VLA Polarimetry of the Active Galaxy 3C 371" 186
Zensus, J. A.: "3C 273: Archetype of Superluminal Sources" 26
Zensus, J. A., and Porcas, R. W.: "Superluminal Motion in a Randomly Oriented Quasar Sample" 126
Zensus, J. A. *see also* Pearson, T. J.

Index of Objects

0016+731 **83–93**
   inverse Compton   91–92
   polarization, optical   263
   spectrum, optical   263, 265
   VLBI   84–86
0026+129
   spectrum, broad-band   220
0026+346   168
0106+130 **(3C 33)**   172
0108+388 **(OC 314)**   237
   optical identification   260
0138+136 **(3C 49)**   172
   EVN+MERLIN   175
0153+744 **83–93**, 169, 236
   inverse Compton   91–92
   motion   89–90
   polarization, optical   263
   spectrum, optical   263, 265
   VLBI   84–86
0212+735 **83–93**, 236
   inverse Compton   91
   polarization, optical   262
   spectrum, optical   262, 265
   superluminal motion   18, 22, 89
   VLBI   84–86
0218+357   168
0224+671 **(4C 67.05)**   236
0229+341 **(3C 68.1)**   117
0235+164   236
   BL Lac object   242
   possibly superluminal   19
   spectrum, broad-band   224
   variability   224
0316+413 **(NGC 1275, 3C 84)**
   76–82, 97, 237
   mm-wavelength observations
     76–82
   subluminal   19, 77–79, 312
   variability   76–78
   VLBI   77–82
0333+321 **(NRAO 140)**   123, 131,
   140, 236
   large-scale structure   130, 132,
     159–160
   spectrum, radio   287

   superluminal motion   18, 290
   VLBI   286
   X-rays   252, 257, 289–290
0355+508 **(NRAO 150)**   236
0415+379 **(3C 111)**   236
   possibly superluminal   19, 22-23
0429+415 **(3C 119)**
   VLBI   176
0430+052 **(3C 120)**   **48–54**, 123, 131,
   214, 236, 298
   accretion disk   270–271
   large-scale structure   51–52, 130,
     159–160
   physical parameters   53
   polarization, optical   247
   power-law variation along jet
     52–53, 315
   spectrum, optical   267, 270–272
   superluminal motion   18, 22,
     49–51, 311–312
   variability, low frequency   201
   variability, optical   225
   variability, radio   225
   VLBI   49–51
   VLBI, polarization   195–196, 315
   X-rays   252, 257
0449−184   266
0454+844 **83–93**, 236
   BL Lac object   242, 260
   inverse Compton   91
   motion   89–90
   spectrum, optical   260
   VLBI   84–86
0459+252 **(3C 133)**
   VLA (depolarization)   105–107
0518+165 **(3C 138)**
   VLBI   175
0538+498 **(3C 147)**   236
   possibly superluminal   19, 22, 176
0605−085 **(OH−010)**   236
   variability, low frequency   201
0615+820 **83–93**, 236
   inverse Compton   91
   motion   89–90
   VLBI   84–86

Index of Objects

0710+439 (**OI 417**)   94–103, 168, 237
   compact double   100–101
   subluminal   100–101
0711+356 (**OI 318**)   237
   possibly superluminal   19, 22
0716+714   **83–93**, 236
   BL Lac object   242
   inverse Compton   91
   motion   89–90
   VLBI   84–86
0723+679 (**3C 179**)   **12–25**, 96–97, 117, 120, 123, 131, 140, 148, 236
   large-scale structure   130, 159–160
   superluminal motion   15–18, 127–128, 313
   VLBI   15–17
0735+178   236
   cosmic conspiracy   219
   spectrum, broad-band   222
   superluminal motion   18, 22
   X-rays   252
0736+017
   spectrum, broad-band   222
0836+710   **83–93**, 236
   inverse Compton   91–92
   superluminal motion   89
   VLBI   84, 87
0838+133 (**3C 207**)
   VLBI   118
0850+581 (**4C 58.17**)   **94–103**, 123, 131, 140, 149, 236
   large-scale structure   98, 130–134
   polarization, optical   262
   spectrum, optical   262, 265
   superluminal motion   18, 22, 98
   VLBI   98
0851+202 (**OJ 287**)   **193–199**, 236
   BL Lac object   242
   spectrum, broad-band   220, 222, 224
   superluminal motion   18, 198
   variability   224
   VLBI, polarization   196–198, 315
0855+143 (**3C 212**)
   VLBI   118
0906+430 (**3C 216**)   **94–103**, 131, 140, 149, 176, 236
   large-scale structure   132, 134, 159–160
   polarization, optical   262
   spectrum, optical   262, 266
   superluminal motion   18, 22, 98–99, 101
   VLBI   98–99

X-rays   253
0923+392 (**4C 39.25**)   **67–71**, 97, 131, 140, 230, 236, 281, 293, 298
   large-scale structure   68, 134
   not contracting   68
   polarization, optical   262
   spectrum, optical   262, 265
   stationary components   67–71
   superluminal motion   18, 22–23, 68–70, 289, 313
   VLBI   67–70
   X-rays   253
0951+699 (**M82**)   97
1003+351 (**3C 236**)   153, 159
1005+077 (**3C 237**)   172
1019+222 (**3C 241**)   172
1038+528A (**OL 564**)   236
   possibly superluminal   19
1039+811   **83–93**, 236
   inverse Compton   91
   motion   89–90
   VLBI   84, 87
1040+123 (**3C 245**)   105, **114–125**, 131, 140, 236
   large-scale structure   122, 134, 159–160
   superluminal motion   18, 22, 118, 120–121
   VLBI   120–121
   X-rays   252
1100+772 (**3C 249.1**)
   VLBI   118
1101+384 (**Mrk 421**)   256
1117+146
   variability, low frequency   203–204
1137+660 (**3C 263**)   **114–125**, **126–128**, 131, 140, 236
   large-scale structure   132, 134, 159–160
   superluminal motion   18, 22, 118–120, 127–128
   VLBI   118–120
   X-rays   252, 257
1150+812   **83–93**, 236
   inverse Compton   91
   superluminal motion   18, 22, 88–89
   VLBI   84, 88
1206+439 (**3C 268.4**)
   VLBI   127
1208+396 (**NGC 4151**)
   spectrum, broad-band   220

1226+023 (**3C 273**) **26–31, 32–33**,
   44, 123, 131, 140, 211, 241,
   236
   blazar properties 233–234
   large-scale structure 28, 134,
      159–160
   millimeter VLBI 32–33
   optical jet (3C 273A) 28
   shock model 215
   superluminal motion 1–3, 12, 18,
      26–30
   UV-bump 239
   variability 1, 227
   VLBI 26–30, 32–33
   wiggles in jet 30
   X-rays 251, 253, 257
1228+127 (**M87**) 237
   black hole 321
   subluminal 19, 23, 312
1253−055 (**3C 279**) **34–39**, 123, 131,
   140, 236, 242, **273–279**
   gravitational bending 3
   inverse Compton 37
   large-scale structure 36, 134,
      159–160
   polarization, optical 34
   polarization, radio 273–279
   shock model 38, 273–279
   spectrum, broad-band 222
   spectrum, optical 267
   superluminal motion 1–3, 12, 15,
      18, 22, 34–38, 311
   variability, low frequency 201
   variability, radio 273–279
   VLBI 34–38
   X-rays 253, 257, 291
1308+326
   spectrum, broad-band 222, 224,
      229
   variability 224
1322−427 (**Cen A**) 237, 322
1328+307 (**3C 286**)
   variability, low frequency 203
1422+202
   VLBI 207–209
1458+718 (**3C 309.1**) 98, 110
1504−167
   variability, low frequency 204
1510−089 236
   variability, low frequency 201,
      204
1518+047 168
1524−136
   variability, low frequency 204
1607+268 (**CTD 93**) **168–170**, 237

spectrum, radio 170
VLBI 170
1611+343 (**DA 406**)
   variability, low frequency 201,
      203–204, 209
   VLBI 207–209
1617−155 (**Sco X-1**) 80
1618+177 (**3C 334**)
   VLBI 118
1624+416
   optical identification 260
1637+574
   polarization, optical 263
   spectrum, optical 263
1637+826 (**NGC 6251**) **162–167**,
   215, 237
   asymmetry 162–166, 241
   large-scale structure 164–165
   subluminal 19, 23, 164, 311
   VLBI 162–164
1638+398 (**NRAO 512**)
   separation from 3C 345 20
1641+399 (**3C 345**) **40–47**, 97, 110,
   123, 131, 140, 211, 236, 241
   acceleration 20–21, 62
   frequency-dependent component
      positions 20–21, 294
   large-scale structure 40, 133–134,
      159–160
   polarization, optical 262
   optical properties 40
   spectrum, broad-band 221–223
   spectrum, optical 262, 266, 267
   stationary core 20, 294
   superluminal motion 18–22,
      40–46, 311, 315
   synchrotron self-Compton model
      40–44
   variability, low frequency 201,
      203
   variability, optical 262, 265
   VLBI 40–46
   VLBI, polarization 194–195, 315
   X-rays 40, 251, 253, 257
1642+690 (**4C 69.21**) **94–103**, 123,
   131, 140, 236
   large-scale structure 133–134,
      159–160
   polarization, optical 262
   spectrum, optical 262, 266
   superluminal motion 18, 22,
      99–100
   VLBI 99–100
1652+398 (**Mrk 501**) 95

Index of Objects

1721+343 (**4C 34.47**) 131, 140, **148–154**, 159–160, 236
   large-scale structure 150, 314
   largest known quasar 134, 150–151
   superluminal motion 18, 22, 151–152
   VLBI 151–152
1732+655
   VLBI 127
1749+701 **83–93**, 236
   BL Lac object 242, 260
   inverse Compton 91
   motion 89–90
   spectrum, optical 260
   VLBI 84, 87
1803+784 **83–93**, 236
   BL Lac object 242, 265
   broadband spectrum 85
   inverse Compton 91–92
   no superluminal motion 89–90
   polarization, optical 263
   spectrum, optical 263, 265–266
   VLBI 84, 87
1807+698 (**3C 371**) 95, **180–185**, **186–192**, 236
   deviations, semi-periodic 181–182, 184
   large-scale structure 187–188
   optical emission 180–181, 186, 242
   polarization, optical 263
   polarization, radio 187–191
   possibly superluminal 19, 22, 181–182, 311
   recollimation, apparent 182–184
   spectrum, optical 263
   VLBI 180–185
1833+326 (**3C 382**)
   spectrum, optical 267
1842+455 (**3C 388**)
   polarization, optical 263
   spectrum, optical 263–266
1845+797 (**3C 390.3**) 236
   accretion disk 270–271
   line profiles 267–272
   polarization, optical 247
   spectrum, optical 267–272
   superluminal motion 18, 22
   variability, optical 268–269
1901+319 (**3C 395**) 68, **72–75**, 131, 140, 169, 236, 298
   anomalous 137, 308, 313
   large-scale structure 72, 133–134, 159–160, 313

   stationary components 72–74
   superluminal motion 18, 22, 73–74, 289
   VLBI 72–74
1909+048 (**SS 433**) 80, 166
1928+738 (**4C 73.18**) **83–93**, **94–103**, 123, 131, 140, **155–161**, 236
   inverse Compton 91
   large-scale structure 133–134, 155–161, 159–160, 314
   polarization, optical 262
   spectrum, optical 262, 264–265
   superluminal motion 18, 22, 89, 99–100, 155
   VLBI 84–85, 87, 99–100
   X-rays 252
1951+498 **126–128**, 236
   superluminal motion 18, 22, 128
   VLBI 128
1957+405 (**Cyg A**) 237
   rotation measure 106
2007+777 **83–93**, 236
   inverse Compton 91
   motion 89–90
   VLBI 84, 87
2016+112 266
2021+614 (**OW 637**) 94–103, 168, 237
   subluminal 19, 101
   compact double 101
2050+364 168
2134+004 (**PHL 61**) 237
   subluminal 19
2147+145
   VLBI 207–208
2155−304 256
2200+420 (**BL Lac**) **60–66**, 97, 122–124, 131, 242, 236
   comparison with 3C 279 38, 275
   deceleration 60–65
   inverse Compton 64
   large-scale structure 134, 159–160
   polarization, optical 262
   shock model 63–64, 273–279
   spectrum, broad-band 224
   spectrum, optical 262, 265–266
   superluminal motion 18, 60–63, 311
   variability 224, 273–279
   variability, low frequency 201, 204
   VLBI 60–63
   X-rays 253
2223−052 (**3C 446**) 236

Index of Objects

  optical emission   242–243
  possibly superluminal   19, 22
  spectrum, optical   267
  UV-bump   239
2230+114 (**CTA 102**)   **206–210**, 236, 301
  anomalous   308
  superluminal motion   18, 22, 209–210
  variability   206, 210, 227, 314
  variability, low frequency   203
  VLBI   207–210
  X-rays   253
2251+158 (**3C 454.3**)   **55–59**, 130, 131, 140, 236, 301
  large-scale structure   56, 134–135
  superluminal brightening   22, 55–58
  superluminal motion   18, 22–23, 55–58, 311
  variability   55, 227
  variability, low frequency   55, 201, 203
  VLBI   55–59
  X-rays   253
2342+821
  optical identification   260

3C 33	see 0106+130
3C 49	see 0138+136
3C 68.1	see 0229+341
3C 84	see 0316+413
3C 111	see 0415+379
3C 119	see 0429+415
3C 120	see 0430+052
3C 133	see 0459+252
3C 138	see 0518+165
3C 147	see 0538+498
3C 179	see 0723+679
3C 207	see 0838+133
3C 212	see 0855+143
3C 216	see 0906+430
3C 236	see 1003+351
3C 237	see 1005+077
3C 241	see 1019+222
3C 245	see 1040+123
3C 249.1	see 1100+772
3C 263	see 1137+660
3C 268.4	see 1206+439
3C 273	see 1226+023
3C 274	see 1228+126
3C 279	see 1253−055
3C 286	see 1328+307
3C 309.1	see 1458+718
3C 334	see 1618+177
3C 345	see 1641+399
3C 371	see 1807+698
3C 382	see 1833+326
3C 388	see 1842+455
3C 390.3	see 1845+797
3C 395	see 1901+319
3C 405	see 1957+405
3C 446	see 2223−052
3C 454.3	see 2251+158
4C 34.47	see 1721+343
4C 39.25	see 0923+392
4C 58.17	see 0850+581
4C 67.05	see 0224+671
4C 69.21	see 1642+690
4C 73.18	see 1928+738
BL Lac	see 2200+420
Cen A	see 1322−427
CTA 102	see 2230+114
CTD 93	see 1607+268
Cyg A	see 1957+405
DA 406	see 1611+343
M82	see 0951+699
M87	see 1228+127
Mrk 421	see 1101+384
Mrk 501	see 1652+398
NGC 1275	see 0316+413
NGC 4151	see 1208−396
NGC 5128	see 1322−427
NGC 6251	see 1637+826
Nova Persei 1901	6
NRAO 140	see 0333+321
NRAO 150	see 0355+508
NRAO 512	see 1638+398
OC 314	see 0108+388
OH−010	see 0605−085
OI 318	see 0711+356
OI 417	see 0710+439
OJ 287	see 0851+202
OL 564	see 1038+528A
OW 637	see 2021+614
PHL 61	see 2134+004
Sco X-1	see 1617−155
SS 433	see 1909+048

Index of Subjects

Accretion disk   270–271, 322–323
Anomalous sources   137, 308, 313
Apparent velocity
   μ–z diagram   **306–309**
   correlation with core strength
      122–123, 137
   correlation with linear size
      115–117, 138
   upper limit   **301–305**
Astrometry, of 3C 345   20
Asymmetric sources   *see* core-jet sources
Asymmetry   162–166, 241
   intrinsic, in NGC 6251   162–166

Bending of jets   *see* Jets, bending
Black holes   321–325
Blazars   218–220, 233–238, 312
BL Lac objects   60, 65, 242, 261, 265
Broad-band spectrum   *see* Spectra, broad-band
Broad-line region, models   270–272
Bulk relativistic motion, evidence for   90–92, 164–166, 184

Christmas trees, computer-controlled   6
Classification of sources   85–88, 94–103, 118, 127
Compact double sources   96–97, 100–101, **168–173**
   evolution   172–173
   relationship to superluminals   168–171
Compact steep-spectrum sources   96, 98–99, 101–102, **174–179**
Component spectra   36–37
Compton brightness limit   1, 206, 314
Compton speed limit   302–305
Contraction   67
Core, motion of   *see* Stationary core
Core-jet sources   26–30, 85, 96
   identification of core   15, 22
Core-to-extended flux ratio (R)   123, 131

   as an indicator of orientation   115–117
   correlation with apparent velocity   120, 122–124, 137
   correlation with linear size   115–117, 138
Cosmic conspiracy   219, 285
Cosmology   5–6, 13–14
Curvature of extended radio sources   115–117, 142–143

Deceleration parameter (q_0)   5–6, 13–14
Depolarization   105–108, 195
Deprojection   143–144, 151–153, 157–158
Deviations, semi-periodic   *see* Wiggles
Doppler boosting   *see* Relativistic beaming
Doppler factor   8, 37, 43–44, 64, 91–92
Doubts, nagging   229–230, 288, 294, *see also* Feelings

Evolution of quasars and radio galaxies   112, 323–325
Extended structure   *see* Large-scale structure

Faraday rotation   187–191
Feelings   70
   gut   230
   uncomfortable   148–154
Flickering   *see* Variability
Frequency-dependent component positions   20–21, 289, 293–294

Grand unified models   **310–326**
Gravitational bending   3
Gravitational lenses   6, 70, 266
h ($H_0/100$ km s^{-1} Mpc^{-1})   6, 14
Host galaxies   111–113, 171, 219, 247–248
Hotspots   70, 74–75, 76, 106, 108, 118, 134, 142–143, 153, 189, 191, 298

Index of Subjects

Hubble constant (H_0) 5–6, 13–14, 105, 143–145, 302, 308–309
Infrared emission 112–113, 218–219, 221, 223, 229, **233–250**, 280, 291, 305, 319, 321–322
Intergalactic medium 106
Interstellar medium 48, 64–65, 165–166, 176, 200–205, 206, 209, 227, 246, 314
Inverse Compton radiation 36–37, 43–44, 64, 77, 91–92, 153, 187, 206, 235–237, 251–259, **280–300**, 321

Jets 26–30, 159, 181–184, 287–288, 294–298
   acceleration mechanisms 301–305
   bending 45, 73–74, 127, 181, 183–184
   continuous or discrete blobs? 180
   curly 110
   emasculated 65
   inhomogeneous 42–44, 315
   instabilities 46, 70, 184, 305
   interaction with interstellar medium 64–65, 127, 165–166, 171
   opening angle 42–43, 294, 315
   optical (3C 273A) 26
   polarization 105–108, 190
   power-law variation along 52–53, 315
   theory 294–298, 301–305, 319–321
   two-sided 159, 164
   viewing geometry 71, 73–74, 110

Kinematics 7–8, 20–22, 44–46, 241, 315

Large-scale structure 26, 28, 36, 40, 51–52, 56, 68, 122, **129–147**, 149–150, 155–160, 162–166, 187–188, 313–314
   beamed? 108
   table of properties 131, (160)
Largest known quasar 134, 150–151
Light-echo 6
Line profiles 267–272
Linear size distribution 137–139, 151
   deprojected 143–144, 151
Linear size problem 10, 138–146, 149–153, 158–159
Low-frequency variability
   see Variability, low-frequency

Magnetic dipole model 6, 298, 309
Misalignment 127, 142–143, 313
Morphology of compact radio sources 85–88, 94–97

Narrow line region 43, 53, 65, 109, 171, 195
Nova Persei 1901 6

Optical emission lines 111, 169, 217, 242–245, 260–266, 267–272
   Hα 267–272
   Hβ 112–113, 240, 264–266, 314

Physical conditions 42–44, 52–53, 80, 315–317
Polarization
   optical 234–237, 262–263
   position angle alignment 244–247
   radio 186–191, 228–229, 273–279
   variation 60–61, 273–279
   VLA imaging 105–108, 186–191, 315
   VLBI imaging **193–199**, 315
Precession 45, 315

Quasars
   are quasars radio galaxies seen end-on? 108–110, 111–113
   cosmological evolution 112
   double-lobed 114–125, 126–128
   infrared emission 112
   optical spectra 111
   underlying galaxies 112

R *see* Core-to-extended flux ratio
Radio galaxies, hidden quasars? 108–110, 111–113
Radio-quiet quasars 9, 217, 219–220, 254, 256, 317–318
Redshifts
   distribution 261
   non-cosmological 1, 6, 13, 307
   origin of 1, 13–14, 306–308
Relativistic beaming model 2, 6–10, 315–319, *see also* Unified schemes
   tests **104–113**, 114–125, 126–128
   why it is not true 152–154, 314, *see also* Linear size problem
   why it is true 105–108, 153, 312–314
Rotation measures 106, 188–191
Rumford Medal 3

Samples
   flux-density limited 83–93, 94–103
   orientation-unbiased 114–125, 126–128
Scintillation 2, 200–205, 206–209, 227, 314
Screen models 6, 153, 298–299
Shocks 38, 58, 63–64, 180, 184, 215, 273–279, 291–298, 316–317
Spectra
   broad-band 85, **217–232**, 237–240
   interpretation 319–321
   dissection 285–288
   optical **260–266**, 267–272,
      see also Optical emission lines
Spectral index (α), sign convention 8
Stationary components 22, 67–71, 72–75
Stationary core 20
Statistics 115–117, 135–146
   core-to-extended flux ratio 176–178
   jet fluxes 104, 109
   jet speeds 8, 104–105, 120, 126, 306–309, 313
   linear sizes 175–177
Subluminal sources 17, 19, 23, 77–79, 89–90, 100–101, 164, 237, 311–313
Superluminal brightening 22, 55–59, 306
Superluminal motion
   acceleration 20–21, 45–46, 62, 315
   between stationary components 67–71, 72–75
   component trajectories 30, 63, 315
   computation of apparent velocity 5–6, 13–14
   correlation with flux outbursts 55, 60–61
   deceleration 60–65, 315
   definition 13
   discovery 1–3, 12
   models 2, 6–10, 306–309
   predicted from X-ray flux 289–290

Superluminal sources
   as blazars 218–221, 233–238, 312
   general properties 15–17, 19–21, 312–315
   identified from non-radio properties 248
   inventory **17–23**, 236–237
   observational problems 17, 97, 311–312
   possible 19, 22–23, 181–182, 236, 311
Surveys 22, **83–93, 94–103**, 114–125, 126–128
Synchro-Compton model 1, 40–44, **280–300**

Ultraviolet emission 237–239
Unified schemes 9–10, 58, 104–113, 135–146, 239–244, 317–319,
   see also Relativistic beaming model
Unresolved sources 88, 96

Variability 224–229, 233–235
   flickering 227, 314
   low-frequency 55, **200–205**, 206, 209, 314
   "occultations" 227
   optical 225–227, 234, 262, 265, 267–270
   radio 1, 55, 76–78, 200–205, 206, 210, 224–227, 273–279, 314
   time scales 202–203, 227
Very Long Baseline Array 4, 212, 311, 326
Very long baseline interferometry
   current capabilities 4, 14, 27, 32–33, 212–215
   early history 1–3, 12–13, 26
   future prospects 211–216
   millimeter wavelengths 32–33, 79–82, 92
   phase referencing 20
   polarization measurement 193–198, 315
Visibility plot, 3C 279 15

Wiggles 30, 71, 181–184, 316

X-rays 37, 40, 77, **251–259**, 280–281, 290–292, 314
   correlations with optical and radio emission 252–257

RAYMOND H. FOGLER LIBRARY

DATE DUE

BOOKS ARE SUBJECT TO RECALL AFTER TWO WEEKS

JAN 15 1988

AUG 26 1987